Michael Schäfer

Computational Engineering – Introduction to Numerical Methods

T0156121

Michael Schäfer

Computational Engineering – Introduction to Numerical Methods

With 204 Figures

 Springer

Professor Dr. rer. nat. Michael Schäfer
Chair of Numerical Methods in Mechanical Engineering
Technische Universität Darmstadt
Petersenstr. 30
64287 Darmstadt
Germany
schaefer@fnb.tu-darmstadt.de

Solutions to the exercises:
www.fnb.tu-darmstadt.de/ceinm/ or www.springer.com/3-540-30686-2

The book is the English edition of the German book: *Numerik im Maschinenbau*

Library of Congress Control Number: 2005938889

ISBN-10 3-540-30685-4   Springer Berlin Heidelberg New York
ISBN-13 978-3-540-30685-6   Springer Berlin Heidelberg New York

Springer is a part of Springer Science+Business Media

springer.com

© Springer-Verlag Berlin Heidelberg 2006
Printed in Germany

Typesetting: Digital data supplied by author
Cover Design: Frido Steinen-Broo, EStudio Calamar, Spain
Production: LE-TeX Jelonek, Schmidt & Vöckler GbR, Leipzig

Printed on acid-free paper   7/3100/YL   5 4 3 2 1 0

Preface

Due to the enormous progress in computer technology and numerical methods that have been achieved in recent years, the use of numerical simulation methods in industry gains more and more importance. In particular, this applies to all engineering disciplines. Numerical computations in many cases offer a cost effective and, therefore, very attractive possibility for the investigation and optimization of products and processes.

Besides the need for developers of corresponding software, there is a strong – and still rapidly growing – demand for qualified specialists who are able to efficiently apply numerical simulation tools to complex industrial problems. The successful and efficient application of such tools requires certain basic knowledge about the underlying numerical methodologies and their possibilities with respect to specific applications. The major concern of this book is the impartation of this knowledge in a comprehensive way.

The text gives a practice oriented introduction in modern numerical methods as they typically are applied in engineering disciplines like mechanical, chemical, or civil engineering. In corresponding applications the by far most frequent tasks are related to problems from heat transfer, structural mechanics, and fluid mechanics, which, therefore, constitute a thematical focus of the text.

The topic must be seen as a strongly interdisciplinary field in which aspects of numerical mathematics, natural sciences, computer science, and the corresponding engineering area are simultaneously important. As a consequence, usually the necessary information is distributed in different textbooks from the individual disciplines. In the present text the subject matter is presented in a comprehensive multidisciplinary way, where aspects from the different fields are treated insofar as it is necessary for general understanding.

Following this concept, the text covers the basics of modeling, discretization, and solution algorithms, whereas an attempt is always made to establish the relationships to the engineering relevant application areas mentioned above. Overarching aspects of the different numerical techniques are emphasized and questions related to accuracy, efficiency, and cost effectiveness, which are most relevant for the practical application, are discussed. The following subjects are addressed in detail:

- *Modelling:* simple field problems, heat transfer, structural mechanics, fluid mechanics.
- *Discretization:* connection to CAD, numerical grids, finite-volume methods, finite-element methods, time discretization, properties of discrete systems.
- *Solution algorithms:* linear systems, non-linear systems, coupling of variables, adaptivity, multi-grid methods, parallelization.
- *Special applications*: finite-element methods for elasto-mechanical problems, finite-volume methods for incompressible flows, simulation of turbulent flows.

The topics are presented in an introductory manner, such that besides basic mathematical standard knowledge in analysis and linear algebra no further prerequisites are necessary. For possible continuative studies hints for corresponding literature with reference to the respective chapter are given.

Important aspects are illustrated by means of application examples. Many exemplary computations done "by hand" help to follow and understand the numerical methods. The exercises for each chapter give the possibility of reviewing the essentials of the methods. Solutions are provided on the web page *www.fnb.tu-darmstadt.de/ceinm/*. The book is suitable either for self-study or as an accompanying textbook for corresponding lectures. It can be useful for students of engineering disciplines, but also for computational engineers in industrial practice. Many of the methods presented are integrated in the flow simulation code FASTEST, which is available from the author.

The text evolved on the basis of several lecture notes for different courses at the *Department of Numerical Methods in Mechanical Engineering* at *Darmstadt University of Technology*. It closely follows the German book *Numerik im Maschinenbau* (Springer, 1999) by the author, but includes several modifications and extensions.

The author would like to thank all members of the department who have supported the preparation of the manuscript. Special thanks are addressed to Patrick Bontoux and the MSNM-GP group of CNRS at Marseille for the warm hospitality at the institute during several visits which helped a lot in completing the text in time. Sincere thanks are given to Rekik Alehegn Mekonnen for proofreading the English text. Last but not least the author would like to thank the Springer-Verlag for the very pleasant cooperation.

Darmstadt
Spring 2006   *Michael Schäfer*

Contents

1

Introduction

In this introductory chapter we elucidate the value of using numerical methods in engineering applications. Also, a brief overview of the historical development of computers is given, which, of course, are a major prerequisite for the successful and efficient use of numerical simulation techniques for solving complex practical problems.

1.1 Usefulness of Numerical Investigations

The functionality or efficiency of technical systems is always determined by certain properties. An ample knowledge of these properties is frequently the key to understanding the systems or a starting point for their optimization. Numerous examples from various engineering branches could be given for this. A few examples, which are listed in Table 1.1, may be sufficient for the motivation.

Table 1.1. Examples for the correlation of properties with functionality and efficiency of technical systems

Property	Functionality/Efficiency
Aerodynamics of vehicles	Fuel consumption
Statics of bridges	Carrying capacity
Crash behavior of vehicles	Chances of passenger survival
Pressure drop in vacuum cleaners	Sucking performance
Pressure distribution in brake pipes	Braking effect
Pollutants in exhaust gases	Environmental burden
Deformation of antennas	Pointing accuracy
Temperature distributions in ovens	Quality of baked products

In engineering disciplines in this context, in particular, solid body and flow properties like

- deformations or stresses,
- flow velocities, pressure or temperature distributions,
- drag or lift forces,
- pressure or energy losses,
- heat or mass transfer rates, ...

play an important role. For engineering tasks the investigation of such properties usually matters in the course of the redevelopment or enhancement of products and processes, where the insights gained can be useful for different purposes. To this respect, exemplarily can be mentioned:

- improvement of efficiency (e.g., performance of solar cells),
- reduction of energy consumption (e.g., current drain of refrigerators),
- increase of yield (e.g., production of video tapes),
- enhancement of safety (e.g., crack propagation in gas pipes, crash behavior of cars),
- improvement of durability (e.g., material fatigue in bridges, corrosion of exhaust systems),
- enhancement of purity (e.g., miniaturization of semi-conductor devices),
- pollutants reduction (e.g., fuel combustion in engines),
- noise reduction (e.g., shaping of vehicle components, material for pavings),
- saving of raw material (e.g., production of packing material),
- understanding of processes, ...

Of course, in the industrial environment in many instances the question of cost reduction, which may arise in one way or another with the above improvements, takes center stage. But it is also often a matter of obtaining a general understanding of processes, which function as a result of long-standing experience and trial and error, but whose actual operating mode is not exactly known. This aspect crops up and becomes a problem particularly if improvements (e.g. as indicated above) should be achieved and the process – under more or less changed basic conditions – does not function anymore or only works in a constricted way (e.g., production of silicon crystals, noise generation of high speed trains, ...).

There are fields of application for the addressed investigations in nearly all branches of engineering and natural sciences. Some important areas are, for instance:

- automotive, aircraft, and ship engineering,
- engine, turbine, and pump engineering,
- reactor and plant construction,
- ventilation, heating, and air conditioning technology,
- coating and deposition techniques,
- combustion and explosion processes,

- production processes in semi-conductor industry,
- energy production and environmental technology,
- medicine, biology, and micro-system technique,
- weather prediction and climate models, ...

Let us turn to the question of what possibilities are available for obtaining knowledge on the properties of systems, since here, compared to alternative investigation methods, the great potential of numerical methods can be seen. In general, the following approaches can be distinguished:

- theoretical methods,
- experimental investigations,
- numerical simulations.

Theoretical methods, i.e., analytical considerations of the equations describing the problems, are only very conditionally applicable for practically relevant problems. The equations, which have to be considered for a realistic description of the processes, are usually so complex (mostly systems of partial differential equations, see Chap. 2) that they are not solvable analytically. Simplifications, which would be necessary in order to allow an analytical solution, often are not valid and lead to inaccurate results (and therefore probably to wrong conclusions). More universally valid approximative formulas, as they are willingly used by engineers, usually cannot be derived from purely analytical considerations for complex systems.

While carrying out experimental investigations one aims to obtain the required system information by means of tests (with models or with real objects) using specialized apparatuses and measuring instruments. In many cases this can cause problems for the following reasons:

- Measurements at real objects often are difficult or even impossible since, for instance, the dimensions are too small or too large (e.g., nano system technique or earth's atmosphere), the processes elapse too slowly or too fast (e.g., corrosion processes or explosions), the objects are not accessible directly (e.g., human body), or the process to be investigated is disturbed during the measurement (e.g., quantuum mechanics).
- Conclusions from model experiments to the real object, e.g., due to different boundary conditions, often are not directly executable (e.g., airplane in wind tunnel and in real flight).
- Experiments are prohibited due to safety or environmental reasons (e.g., impact of a tanker ship accident or an accident in a nuclear reactor).
- Experiments are often very expensive and time consuming (e.g., crash tests, wind tunnel costs, model fabrication, parameter variations, not all interesting quantities can be measured at the same time).

Besides (or rather between) theoretical and experimental approaches, in recent years numerical simulation techniques have become established as a widely self-contained scientific discipline. Here, investigations are performed by means of numerical methods on computers. The advantages of numerical simulations compared to purely experimental investigations are quite obvious:

- Numerical results often can be obtained faster and at lower costs.
- Parameter variations on the computer usually are easily realizable (e.g., aerodynamics of different car bodies).
- A numerical simulation often gives more comprehensive information due to the global and simultaneous computation of different problem-relevant quantities (e.g., temperature, pressure, humidity, and wind for weather forecast).

An important prerequisite for exploiting these advantages is, of course, the reliability of the computations. The possibilities for this have significantly improved in recent years due developments which have contributed a great deal to the "booming" of numerical simulation techniques (this will be briefly sketched in the next section). However, this does *not* mean that experimental investigations are (or will become) superfluous. Numerical computations surely will never completely replace experiments and measurements. Complex physical and chemical processes, like turbulence, combustion, etc., or non-linear material properties have to be modelled realistically, for which as near to exact and detailed measuring data are indispensable. Thus, both areas, numerics *and* experiments, must be further developed and ideally used in a complementary way to achieve optimal solutions for the different requirements.

1.2 Development of Numerical Methods

The possibility of obtaining approximative solutions via the application of finite-difference methods to the partial differential equations, as they typically arise in the engineering problems of interest here, was already known in the 19th century (the mathematicians Gauß and Euler should be mentioned as pioneers). However, these methods could not be exploited reasonably due to the too high number of required arithmetic operations and the lack of computers. It was with the development of electronic computers that these numerical approaches gained importance. This development was (and is) very fast-paced, as can be well recognized from the maximally possible number of floating point operations per second (Flops) achieved by the computers which is indicated in Table 1.2. Comparable rates of improvement can be observed for the available memory capacity (also see Table 1.2).

However, not only the advances in computer technology have had a crucial influence on the possibilities of numerical simulation methods, but also the continuous further development of the numerical algorithms has contributed significantly to this. This becomes apparent when one contrasts the developments in both areas in recent years as indicated in Fig. 1.1. The improved

Table 1.2. Development of computing power and memory capacity of electronic computers

Year	Computer	Floating point operations per second (Flops)	Memory space in Bytes
1949	EDSAC 1	$1 \cdot 10^2$	$2 \cdot 10^3$
1964	CDC 6600	$3 \cdot 10^6$	$9 \cdot 10^5$
1976	CRAY 1	$8 \cdot 10^7$	$3 \cdot 10^7$
1985	CRAY 2	$1 \cdot 10^9$	$4 \cdot 10^9$
1997	Intel ASCI	$1 \cdot 10^{12}$	$3 \cdot 10^{11}$
2002	NEC Earth Simulator	$4 \cdot 10^{13}$	$1 \cdot 10^{13}$
2005	IBM Blue Gene/L	$3 \cdot 10^{14}$	$5 \cdot 10^{13}$
2009	IBM Blue Gene/Q	$3 \cdot 10^{15}$	$5 \cdot 10^{14}$

capabilities with respect to a realistic modeling of the processes to be investigated also have to be mentioned in this context. An end to these developments is not yet in sight and the following trends are on the horizon for the future:

- Computers will become ever faster (higher integrated chips, higher clock rates, parallel computers) and the memory capacity will simultaneously increase.
- The numerical algorithms will become more and more efficient (e.g., by adaptivity concepts).
- The possibilities of a realistic modeling will be further improved by the allocation of more exact and detailed measurement data.

One can thus assume that the capabilities of numerical simulation techniques will greatly increase in the future.

Along with the achieved advances, the application of numerical simulation methods in industry increases rapidly. It can be expected that this trend will be even more pronounced in the future. However, with the increased possibilities the demand for simulations of more and more complex tasks also rises. This in turn means that the complexity of the numerical methods and the corresponding software further increases. Therefore, as is already the case in recent years, the field will be an area of active research and development in the foreseeable future. An important aspect in this context is that developments frequently undertaken at universities are rapidly made available for efficient use in industrial practice.

Based on the aforementioned developments, it can be assumed that in the future there will be a continuously increasing demand for qualified specialists, who are able to apply numerical methods in an efficient way for complex industrial problems. An important aspect here is that the possibilities *and also* the limitations of numerical methods and the corresponding computer software for the respective application area are properly assessed.

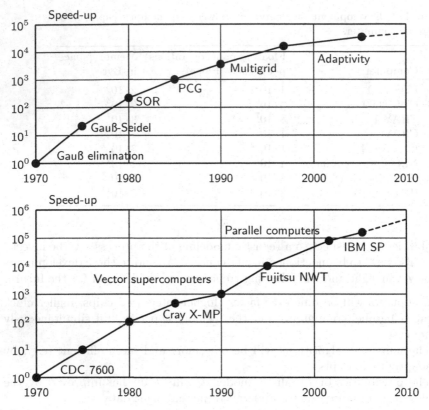

Fig. 1.1. Developments in computer technology (bottom) and numerical methods (top)

1.3 Characterization of Numerical Methods

To illustrate the different aspects that play a role when employing numerical simulation techniques for the solution of engineering problems, the general procedure is represented schematically in Fig. 1.2.

The first step consists in the appropriate mathematical modeling of the processes to be investigated or, in the case when an existing program package is used, in the choice of the model which is best adapted to the concrete problem. This aspect, which we will consider in more detail in Chap. 2, must be considered as crucial, since the simulation usually will not yield any valuable results if it is not based on an adequate model.

The continuous problem that result from the modeling – usually systems of differential or integral equations derived in the framework of continuum mechanics – must then be suitably approximated by a discrete problem, i.e., the unknown quantities to be computed have to be represented by a finite

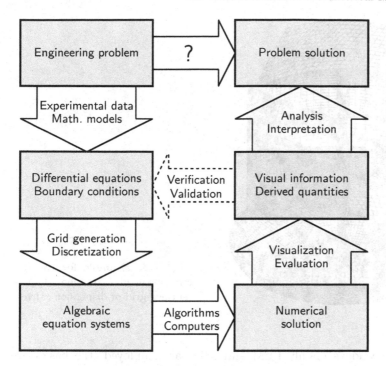

Fig. 1.2. Procedure for the application of numerical simulation techniques for the solution of engineering problems

number of values. This process, which is called *discretization*, mainly involves two tasks:

- the discretization of the problem domain,
- the discretization of the equations.

The discretization of the problem domain, which is addressed in Chap. 3, approximates the continuous domain (in space and time) by a finite number of subdomains (see Fig. 1.3), in which then numerical values for the unknown quantities are determined. The set of relations for the computation of these values are obtained by the discretization of the equations, which approximates the continuous systems by discrete ones. In contrast to an analytical solution, the numerical solution thus yields a set of values related to the discretized problem domain from which the approximation of the solution can be constructed.

There are primarily three different approaches available for the discretization procedure:

- the finite-difference method (FDM),
- the finite-volume method (FVM),
- the finite-element method (FEM).

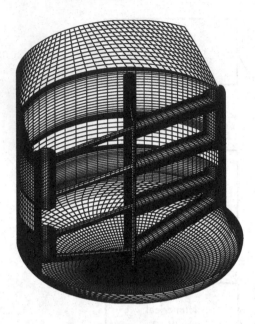

Fig. 1.3. Example for the discretization of a problem domain (surface grid of dispersion stirrer)

In practice nowadays mainly FEM and FVM are employed (the basics are addressed in detail in Chaps. 4 and 5). While FEM is predominantly used in the area of structural mechanics, FVM dominates in the flow mechanical area. Because of the importance of these two application areas in combination with the corresponding discretization technique, we will deal with them separately in Chaps. 9 and 10. For special pupuses, e.g., for the time discretization, which is the topic of Chap. 6, or for special approximations in the course of FVM and FEM, FDM is often also applied (the corresponding basics are recalled where needed). It should be noted that there are other discretization methods, e.g., spectral methods or meshless methods, which are used for special purposes. However, since these currently are not in widespread use we do not consider them further here.

The next step in the course of the simulation consists in the solution of the algebraic equation systems (the actual computation), where one frequently is faced with equations with several millions of unknowns (the more unknowns, the more accurate the numerical result will be). Here, algorithmic questions and, of course, computers come into play. The most relevant aspects in this regard are treated in Chaps. 7 and 12.

The computation in the first instance results in a usually huge amount of numbers, which normally are not intuitively understood. Therefore, for the evaluation of the computed results a suitable visualization of the results is important. For this purpose special software packages are available, which meanwhile have reached a relatively high standard. We do not address this topic further here.

After the results are available in an interpretable form, it is essential to inspect them with respect to their quality. During all prior steps, errors are inevitably introduced, and it is necessary to get clarity about their quantity (e.g., reference experiments for model error, systematic computations for numerical errors). Here, two questions have to be distinguished:

- *Validation:* Are the proper equations solved?
- *Verification:* Are the equations solved properly?

Often, after the validation and verification it is necessary to either adapt the model or to repeat the computation with a better discretization accuracy. These crucial questions, which also are closely linked to the properties of the model equations and the discretization techniques, are discussed in detail in Chap. 8.

In summary, it can be stated that related to the application of numerical methods for engineering problems, the following areas are of particular importance:

- Mathematical modelling of continuum mechanical processes.
- Development and analysis of numerical algorithms.
- Implementation of numerical methods into computer codes.
- Adaption and application of numerical methods to concrete problems.
- Validation, verification, evaluation and interpretation of numerical results.

The corresponding requirements and their interdependencies are indicated schematically in Fig. 1.4.

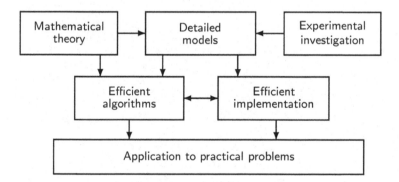

Fig. 1.4. Requirements and interdependencies for the numerical simulation of practical engineering problems

Regarding the above considerations, one can say that one is faced with a strongly interdisciplinary field, in which aspects from engineering science, natural sciences, numerical mathematics, and computer science (see Fig. 1.5) are involved. An important prerequisite for the successful and efficient use of

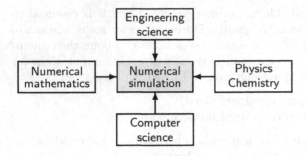

Fig. 1.5. Interdisciplinarity of numerical simulation of engineering problems

numerical simulation methods is, in particular, the efficient interaction of the different methodologies from the different areas.

2

Modeling of Continuum Mechanical Problems

A very important aspect when applying numerical simulation techniques is the "proper" mathematical modeling of the processes to be investigated. If there is no adequate underlying model, even a perfect numerical method will not yield reasonable results. Another essential issue related to modeling is that frequently it is possible to significantly reduce the computational effort by certain simplifications in the model. In general, the modeling should follow the principle already formulated by Albert Einstein: *as simple as possible, but not simpler*. Because of the high relevance of the topic in the context of the practical use of numerical simulation methods, we will discuss here the most essential basics for the modeling of continuum mechanical problems as they primarily occur in engineering applications. We will dwell on continuum mechanics only to the extent as it is necessary for a basic understanding of the models.

2.1 Kinematics

For further considerations some notation conventions are required, which we will introduce first. In the Euclidian space $\mathbb{R}^3$ we consider a Cartesian coordinate system with the basis unit vectors $\mathbf{e}_1$, $\mathbf{e}_2$, and $\mathbf{e}_3$ (see Fig. 2.1). The continuum mechanical quantities of interest are *scalars (zeroth-order tensors)*, *vectors (first-order tensors)*, and *dyads (second-order tensors)*, for which we will use the following notations:

- scalars with letters in normal font: a, b, ..., A, B, ..., α, β, ...,
- vectors with bold face lower case letters: $\mathbf{a}$, $\mathbf{b}$, ...,
- dyads with bold face upper case letters: $\mathbf{A}$, $\mathbf{B}$, ...

The different notations of the tensors are summarized in Table 2.1. We denote the coordinates of vectors and dyads with the corresponding letters in normal font (with the associated indexing). We mainly use the coordinate notation, which usually also constitutes the basis for the realization of a model within a computer program. To simplify the notation, *Einstein's summation convention* is employed, i.e., a summation over double indices is implied. For the basic conception of tensor calculus, which we need in some instances, we refer to the corresponding literature (see, e.g., [19]).

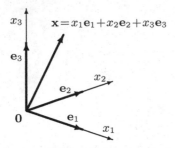

Fig. 2.1. Cartesian coordinate system with unit basis vectors e_1, e_2, and e_3

Table 2.1. Notations for Cartesian tensors

Order	Name	Notation	
0	Scalar	ϕ	
1	Vector	$\mathbf{v} = v_i \mathbf{e}_i$	(symbolic)
		v_i	(components, coordinates)
2	Dyad	$\mathbf{A} = A_{ij} \mathbf{e}_i \mathbf{e}_j$	(symbolic)
		A_{ij}	(components, coordinates)

Movements of bodies are described by the movement of their material points. The material points are identified by mapping them to points in $\mathbb{R}^3$ and a spatially fixed reference point **0**. Then, the position of a material point at every point in time t is determined by the *position vector* $\mathbf{x}(t)$. To distinguish the material points, one selects a *reference configuration* for a point in time t_0, at which the material point possesses the position vector $\mathbf{x}(t_0) = \mathbf{a}$. Thus, the position vector **a** is assigned to the material point as a marker. Normally, t_0 is related to an initial configuration, whose modifications have to be computed (often $t_0 = 0$). With the Cartesian coordinate system already introduced, one has the representations $\mathbf{x} = x_i \mathbf{e}_i$ and $\mathbf{a} = a_i \mathbf{e}_i$, and for the motion of the material point with the marker **a** one obtains the relations (see also Fig. 2.2):

$$x_i = x_i(\mathbf{a}, t) \quad \text{pathline of } \mathbf{a},$$
$$a_i = a_i(\mathbf{x}, t) \quad \text{material point } \mathbf{a} \text{ at time } t \text{ at position } \mathbf{x}.$$

x_i are denoted as *spatial coordinates* (or *local coordinates*) and a_i as *material* or *substantial coordinates*. If the assignment

$$x_i(a_j, t) \Leftrightarrow a_i(x_j, t)$$

is reversably unique, it defines a *configuration* of the body. This is exactly the case if the *Jacobi determinant* J of the mapping does not vanish, i.e.,

$$J = \det\left(\frac{\partial x_i}{\partial a_j}\right) \neq 0,$$

where the determinant $\det(\mathbf{A})$ of a dyad $\mathbf{A}$ is defined by

$$\det(\mathbf{A}) = \epsilon_{ijk} A_{i1} A_{j2} A_{k3}$$

with the *Levi-Civita symbol* (or *permutation symbol*)

$$\epsilon_{ijk} = \begin{cases} 1 & \text{for} \quad (i,j,k) = (1,2,3), (2,3,1), (3,1,2), \\ -1 & \text{for} \quad (i,j,k) = (1,3,2), (3,2,1), (2,1,3), \\ 0 & \text{for} \quad i = j \text{ or } i = k \text{ or } j = k. \end{cases}$$

The sequence of configurations $\mathbf{x} = \mathbf{x}(\mathbf{a}, t)$, with the time t as parameter, is called *deformation* (or *movement*) of the body.

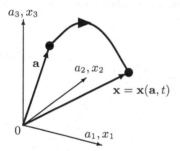

Fig. 2.2. Pathline of a material point $\mathbf{a}$ in a Cartesian coordinate system

For the description of the properties of material points, which usually vary with their movement (i.e., with the time), one distinguishes between the *Lagrangian* and the *Eulerian descriptions*. These can be characterized as follows:

- *Lagrangian description*: Formulation of the properties as functions of $\mathbf{a}$ and t. An observer is linked with the material point and measures the change in its properties.
- *Eulerian description*: Formulation of the properties as functions of $\mathbf{x}$ and t. An observer is located at position $\mathbf{x}$ and measures the changes there, which occur due to the fact that at different times t different material points $\mathbf{a}$ are at position $\mathbf{x}$.

The Lagrangian description is also called *material, substantial,* or *reference-based description*, whereas the Eulerian one is known as *spatial* or *local description*.

In solid mechanics mainly the Langrangian description is employed since usually a deformed state has to be determined from a known reference configuration, which naturally can be done by tracking the corresponding material points. In fluid mechanics mainly the Eulerian description is employed since usually the physical properties (e.g., pressure, velocity, etc.) at a specific location of the problem domain are of interest.

According to the two different descriptions one defines two different time derivatives: the *local time derivative*

$$\frac{\partial \phi}{\partial t} = \left.\frac{\partial \phi(\mathbf{x}, t)}{\partial t}\right|_{\mathbf{x}\ \text{fixed}},$$

which corresponds to the temporal variation of ϕ, which an observer measures at a fixed position $\mathbf{x}$, and the *material time derivative*

$$\frac{D\phi}{Dt} = \left.\frac{\partial \phi(\mathbf{a}, t)}{\partial t}\right|_{\mathbf{a}\ \text{fixed}},$$

which corresponds to the temporal variation of ϕ, which an observer linked to the material point $\mathbf{a}$ measures. In the literature, the material time derivative often is also denoted as $\dot{\phi}$. Between the two time derivatives there exists the following relationship:

$$\underbrace{\frac{D\phi}{Dt}}_{\text{material}} = \underbrace{\left.\frac{\partial \phi}{\partial t}\right|_{\mathbf{x}\ \text{fixed}}}_{\text{local}} + \underbrace{v_i \left.\frac{\partial \phi}{\partial x_i}\right|_{\mathbf{a}\ \text{fixed}}}_{\text{convective}}, \tag{2.1}$$

where

$$v_i = \frac{Dx_i}{Dt}$$

are the (Cartesian) coordinates of the *velocity vector* $\mathbf{v}$.

In solid mechanics, one usually works with displacements instead of deformations. The displacement $\mathbf{u} = u_i \mathbf{e}_i$ (in Lagrangian description) is defined by

$$u_i(\mathbf{a}, t) = x_i(\mathbf{a}, t) - a_i. \tag{2.2}$$

Using the displacements, strain tensors can be introduced as a measure for the deformation (strain) of a body. Strain tensors quantify the deviation of a deformation of a deformable body from that of a rigid body. There are various ways of defining such strain tensors. The most usual one is the *Green-Lagrange strain tensor* $\mathbf{G}$ with the coordinates (in Lagrangian description):

$$G_{ij} = \frac{1}{2}\left(\frac{\partial u_i}{\partial a_j} + \frac{\partial u_j}{\partial a_i} + \frac{\partial u_k}{\partial a_i}\frac{\partial u_k}{\partial a_j}\right).$$

This definition of **G** is the starting point for a frequently employed *geometrical linearization* of the kinematic equations, which is valid in the case of "small" displacements (details can be found, e.g., in [19]), i.e.,

$$\left| \frac{\partial u_i}{\partial a_j} \right| = \sqrt{\frac{\partial u_i}{\partial a_j} \frac{\partial u_j}{\partial a_i}} \ll 1 . \tag{2.3}$$

In this case the non-linear part of **G** is neglected, leading to the linearized strain tensor called *Green-Cauchy* (or also *linear* or *infinitesimal*) *strain tensor*:

$$\varepsilon_{ij} = \frac{1}{2} \left(\frac{\partial u_i}{\partial a_j} + \frac{\partial u_j}{\partial a_i} \right) . \tag{2.4}$$

In a geometrically linear theory there is no need to distinguish between the Lagrangian and Eulerian description. Due to the assumption (2.3) one has

$$\frac{\partial u_i}{\partial a_j} = \frac{\partial x_i}{\partial a_j} - \delta_{ij} \approx 0 ,$$

where

$$\delta_{ij} = \begin{cases} 1 & \text{for} \quad i = j , \\ 0 & \text{for} \quad i \neq j \end{cases}$$

denotes the *Kronecker symbol*. Thus, one has

$$\frac{\partial x_i}{\partial a_j} \approx \delta_{ij} \quad \text{or} \quad \frac{\partial}{\partial a_i} \approx \frac{\partial}{\partial x_i} ,$$

which means that the derivatives with respect to **a** and **x** can be interpreted to be identical.

2.2 Basic Conservation Equations

The mathematical models, on which numerical simulation methods for most engineering applications are based, are derived from the fundamental conservation laws of continuum mechanics for mass, momentum, moment of momentum, and energy. Together with problem specific material laws and suitable initial and boundary conditions, these give the basic (differential or integral) equations, which can be solved numerically. In the following we briefly describe the conservation laws, where we also discuss different formulations, as they constitute the starting point for the application of the different discretization techniques. The material theory will not be addressed explicitly, but in Sects. 2.3, 2.4, and 2.5 we will provide examples of a couple of material laws as they are frequently employed in engineering applications. For a detailed description of the continuum mechanical basics of the formulations we refer to the corresponding literature (e.g., [19, 23]).

Continuum mechanical conservation quantities of a body, let them be denoted generally by $\psi = \psi(t)$, can be defined as (spatial) integrals of a field quantity $\phi = \phi(\mathbf{x}, t)$ over the (temporally varying) volume $V = V(t)$ that the body occupies in its actual configuration at time t:

$$\psi(t) = \int_{V(t)} \phi(\mathbf{x}, t)\, dV\,.$$

Here ψ can depend on the time either via the integrand ϕ or via the integration range V. Therefore, the following relation for the temporal change of material volume integrals over a temporally varying spatial integration domain is important for the derivation of the balance equations (see, e.g., [23]):

$$\frac{D}{Dt} \int_{V(t)} \phi(\mathbf{x}, t)\, dV = \int_{V(t)} \left[\frac{D\phi(\mathbf{x}, t)}{Dt} + \phi(\mathbf{x}, t)\frac{\partial v_i(\mathbf{x}, t)}{\partial x_i} \right] dV\,. \tag{2.5}$$

Due to the relation between the material and local time derivatives given by (2.1), one has further:

$$\int_V \left(\frac{D\phi}{Dt} + \phi\frac{\partial v_i}{\partial x_i} \right) dV = \int_V \left[\frac{\partial \phi}{\partial t} + \frac{\partial (\phi v_i)}{\partial x_i} \right] dV\,. \tag{2.6}$$

For a more compact notation we have skipped the corresponding dependence of the quantities from space and time, and we will frequently also do so in the following. Equation (2.5) (sometimes also (2.6)) is called *Reynolds transport theorem*.

2.2.1 Mass Conservation

The *mass m* of an arbitrary volume V is defined by

$$m(t) = \int_V \rho(\mathbf{x}, t)\, dV$$

with the *density ρ*. The mass conservation theorem states that if there are no mass sources or sinks, the total mass of a body remains constant for all times:

$$\frac{D}{Dt} \int_V \rho\, dV = 0\,. \tag{2.7}$$

For the mass before and after a deformation we have:

$$\int_{V_0} \rho_0(\mathbf{a}, t)\, dV_0 = \int_V \rho(\mathbf{x}, t)\, dV\,,$$

where $\rho_0 = \rho(t_0)$ and $V_0 = V(t_0)$ denote the density and the volume, respectively, before the deformation (i.e., in the reference configuration). Thus, during a deformation the volume and the density can change, but not the mass. The following relations are valid:

$$\frac{dV}{dV_0} = \frac{\rho_0}{\rho} = \det\left(\frac{\partial x_i}{\partial a_j}\right).$$

Using the relations (2.5) and (2.6) and applying the Gauß integral theorem (e.g., [19]) one obtains from (2.7):

$$\int_V \frac{\partial \rho}{\partial t}\, dV + \int_S \rho v_i n_i\, dS = 0,$$

where $\mathbf{n} = n_i \mathbf{e}_i$ is the outward unit normal vector at the closed surface S of the volume V (see Fig. 2.3). This representation of the mass conservation allows the physical interpretation that the temporal change of the mass contained in the volume V equals the inflowing and outflowing mass through the surface. In differential (conservative) form the mass balance reads:

$$\frac{\partial \rho}{\partial t} + \frac{\partial(\rho v_i)}{\partial x_i} = 0. \tag{2.8}$$

This equation is also called *continuity equation*.

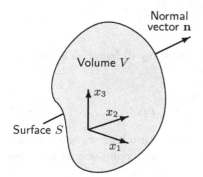

Fig. 2.3. Notations for application of Gauß integral theorem

For an *incompressible* material one has:

$$\det\left(\frac{\partial x_i}{\partial a_j}\right) = 1 \quad \text{and} \quad \frac{D\rho}{Dt} = \frac{\partial v_i}{\partial x_i} = 0,$$

i.e., the velocity field in this case is divergence-free.

2.2.2 Momentum Conservation

The *momentum vector* $\mathbf{p} = p_i \mathbf{e}_i$ of a body is defined by

$$p_i(t) = \int_V \rho(\mathbf{x}, t) v_i(\mathbf{x}, t) \, dV \,.$$

The principle of balance of momentum states that the temporal change of the momentum of a body equals the sum of all body and surface forces acting on the body. This can be expressed as follows:

$$\underbrace{\frac{D}{Dt} \int_V \rho v_i \, dV}_{\text{change of momentum}} = \underbrace{\int_S T_{ij} n_j \, dS}_{\text{surface forces}} + \underbrace{\int_V \rho f_i \, dV}_{\text{volume forces}} \,, \qquad (2.9)$$

where $\mathbf{f} = f_i \mathbf{e}_i$ are the volume forces per mass unit. T_{ij} are the components of the *Cauchy stress tensor* $\mathbf{T}$, which describes the state of stress of the body in each point (a measure for the internal force in the body). The components with $i = j$ are called *normal stresses* and the components with $i \neq j$ are called *shear stresses*. (In the framework of structural mechanics $\mathbf{T}$ is usually denoted as $\boldsymbol{\sigma}$.)

Applying the Gauß integral theorem to the surface integral in (2.9) one gets:

$$\frac{D}{Dt} \int_V \rho v_i \, dV = \int_V \frac{\partial T_{ij}}{\partial x_j} \, dV + \int_V \rho f_i \, dV \,.$$

Using the relations (2.5) and (2.6) yields the following differential form of the momentum balance in Eulerian description:

$$\frac{\partial(\rho v_i)}{\partial t} + \frac{\partial(\rho v_i v_j)}{\partial x_j} = \frac{\partial T_{ij}}{\partial x_j} + \rho f_i \,. \qquad (2.10)$$

For the Lagrangian representation of the momentum balance one normally uses the *second Piola-Kirchhoff stress tensor* $\mathbf{P}$, whose components are given by:

$$P_{ij} = \frac{\rho_0}{\rho} \frac{\partial a_i}{\partial x_k} T_{kl} \frac{\partial a_j}{\partial x_l} \,.$$

With this, the Lagrangian formulation of the momentum balance reads in differential form:

$$\rho_0 \frac{D^2 x_i}{Dt^2} = \frac{\partial}{\partial a_j} \left(P_{jk} \frac{\partial x_i}{\partial a_k} \right) + \rho_0 f_i \,. \qquad (2.11)$$

2.2.3 Moment of Momentum Conservation

The *moment of momentum vector* $\mathbf{d} = d_i \mathbf{e}_i$ of a body is defined by

$$\mathbf{d}(t) = \int_V \mathbf{x} \times \rho(\mathbf{x}, t) \mathbf{v}(\mathbf{x}, t) \, dV \,,$$

where "$\times$" denotes the usual vector product, which for two vectors $\mathbf{a} = a_i \mathbf{e}_i$ and $\mathbf{b} = b_j \mathbf{e}_j$ is defined by $\mathbf{a} \times \mathbf{b} = a_i b_j \epsilon_{ijk} \mathbf{e}_k$. The principle of balance of moment of momentum states that the temporal change of the total moment of momentum of a body equals the toal moment of all body and surface forces acting on the body. This can be expressed as follows:

$$\underbrace{\frac{D}{Dt} \int_V (\mathbf{x} \times \rho \mathbf{v}) \, dV}_{\substack{\text{change of moment} \\ \text{of momentum}}} = \underbrace{\int_V (\mathbf{x} \times \rho \mathbf{f}) \, dV}_{\substack{\text{moment of} \\ \text{volume forces}}} + \underbrace{\int_S (\mathbf{x} \times \mathbf{T}\mathbf{n}) \, dS.}_{\substack{\text{moment of} \\ \text{surface forces}}} \qquad (2.12)$$

Applying the Gauß integral theorem and using the mass and momentum conservation as well as the relations (2.5) and (2.6), the balance of moment of momentum can be put into the following simple form (Exercise 2.2):

$$T_{ij} = T_{ji} \,,$$

i.e., the conservation of the moment of momentum is expressed by the symmetry of the Cauchy stress tensor.

2.2.4 Energy Conservation

The *total energy* W of a body is defined by

$$W(t) = \underbrace{\int_V \rho e \, dV}_{\substack{\text{internal} \\ \text{energy}}} + \underbrace{\frac{1}{2} \int_V \rho v_i v_i \, dV}_{\substack{\text{kinetic} \\ \text{energy}}}$$

with the *specific internal energy* e. The *power of external forces* P_a (surface and volume forces) is given by

$$P_a(t) = \underbrace{\int_S T_{ij} v_j n_i \, dS}_{\substack{\text{power of} \\ \text{surface forces}}} + \underbrace{\int_V \rho f_i v_i \, dV}_{\substack{\text{power of} \\ \text{volume forces}}}$$

and for the *power of heat supply* Q one has

$$Q(t) \;=\; \underbrace{\int_V \rho q \, dV}_{\substack{\text{power of} \\ \text{heat sources}}} \;-\; \underbrace{\int_S h_i n_i \, dS}_{\substack{\text{power of} \\ \text{surface supply}}} \;,$$

where q denotes (scalar) heat sources and $\mathbf{h} = h_i \mathbf{e}_i$ denotes the heat flux vector per unit area. The principle of energy conservation states that the temporal change of the total energy W equals the total external energy supply $P_a + Q$:

$$\frac{DW}{Dt} = P_a + Q .$$

This theorem is also known as the *first law of thermodynamics*.

Using the above definitions the energy conservation law can be written as follows:

$$\frac{D}{Dt} \int_V \rho(e + \tfrac{1}{2} v_i v_i) \, dV = \int_S (T_{ij} v_j - h_i) n_i \, dS + \int_V \rho(f_i v_i + q) \, dV . \qquad (2.13)$$

After some transformations (using (2.5) and (2.6)), application of the Gauß integral theorem, and using the momentum conservation law (2.10), one obtains for the energy balance the following differential form (Exercise 2.3):

$$\frac{\partial(\rho e)}{\partial t} + \frac{\partial(\rho v_i e)}{\partial x_i} = T_{ij} \frac{\partial v_j}{\partial x_i} - \frac{\partial h_i}{\partial x_i} + \rho q . \qquad (2.14)$$

2.2.5 Material Laws

The unknown physical quantities that appear in the balance equations of the previous sections in Table 2.2 are presented alongside with the number of equations available for their computation. Since there are more unknowns than equations, it is necessary to involve additional problem specific equations, which are called *constitutive* or *material laws*, that suitably relate the unknowns to each other. These can be algebraic relations, differential equations, or integral equations. As already indicated, we will not go into the details of material theory, but in the next sections we will give examples of continuum mechanics problem formulations as they result from special material laws which are of high relevance in engineering applications.

2.3 Scalar Problems

A number of practically relevant engineering tasks can be described by a single (partial) differential equation. In the following some representative examples that frequently appear in practice are given.

Table 2.2. Unknown physical quantities and conservation laws

Unknown	No.	Equation	No.
Density ρ	1	Mass conservation	1
Velocity v_i	3	Momentum conservation	3
Stress tensor T_{ij}	9	Moment of momentum conservation	3
Internal energy e	1	Energy conservation	1
Heat flux h_i	3		
Sum 17		Sum 8	

2.3.1 Simple Field Problems

Some simple continuum mechanical problems can be described by a differential equation of the form

$$-\frac{\partial}{\partial x_i}\left(a\frac{\partial \phi}{\partial x_i}\right) = g, \tag{2.15}$$

which has to be valid in a problem domain Ω. An unknown scalar function $\phi = \phi(\mathbf{x})$ is searched for. The coefficient function $a = a(\mathbf{x})$ and the right hand side $g = g(\mathbf{x})$ are prescribed. In the case $a = 1$, (2.15) is called *Poisson equation*. If, additionally, $g = 0$, one speaks of a *Laplace equation*.

In order to fully define a problem governed by (2.15), boundary conditions for ϕ have to be prescribed at the whole boundary Γ of the problem domain Ω. Here, the following three types of conditions are the most important ones:

– Dirichlet condition: $\phi = \phi_{\mathrm{b}}$,

– Neumann condition: $a\dfrac{\partial \phi}{\partial x_i}n_i = b_{\mathrm{b}}$,

– Cauchy condition: $c_{\mathrm{b}}\phi + a\dfrac{\partial \phi}{\partial x_i}n_i = b_{\mathrm{b}}$.

ϕ_{b}, b_{b}, and c_{b} are prescribed functions on the boundary Γ and n_i are the components of the outward unit normal vector to Γ. The different boundary condition types can occur for one problem at different parts of the boundary (mixed boundary value problems).

The problems described by (2.15) do not involve time dependence. Thus, one speaks of *stationary* or *steady state* field problems. In the time-dependent (unsteady) case, in addition to the dependence on the spatial coordinate $\mathbf{x}$, all quantities may also depend on the time t. The corresponding differential equation for the description of unsteady field problems reads:

$$\frac{\partial \phi}{\partial t} - \frac{\partial}{\partial x_i}\left(a\frac{\partial \phi}{\partial x_i}\right) = g \tag{2.16}$$

for the unknown scalar function $\phi = \phi(\mathbf{x}, t)$. For unsteady problems, in addition to the boundary conditions (that in this case also may depend on the time), an initial condition $\phi(\mathbf{x}, t_0) = \phi_0(\mathbf{x})$ has to be prescribed to complete the problem definition.

Examples of physical problems that are described by equations of the types (2.15) or (2.16) are:

- temperature for heat conduction problems,
- electric field strength in electro-static fields,
- pressure for flows in porous media,
- stress function for torsion problems,
- velocity potential for irrotational flows,
- cord line for sagging cables,
- deflection of elastic strings or membranes.

In the following we give two examples for such applications, where we only consider the steady problem. The corresponding unsteady problem formulations can be obtained analogously as the transition from (2.15) to (2.16).

Interpreting ϕ as the deflection u of a homogeneous *elastic membrane*, (2.15) describes its deformation under an external load ($i = 1, 2$):

$$-\frac{\partial}{\partial x_i} \left(\tau \frac{\partial u}{\partial x_i} \right) = f \tag{2.17}$$

with the *stiffness* τ and the *force density* f (see Fig. 2.4). Under certain assumptions which will not be detailed here, (2.17) can be derived from the momentum balance (2.10).

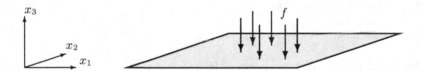

Fig. 2.4. Deformation of an elastic membrane under external load

As boundary conditions Dirichlet or Neumann conditions are possible, which in this context have the following meaning:

– Prescribed deflection (Dirichlet condition): $u = u_{\mathrm{b}}$,

– Prescribed stress (Neumann condition): $\tau \dfrac{\partial u}{\partial x_i} n_i = t_{\mathrm{b}}$.

As a second example we consider *incompressible potential flows*. For an irrotational flow, i.e., if the flow velocity fulfills the relations

$$\frac{\partial v_j}{\partial x_i} \epsilon_{ijk} = 0,$$

there exists a velocity potential ψ, which is defined by

$$v_i = \frac{\partial \psi}{\partial x_i} \,. \tag{2.18}$$

Inserting the relation (2.18) into the mass conservation equation (2.8), under the additional assumption of an incompressible flow (i.e., $D\rho/Dt = 0$), the following equation for the determination of ψ results:

$$\frac{\partial^2 \psi}{\partial x_i^2} = 0 \,. \tag{2.19}$$

This equation corresponds to (2.15) with $f = 0$ and $a = 1$.

The assumptions of a potential flow are frequently employed for the investigation of the flow around bodies, e.g., for aerodynamical investigations of vehicles or airplanes. In the case of fluids with small viscosity (e.g., air) flowing at relatively high velocities, these assumptions are justifiable. In regions where the flow accelerates (outside of boundary layers), one obtains a comparably good approximation for the real flow situation. As an example for a potential flow, Fig. 2.5 shows the *streamlines* (i.e., lines with $\psi = const.$) for the flow around a circular cylinder.

As boundary conditions at the body one has the following Neumann condition (kinematic boundary condition):

$$\frac{\partial \psi}{\partial x_i} n_i = v_{\mathrm{b}i} n_i \,,$$

where $\mathbf{v}_\mathrm{b} = v_{\mathrm{b}i} \mathbf{e}_i$ is the velocity with which the body moves. Having computed ψ in this way, one obtains v_i from (2.18). The pressure p, which is uniquely determined only up to an additive constant C, can then be determined from the *Bernoulli equation* (see [23])

$$p = C - \rho \frac{\partial \psi}{\partial t} - \frac{1}{2} \rho v_i v_i \,.$$

2.3.2 Heat Transfer Problems

A very important class of problems for engineering applications are heat transfer problems in solids or fluids. Here, usually one is interested in temperature

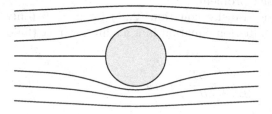

Fig. 2.5. Streamlines for potential flow around circular cylinder

distributions, which result due to diffusive, convective, and/or radiative heat transport processes under certain boundary conditions. In simple cases such problems can be described by a single scalar transport equation for the temperature T (diffusion in solids, diffusion and convection in fluids).

Let us consider first the more general case of the heat transfer in a fluid. The heat conduction in solids then results from this as a special case. We will not address the details of the derivation of the corresponding differential equations, which can be obtained under certain assumptions from the energy conservation equation (2.14).

We consider a flow with the (known) velocity $\mathbf{v} = v_i \mathbf{e}_i$. As constitutive relation for the heat flux vector we employ Fourier's law (for isotropic materials)

$$h_i = -\kappa \frac{\partial T}{\partial x_i} \tag{2.20}$$

with the *heat conductivity* κ. This assumption is valid for nearly all relevant applications. Assuming in addition that the specific heat capacity of the fluid is constant, and that the work done by pressure and friction forces can be neglected, the following convection-diffusion equation for the temperature T can be derived from the energy balance (2.14) (see also Sect. 2.5.1):

$$\frac{\partial(\rho c_p T)}{\partial t} + \frac{\partial}{\partial x_i}\left(\rho c_p v_i T - \kappa \frac{\partial T}{\partial x_i}\right) = \rho q \tag{2.21}$$

with possibly present heat sources or sinks q and the specific heat capacity c_p (at constant pressure).

The most frequently occuring boundary conditions are again of Dirichlet, Neumann, or Cauchy type, which in this context have the following meaning:

- Prescribed temperature: $T = T_b$,

- Prescribed heat flux: $\kappa \dfrac{\partial T}{\partial x_i} n_i = h_b$,

- Heat flux proportional to heat transport: $\kappa \dfrac{\partial T}{\partial x_i} n_i = \tilde{\alpha}(T_b - T)$.

Here, T_b and h_b are prescribed values at the problem domain boundary Γ for the temperature and the heat flux in normal direction, respectively, and $\tilde{\alpha}$ is the *heat transfer coefficient*. In Fig. 2.6 the configuration of a plate heat exchanger is given together with the corresponding boundary conditions as a typical example for a heat transfer problem.

As a special case of the heat transfer equation (2.21) for $v_i = 0$ (only diffusion) we obtain the heat conduction equation in a medium at rest (fluid or solid):

$$\frac{\partial(\rho c_p T)}{\partial t} - \frac{\partial}{\partial x_i}\left(\kappa \frac{\partial T}{\partial x_i}\right) = \rho q . \tag{2.22}$$

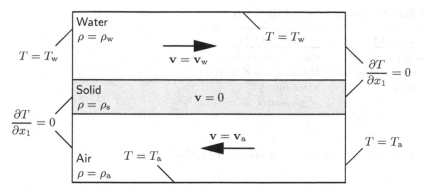

Fig. 2.6. Heat transfer problem in a plate heat exchanger with corresponding boundary conditions

The corresponding equations for steady heat transfer are obtained from (2.21) and (2.22) by simply dropping the term with the time derivative.

Besides conduction and convection, *thermal radiation* is another heat transfer mechanism playing an important role in technical applications, in particular at high absolute temperature levels (e.g., in furnaces, combustion chambers, ...). Usually highly non-linear effects are related to radiation phenomena, which have to be considered by additional terms in the differential equations and/or boundary conditions. For this topic we refer to [22].

An equation completely analogous to (2.21) can be derived for the species transport in a fluid. Instead of the temperature in this case one has the species *concentration c* as the unknown variable. The heat conductivity corresponds to the diffusion coefficient D and the heat source q has to be replaced by a mass source R. The material law corresponding to Fourier's law (2.20)

$$j_i = -D\frac{\partial c}{\partial x_i}$$

for the mass flux $\mathbf{j} = j_i \mathbf{e}_i$ is known as *Fick's law*. With this, the corresponding equation for the species transport reads:

$$\frac{\partial(\rho c)}{\partial t} + \frac{\partial}{\partial x_i}\left(\rho v_i c - D\frac{\partial c}{\partial x_i}\right) = R. \tag{2.23}$$

The types of boundary conditions and their meaning for species transport problems are fully analogous to that for the heat transport. In Table 2.3 the analogy between heat and species transport is summarized.

An equation of the type (2.21) or (2.23) will be used in the following frequently for different purposes as an exemplary model equation. For this we employ the general form

$$\frac{\partial(\rho\phi)}{\partial t} + \frac{\partial}{\partial x_i}\left(\rho v_i \phi - \alpha\frac{\partial\phi}{\partial x_i}\right) = f, \tag{2.24}$$

Table 2.3. Analogy of heat and species transport

Heat transport	Species transport
Temperature T	Concentration c
Heat conductivity κ	Diffusion coefficient D
Heat flux $\mathbf{h}$	Mass flux $\mathbf{j}$
Heat source q	Mass source R

which is called *general scalar transport equation*.

2.4 Structural Mechanics Problems

In structural mechanics problems, in general, the task is to determine deformations of solid bodies, which arise due to the action of various kinds of forces.
From this, for instance, stresses in the body can be determined, which are of
great importance for many applications. (It is also possible to directly formulate equations for the stresses, but we will not consider this here.) For the
different material properties there exist a large number of material laws, which
together with the balance equations (see Sect. 2.2) lead to diversified complex
equation systems for the determination of deformations (or displacements).

In principle, for structural mechanics problems one distinguishes between
linear and non-linear models, where the non-linearity can be of geometrical
and/or physical nature. Geometrically linear problems are characterized by
the linear strain-displacements relation (see Sect. 2.1)

$$\varepsilon_{ij} = \frac{1}{2} \left(\frac{\partial u_i}{\partial x_j} + \frac{\partial u_j}{\partial x_i} \right) , \qquad (2.25)$$

whereas physically linear problems are based on a material law involving a
linear relation between strains and stresses. In Table 2.4 the different model
classes are summarized.

We restrict ourselves to the formulation of the equations for two simpler
linear model classes, i.e., the linear elasticity theory and the linear thermo-

Table 2.4. Model classes for structural mechanics problems

	Geometrically linear	Geometrically non-linear
Physically linear	small displacements small strains	large displacements small strains
Physically non-linear	small displacements large strains	large displacements large strains

elasticity, which can be used for many typical engineering applications. Furthermore, we briefly address hyperelasticity as an example of a non-linear model class. For other classes, i.e., elasto-plastic, visco-elastic, or visco-plastic materials, we refer to the corresponding literature (e.g., [14]).

2.4.1 Linear Elasticity

The theory of linear elasticity is a geometrically and physically linear one. As already outlined in Sect. 2.1, there is no need to distinguish between Eulerian and Lagrangian description for a geometrically linear theory. In the following the spatial coordinates are denoted by x_i.

The equations of the linear elasticity theory are obtained from the linearized strain-displacement relations (2.25), the momentum conservation law (2.10) formulated for the displacements (in the framework of structural mechanics this often also is denoted as *equation of motion*)

$$\rho\frac{D^2 u_i}{Dt^2} = \frac{\partial T_{ij}}{\partial x_j} + \rho f_i \,, \tag{2.26}$$

and the assumption of a linear elastic material behavior, which is characterized by the constitutive equation

$$T_{ij} = \lambda \varepsilon_{kk} \delta_{ij} + 2\mu \varepsilon_{ij} \,. \tag{2.27}$$

Equation (2.27) is known as *Hooke's law*. λ and μ are the *Lamé constants*, which depend on the corresponding material (μ is also known as *bulk modulus*). The *elasticity modulus* (or *Young modulus*) E and the *Poisson ratio* ν are often employed instead of the Lamé constants. The relations between these quantities are:

$$\lambda = \frac{E\nu}{(1+\nu)(1-2\nu)} \quad \text{and} \quad \mu = \frac{E}{2(1+\nu)} \,. \tag{2.28}$$

Hooke's material law (2.27) is applicable for a large number of applications for different materials (e.g., steel, glass, stone, wood,...). Necessary prerequisites are that the stresses are not "too big", and that the deformation happens within the elastic range of the material (see Fig. 2.7).

The material law for the stress tensor frequently is also given in the following notation:

$$\begin{bmatrix} T_{11} \\ T_{22} \\ T_{33} \\ T_{12} \\ T_{13} \\ T_{23} \end{bmatrix} = \underbrace{\frac{E}{(1+\nu)(1-2\nu)} \begin{bmatrix} 1-\nu & \nu & \nu & 0 & 0 & 0 \\ \nu & 1-\nu & \nu & 0 & 0 & 0 \\ \nu & \nu & 1-\nu & 0 & 0 & 0 \\ 0 & 0 & 0 & 1-2\nu & 0 & 0 \\ 0 & 0 & 0 & 0 & 1-2\nu & 0 \\ 0 & 0 & 0 & 0 & 0 & 1-2\nu \end{bmatrix}}_{C} \begin{bmatrix} \varepsilon_{11} \\ \varepsilon_{22} \\ \varepsilon_{33} \\ \varepsilon_{12} \\ \varepsilon_{13} \\ \varepsilon_{23} \end{bmatrix} .$$

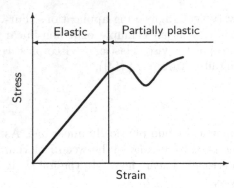

Fig. 2.7. Qualitative strain-stress relation of real materials with linear elastic range

Due to the principle of balance of moment of momentum, $\mathbf{T}$ has to be symmetric, such that only the given 6 components are necessary in order to fully describe $\mathbf{T}$. The matrix $\mathbf{C}$ is called *material matrix*. Putting the material law in the general form

$$T_{ij} = E_{ijkl}\varepsilon_{kl}\,,$$

the fourth order tensor $\mathbf{E}$ with the components E_{ijkl} is called the *elasticity tensor* (of course, the entries in the matrix $\mathbf{C}$ and the corresponding components of $\mathbf{E}$ match).

Finally, one obtains from (2.25), (2.26), and (2.27) by eliminating ε_{ij} and T_{ij} the following system of differential equations for the displacements u_i:

$$\rho\frac{D^2 u_i}{Dt^2} = (\lambda + \mu)\frac{\partial^2 u_j}{\partial x_i \partial x_j} + \mu\frac{\partial^2 u_i}{\partial x_j \partial x_j} + \rho f_i\,. \tag{2.29}$$

These equations are called (unsteady) *Navier-Cauchy equations of linear elasticity theory.* For steady problems correspondingly one has:

$$(\lambda + \mu)\frac{\partial^2 u_j}{\partial x_i \partial x_j} + \mu\frac{\partial^2 u_i}{\partial x_j \partial x_j} + \rho f_i = 0\,. \tag{2.30}$$

Possible boundary conditions for linear elasticity problems are:

– Prescribed displacements: $u_i = u_{\mathrm{b}i}$ on Γ_1,

– Prescribed stresses: $T_{ij}n_j = t_{\mathrm{b}i}$ on Γ_2.

The boundary parts Γ_1 and Γ_2 should be disjoint and should cover the full problem domain boundary Γ, i.e., $\Gamma_1 \cap \Gamma_2 = \emptyset$ and $\Gamma_1 \cup \Gamma_2 = \Gamma$.

Besides the formulation given by (2.29) or (2.30) as a system of partial differential equations, there are other equivalent formulations for linear elasticity problems. We will give here two other ones that are important in connection with different numerical methods. We restrict ourselves to the steady case.

(Scalar) multiplication of the differential equation system (2.30) with a test function $\varphi = \varphi_i \mathbf{e}_i$, which vanishes at the boundary part Γ_1, and integration over the problem domain Ω yields:

$$\int_\Omega \left[(\lambda+\mu) \frac{\partial^2 u_j}{\partial x_i \partial x_j} + \mu \frac{\partial^2 u_i}{\partial x_j \partial x_j} \right] \varphi_i \, d\Omega + \int_\Omega \rho f_i \varphi_i \, d\Omega = 0. \tag{2.31}$$

By integration by parts of the first integral in (2.31) one gets:

$$\int_\Omega \left[(\lambda+\mu) \frac{\partial u_j}{\partial x_i} + \mu \frac{\partial u_i}{\partial x_j} \right] \frac{\partial \varphi_i}{\partial x_j} \, d\Omega = \int_\Gamma T_{ij} n_j \varphi_i \, d\Gamma + \int_\Omega \rho f_i \varphi_i \, d\Omega. \tag{2.32}$$

Since $\varphi_i = 0$ on Γ_1 in (2.32) the corresponding part in the surface integral vanishes and in the remaining part over Γ_2 for $T_{ij} n_j$ the prescribed stress t_{bi} can be inserted. Thus, one obtains:

$$\int_\Omega \left[(\lambda+\mu) \frac{\partial u_j}{\partial x_i} + \mu \frac{\partial u_i}{\partial x_j} \right] \frac{\partial \varphi_i}{\partial x_j} \, d\Omega = \int_{\Gamma_2} t_{bi} \varphi_i \, d\Gamma + \int_\Omega \rho f_i \varphi_i \, d\Omega. \tag{2.33}$$

The requirement that the relation (2.33) is fulfilled for a suitable class of test functions (let this be denoted by $\mathcal{H}$) results in a formulation of the linear elasticity problem as a variational problem:

Find $\mathbf{u} = u_i \mathbf{e}_i$ with $u_i = u_{bi}$ on Γ_1, such that

$$\int_\Omega \left[(\lambda+\mu) \frac{\partial u_j}{\partial x_i} + \mu \frac{\partial u_i}{\partial x_j} \right] \frac{\partial \varphi_i}{\partial x_j} \, d\Omega = \int_\Omega \rho f_i \varphi_i \, d\Omega + \int_{\Gamma_2} t_{bi} \varphi_i \, d\Gamma \tag{2.34}$$

for all $\varphi = \varphi_i \mathbf{e}_i$ in $\mathcal{H}$.

The question remains of which functions should be contained in the function space $\mathcal{H}$. Since this is not essential for the following, we will not provide an exact definition (this can be found, for instance, in [3]). It is important that the test functions φ vanish on the boundary part Γ_1. Further requirements mainly concern the integrability and differentiability properties of the functions (all appearing terms must be defined).

The formulation (2.34) is called *weak formulation*, where the term "weak" relates to the differentiability of the functions involved (there are only first derivatives, in contrast to the second derivatives in the differential formulation (2.30)). Frequently, in the engineering literature, the formulation (2.34) is also called *principle of virtual work* (or *principle of virtual displacements*). The test functions in this context are called *virtual displacements*.

Another alternative formulation of the linear elasticity problem is obtained starting from the expression for the *potential energy* $P = P(\mathbf{u})$ of the body dependent on the displacements:

$$P(\mathbf{u}) = \frac{1}{2} \int_{\Omega} \left[(\lambda + \mu) \frac{\partial u_j}{\partial x_i} + \mu \frac{\partial u_i}{\partial x_j} \right] \frac{\partial u_i}{\partial x_j} \, d\Omega - \int_{\Omega} \rho f_i u_i \, d\Omega - \int_{\Gamma_2} t_{bi} u_i \, d\Gamma . \quad (2.35)$$

One gets the solution by looking among all possible displacements, which fulfill the boundary condition $u_i = u_{bi}$ on Γ_1, for the one at which the potential energy takes its minimum. The relationship of this formulation, called the *principle of minimum of potential energy*, with the weak formulation (2.34) becomes apparent if one considers the derivative of P with respect to $\mathbf{u}$ (in a suitable sense). The minimum of the potential energy is taken if the *first variation* of P, i.e., a derivative in a functional analytic sense, vanishes (analogous to the usual differential calculus), which corresponds to the validity of (2.33).

Contrary to the differential formulation, in the weak formulation (2.34) and the energy formulation (2.35) the stress boundary condition $T_{ij} n_j = t_{bi}$ on Γ_2 is not enforced explicitly, but is implicitly contained in the corresponding boundary integral over Γ_2. The solutions fulfill this boundary condition automatically, albeit only in a weak (integral) sense. With respect to the construction of a numerical method, this can be considered as an advantage since only (the more simple) displacement boundary conditions $u_i = u_{bi}$ on Γ_1 have to be considered explicitly. In this context, the stress boundary conditions are also called *natural boundary conditions*, whereas in the case of displacements boundary conditions one speaks about *essential* or *geometric boundary conditions*.

It should be emphasized that the different formulations basically all describe one and the same problem, but with different approaches. However, the proof that the formulations from a rigorous mathematical point of view in fact are equivalent (or rather which conditions have to be fulfilled for this) requires advanced functional analytic methods and is relatively difficult. Since this is not essential for the following, we will not go into detail on this matter (see, e.g., [3]).

So far, we have considered the general linear elasticity equations for three-dimensional problems. In practice, very often these can be simplified by suitable problem specific assumptions, in particular with respect to the spatial dimension. In the following we will consider some of these special cases, which often can be found in applications.

2.4.2 Bars and Beams

The simplest special case of a linear elasticity problem results for a *tensile bar*. We consider a bar with length L and cross-sectional area $A = A(x_1)$ as shown in Fig. 2.8.

The equations for the bar can be used for the problem description if the following requirements are fulfilled:

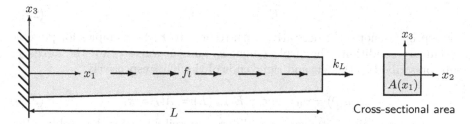

Fig. 2.8. Tensile bar under load in longitudinal direction

- forces only act in x_1-direction,
- the cross-section remains plane and moves only in x_1-direction.

Under these assumptions we have

$$u_2 = u_3 = 0$$

and the unknown displacement u_1 only depends from x_1:

$$u_1 = u_1(x_1).$$

In the strain tensor only the component

$$\varepsilon_{11} = \frac{\partial u_1}{\partial x_1}$$

is different from zero. Furthermore, there is only normal stress acting in x_1-direction, such that in the stress tensor only the component T_{11} is non-zero. The equation of motion for the bar reads

$$\frac{\partial(AT_{11})}{\partial x_1} + f_1 = 0, \tag{2.36}$$

where $f_1 = f_1(x_1)$ denotes the continuous longitudinal load of the bar in x_1-direction. If, for instance, the self-weight of the bar should be considered when the acceleration of gravity g acts in x_1-direction, we have $f_1 = \rho A g$. The derivation of the bar equation (2.36) can be carried out via the integral momentum balance (the cross-sectional area A shows up by carrying out the integration in x_2- and x_3-direction). Hooke's law becomes:

$$T_{11} = E\varepsilon_{11}. \tag{2.37}$$

In summary, one is faced with a one-dimensional problem only. To avoid redundant indices we write $u = u_1$ and $x = x_1$. Inserting the material law (2.37) in the equation of motion (2.36) finally yields the following (ordinary) differential equation for the unknown displacement u:

$$(EAu')' + f_1 = 0 . \tag{2.38}$$

The prime denotes the derivative with respect to x. As examples for possible boundary conditions, the displacement u_0 and the stress t_L (or the force k_L) are prescribed at the left and right ends of the bar, respectively:

$$u(0) = u_0 \quad \text{and} \quad EAu'(L) = At_L = k_L . \tag{2.39}$$

In order to illustrate again the different possibilities for the problem formulation using a simple example, we also give the weak and potential energy formulations for the tensile bar with the boundary conditions (2.39). The principle of minimum potential energy in this case reads:

$$P(u) = \frac{1}{2} \int_0^L EA(u')^2 \, dx - \int_0^L f_1 u \, dx - u(L)k_L \quad \rightarrow \quad \text{Minimum} ,$$

where the minimum is sought among all displacements for which $u(0) = u_0$. The "derivative" (first variation) of the potential energy with respect to u is given by

$$\lim_{\alpha \to 0} \frac{dP(u + \alpha \varphi)}{d\alpha} = \int_0^L EAu'\varphi' \, dx - \int_0^L f_1 \varphi \, dx - \varphi(L)k_L .$$

The principle of virtual work for the tensile bar thus reads:

$$\int_0^L EAu'\varphi' \, dx = \int_0^L f_1 \varphi \, dx + \varphi(L)k_L \tag{2.40}$$

for all virtual displacements φ with $\varphi(0) = 0$.

The relationship of this weak formulation with the differential formulation (2.38) with the boundary conditions (2.39) becomes apparent if one integrates the integral on the left hand side of (2.40) by parts:

$$\int_0^L (EAu')' \, \varphi \, dx + [EAu'\varphi]_0^L - \int_0^L f_1 \varphi \, dx - \varphi(L)k_L =$$

$$\int_0^L \left[-(EAu')' - f_1 \right] \varphi \, dx + [EAu'(L) - k_L] \varphi(L) = 0 .$$

The last equation obviously is fulfilled if u is a solution of the differential equation (2.38) that satisfies the boundary conditions (2.39). Therefore, the principles of virtual work and of minimum potential energy are fulfilled. The reverse conclusion is not that clear, since there can be displacements satisfying the principles of virtual work and minimum potential energy, but not the differential equation (2.38) (in the classical sense). For this, additional differentiability properties of the displacements are necessary, which, however, we will not elaborate here.

Another special case of linear elasticity theory, which can be described by a one-dimensional equation, is beam bending (see Fig. 2.9). We will focus here on the *shear-rigid beam* (or *Bernoulli beam*). This approximation is based on the assumption that during the bending along one main direction, plane cross-sections remain plane and normals to the neutral axis (x_1-axis in Fig. 2.9) remain normal to this axis also in the deformed state. Omitting the latter assumption one obtains the *shear-elastic beam* (or *Timoshenko beam*).

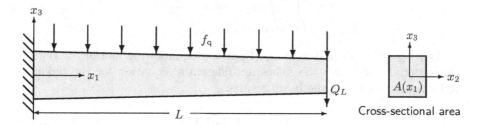

Fig. 2.9. Beam under vertical load

Under the assumptions for the shear-rigid beam, the displacement u_1 can be expressed by the inclination of the bending line u_3 (deflection parallel to the x_3-axis):

$$u_1 = -x_3 \frac{\partial u_3}{\partial x_1} \, .$$

In the strain tensor, just as for the tensile bar, only the component

$$\varepsilon_{11} = \frac{\partial u_1}{\partial x_1} = -x_3 \frac{\partial^2 u_3}{\partial x_1^2}$$

is different from zero. The equation of motion for the shear-rigid beam reads

$$\frac{\partial^2 T_{11}}{\partial x_1^2} + f_q = 0 \, , \tag{2.41}$$

where $f_q = f_q(x_1)$ denotes the continuous lateral load (uniform load) of the beam in x_3-direction. For instance, to take into account the self-weight of the beam one has again $f_q = \rho A g$, where the acceleration of gravity g now is acting in x_3-direction. $A = A(x_1)$ is the cross-sectional area of the beam. For the normal stresses in x_1-direction (all other stresses are zero) one has the material law

$$T_{11} = B\varepsilon_{11} , \tag{2.42}$$

where

$$B = EI \quad \text{with} \quad I = \int_A x_3^2 \, dx_2 dx_3$$

is the *flexural stiffness* of the beam. I is called *axial geometric moment of inertia*. In the case of a rectangular cross-section with width b and height h, for instance, one has

$$I = \int_{-h/2}^{h/2} \int_{-b/2}^{b/2} x_3^2 \, dx_2 dx_3 = \frac{1}{12} bh^3 .$$

Writing $w = u_3$ and $x = x_1$, from the equation of motion (2.41) and the material law (2.42) the following differential equation for the unknown deflection $w = w(x)$ of the beam results:

$$(Bw'')'' + f_q = 0 . \tag{2.43}$$

Thus, the beam equation (2.43) is an ordinary differential equation of fourth order. This also has consequences with respect to the boundary conditions that have to be prescribed. For problems of the considered type there is the rule that the number of boundary conditions should be half the order of the differential equation. Thus for the beam two conditions at each of the interval boundaries have to be prescribed.

Concerning the combination of the boundary conditions there are different possibilities in prescribing two of the following quantities: the deflection, its derivative, the bending moment

$$M = Bw'' ,$$

or the transverse force

$$Q = Bw''' .$$

The prescription of the two latter quantities corresponds to the natural boundary conditions. For instance, if the beam is clamped at the left end $x = 0$ and free at the right end $x = L$, one has the boundary conditions (see Fig. 2.10, left)

$$w(0) = 0 \quad \text{and} \quad w''(0) = 0 \tag{2.44}$$

as well as

$$M(L) = Bw''(L) = 0 \quad \text{and} \quad Q(L) = Bw'''(L) = 0 . \tag{2.45}$$

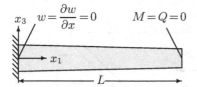 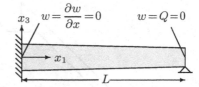

Fig. 2.10. Boundary conditions for beams with clamped left and free (left) and simply supported (right) right end

For a simply supported right end one would have (see Fig. 2.10, right):

$$w(L) = 0 \quad \text{and} \quad M(L) = Bw''(L) = 0.$$

The potential energy of a shear-rigid beam, for instance, with clamped left end and prescription of M und Q at the right end, is given by:

$$P(w) = \frac{B}{2} \int_0^L (w'')^2 \mathrm{d}x - \int_0^L \rho A f_q w \, \mathrm{d}x - w(L)Q_L - w'(L)M(L). \qquad (2.46)$$

The unknown deflection is defined as the minimum of this potential energy among all deflections w satisfying the boundary conditions (2.44).

2.4.3 Disks and Plates

A further special case of the general linear elasticity equations are problems with *plane stress state*. The essential assumptions for this case are that the displacements, strains, and stresses only depend on two spatial dimensions (e.g., x_1 and x_2) and that the problem can be treated as two-dimensional. For thin disks loaded by forces in their planes, these assumptions, for instance, are fulfilled in good approximation (see Fig. 2.11).

For problems with plane stress state for the stresses one has

$$T_{13} = T_{23} = T_{33} = 0.$$

For the strains in the x_1-x_2-plane from the general relation (2.25) it follows:

$$\varepsilon_{11} = \frac{\partial u_1}{\partial x_1}, \quad \varepsilon_{22} = \frac{\partial u_2}{\partial x_2} \quad \text{and} \quad \varepsilon_{12} = \frac{1}{2}\left(\frac{\partial u_1}{\partial x_2} + \frac{\partial u_2}{\partial x_1}\right).$$

Instead of (2.28) for the Lamè constant λ the relation

$$\lambda = \frac{E\nu}{1 - \nu^2}$$

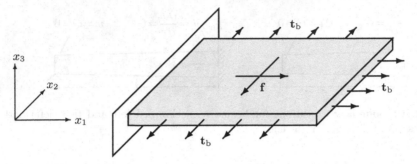

Fig. 2.11. Thin disk in plane stress state

has to be employed (see, e.g., [25] for a motivation of this). With this, Hooke's law for the plane stress state is expressed by the following stress-strain relation (in matrix notation):

$$\begin{bmatrix} T_{11} \\ T_{22} \\ T_{12} \end{bmatrix} = \frac{E}{1-\nu^2} \begin{bmatrix} 1 & \nu & 0 \\ \nu & 1 & 0 \\ 0 & 0 & 1-\nu \end{bmatrix} \begin{bmatrix} \varepsilon_{11} \\ \varepsilon_{22} \\ \varepsilon_{12} \end{bmatrix}. \tag{2.47}$$

Note that in general a strain in x_3-direction can also appear, which is given by

$$\varepsilon_{33} = -\frac{\nu}{E}(T_{11}+T_{22}).$$

From the matarial law (2.47) and the (steady) Navier-Cauchy equations (2.30) finally the following system of differential equations for the two displacements $u_1 = u_1(x_1, x_2)$ and $u_2 = u_2(x_1, x_2)$ results:

$$(\lambda+\mu)\left(\frac{\partial^2 u_1}{\partial x_1^2} + \frac{\partial^2 u_2}{\partial x_1 \partial x_2}\right) + \mu\left(\frac{\partial^2 u_1}{\partial x_1^2} + \frac{\partial^2 u_1}{\partial x_2^2}\right) + \rho f_1 = 0, \tag{2.48}$$

$$(\lambda+\mu)\left(\frac{\partial^2 u_1}{\partial x_2 \partial x_1} + \frac{\partial^2 u_2}{\partial x_2^2}\right) + \mu\left(\frac{\partial^2 u_2}{\partial x_1^2} + \frac{\partial^2 u_2}{\partial x_2^2}\right) + \rho f_2 = 0. \tag{2.49}$$

The displacement boundary conditions are

$$u_1 = u_{b1} \text{ and } u_2 = u_{b2} \text{ on } \Gamma_1,$$

whereas the stress boundary conditions on Γ_2 read

$$\frac{E}{1-\nu^2}\left(\frac{\partial u_1}{\partial x_1} + \nu\frac{\partial u_2}{\partial x_2}\right) n_1 + \frac{E}{2(1+\nu)}\left(\frac{\partial u_1}{\partial x_2} + \frac{\partial u_2}{\partial x_1}\right) n_2 = t_{b1},$$

$$\frac{E}{2(1+\nu)}\left(\frac{\partial u_1}{\partial x_2} + \frac{\partial u_2}{\partial x_1}\right) n_1 + \frac{E}{1-\nu^2}\left(\nu\frac{\partial u_1}{\partial x_1} - \frac{\partial u_2}{\partial x_2}\right) n_2 = t_{b2}.$$

Analogous to disks, also problems involving long bodies, whose geometries and loads do not change in longitudinal direction, can be reduced to two spatial dimensions (see Fig. 2.12). In this case one speaks of problems with *plane strain state*. Again, the displacements, strains, and stresses only depend on two spatial directions (again denoted by x_1 and x_2).

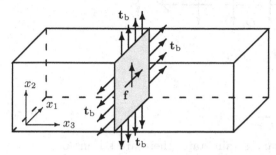

Fig. 2.12. Disk in plane strain state

The plane strain state is characterized by

$$\varepsilon_{13} = \varepsilon_{23} = \varepsilon_{33} = 0\,.$$

The normal stress in x_3-direction T_{33} does not necessarily vanish in this case. The essential difference from a disk in plane stress state is the different strain-stress relation. For the plane strain state this reads

$$\begin{bmatrix} T_{11} \\ T_{22} \\ T_{12} \end{bmatrix} = \frac{E}{(1+\nu)(1-2\nu)} \begin{bmatrix} 1-\nu & \nu & 0 \\ \nu & 1-\nu & 0 \\ 0 & 0 & 1-2\nu \end{bmatrix} \begin{bmatrix} \varepsilon_{11} \\ \varepsilon_{22} \\ \varepsilon_{12} \end{bmatrix},$$

where again the original Lamé constant λ given by (2.28) must be used. With this a two-dimensional differential equation system for the unknown displacements similar to (2.48) and (2.49) results.

The deformation of a thin plate, which is subjected to a vertical load (see Fig. 2.13), can under certain conditions also be formulated as a two-dimensional problem. The corresponding assumptions are known as *Kirchhoff hypotheses*:

- the plate thickness is small compared to the dimensions in the other two spatial directions,
- the vertical deflection u_3 of the midplane and its derivatives are small,
- the normals to the midplane remain straight and normal to the midplane during the deformation,
- the stresses normal to the midplane are negligible.

This case is denoted as *shear-rigid plate* (or also *Kirchhoff plate*), and is a special case of a more general plate theory that will not be addressed here (e.g., [6]).

Under the above assumptions one has the following relations for the displacements:

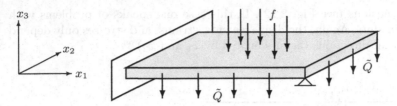

Fig. 2.13. Thin plate under vertical load

$$u_1 = -x_3 \frac{\partial u_3}{\partial x_1} \quad \text{and} \quad u_2 = -x_3 \frac{\partial u_3}{\partial x_2}. \tag{2.50}$$

As in the case of a disk in plane strain state, there are strains only in x_1- and x_2-direction and shear acts only in the x_1-x_2-plane. The corresponding components of the strain tensor read

$$\varepsilon_{11} = -x_3 \frac{\partial^2 u_3}{\partial x_1^2}, \quad \varepsilon_{22} = -x_3 \frac{\partial^2 u_3}{\partial x_2^2} \quad \text{and} \quad \varepsilon_{12} = -x_3 \frac{\partial^2 u_3}{\partial x_1 \partial x_2}.$$

All other components vanish. For the stresses one has

$$T_{13} = T_{23} = T_{33} = 0.$$

Due to the assumption that the stresses normal to the midplane are very small compared to those due to bending moments and thus can be neglected, the stress-displacement relation (2.47) for the plane stress state also can be employed here. Together with the equation of motion – after introduction of stress resultants (for details we refer to the corresponding literature, e.g., [6]) – the following differential equation for the unknown deflection $w = w(x_1, x_2)$ (we again write $w = u_3$) results:

$$K \left(\frac{\partial^4 w}{\partial x_1^4} + 2 \frac{\partial^4 w}{\partial x_1^2 \partial x_2^2} + \frac{\partial^4 w}{\partial x_2^4} \right) = \rho f. \tag{2.51}$$

The coefficient

$$K = \frac{E}{1 - \nu^2} \int_{-d/2}^{d/2} x_3^2 \, dx_3 = \frac{E d^3}{12(1 - \nu^2)},$$

where d is the plate thickness, is called *plate stiffness*. As in the case of a beam, the Kirchhoff plate theory results in a differential equation, albeit a partial one, of fourth order. Equation (2.51) also is called *biharmonic equation*.

As boundary conditions for plate problems one has, for instance, for a clamped boundary

$$w = 0 \quad \text{and} \quad \frac{\partial w}{\partial x_1} n_1 + \frac{\partial w}{\partial x_2} n_2 = 0, \tag{2.52}$$

and for a simply supported boundary

$$w = 0 \quad \text{and} \quad \frac{\partial^2 w}{\partial x_1^2} + \frac{\partial^2 w}{\partial x_2^2} = 0. \tag{2.53}$$

At a free boundary the bending moment as well as the sum of the transverse shear forces and the torsional moments have to vanish (for this we also refer to the literature, e.g., [6]). As usual, the different boundary conditions again can be prescribed at different parts of the boundary Γ.

After having computed the deflection w on the basis of (2.51) and the corresponding boundary conditions, the displacements u_1 and u_2 result from the expressions (2.50).

2.4.4 Linear Thermo-Elasticity

A frequently occuring problem in structural mechanics applications is deformations, for which also thermal effects play an essential role. For geometrically and physically linear problems such tasks can be described in the framework of the *linear thermo-elasticity*. The corresponding equations result from the momentum balance (2.10) and the energy balance (2.14) using the linearized strain-displacement relation (2.25) (geometrically linear theory) and the assumption of a simple linear thermo-elastic material (physically linear theory).

We briefly sketch the basic ideas for the derivation of the equations (details can be found, e.g., in [19]). For this, we first introduce the *specific dissipation function* ψ defined by

$$\psi = T_{ij} \frac{D\varepsilon_{ji}}{Dt} - \rho \frac{D}{Dt}(e - Ts) + \rho s \frac{DT}{Dt},$$

which is a measure for the energy dissipation in the continuum. Here, s is the *specific internal entropy*. An assumption for a simple thermo-elastic material is that there is no energy dissipation, i.e., one has

$$T_{ij} \frac{D\varepsilon_{ji}}{Dt} - \rho \frac{D}{Dt}(e - Ts) + \rho s \frac{DT}{Dt} = 0. \tag{2.54}$$

Together with the energy conservation equation (2.14) it follows from (2.54):

$$T\rho \frac{Ds}{Dt} = \frac{\partial h_i}{\partial x_i} + \rho q.$$

Assuming, as usual, the validity of Fourier's law (2.20) yields:

$$T\rho \frac{Ds}{Dt} = -\frac{\partial}{\partial x_i}\left(\kappa \frac{\partial T}{\partial x_i}\right) + \rho q. \tag{2.55}$$

Under the assumption of small temperature variations it is now possible to introduce a linearization, i.e., one considers the deviation $\theta = T - \bar{T}$ of the temperature from a (mean) reference temperature $\bar{T}$. For a simple linear thermo-elastic material one has the constitutive equations

$$T_{ij} = (\lambda \varepsilon_{kk} - \alpha\theta)\,\delta_{ij} + 2\mu\varepsilon_{ij}\,, \tag{2.56}$$

$$\rho s = \alpha\varepsilon_{ii} + c_p\theta \tag{2.57}$$

for the stress tensor and the entropy. Here, α is the *thermal expansion coefficient*. The relation (2.56) is known as *Duhamel-Neumann equation*. The material equations (2.56) and (2.57) provide linear relations between the stresses, strains, the temperature, and the entropy.

Together with the momentum conservation equation (2.26) one finally obtains from (2.55), (2.56), and (2.57) the following system of differential equations for the displacements u_i and the temperature variation θ:

$$\rho\frac{D^2 u_i}{Dt^2} + \alpha\frac{\partial\theta}{\partial x_i} - (\lambda + \mu)\frac{\partial^2 u_j}{\partial x_i \partial x_j} - \mu\frac{\partial^2 u_i}{\partial x_j \partial x_j} = \rho f_i\,, \tag{2.58}$$

$$-\frac{\partial}{\partial x_i}\left(\kappa\frac{\partial\theta}{\partial x_i}\right) + c_p\bar{T}\frac{D\theta}{Dt} + \alpha\bar{T}\frac{D\varepsilon_{kk}}{Dt} = \rho q\,. \tag{2.59}$$

For steady problems the equations simplify to:

$$\alpha\frac{\partial\theta}{\partial x_i} - (\lambda + \mu)\frac{\partial^2 u_j}{\partial x_i \partial x_j} - \mu\frac{\partial^2 u_i}{\partial x_j \partial x_j} = \rho f_i\,, \tag{2.60}$$

$$-\frac{\partial}{\partial x_i}\left(\kappa\frac{\partial\theta}{\partial x_i}\right) = \rho q\,. \tag{2.61}$$

The boundary conditions for the displacements are analogous to that of the linear elasticity theory (see Sect. 2.4.1). For the temperature variations the same type of conditions as for scalar heat transfer problems can be employed (see Sect. 2.3.2).

An exemplary thermo-elastic problem with boundary conditions is shown in Fig. 2.14. The configuration (in plane strain state) consists in a concentric pipe wall, which at the outer side is fixed and isolated, and at the inner side is supplied with a temperature θ_{f} and a stress $t_{\mathrm{f}i}$ (e.g., by a hot fluid).

The equations of the linear elasticity theory from Sect. 2.4.1 can be viewed as a special case of the linear thermo-elasticity, if one assumes that heat changes are so slow so as not to cause inertia forces. As similarly outlined in the previous sections for bars, beams, disks, or plates, special cases of the linear thermo-elasticity can be derived allowing for consideration of thermal effects for the corresponding problem classes.

2.4.5 Hyperelasticity

As an example for a geometrically and physically non-linear theory, we will give the equations for large deformations of hyperelastic materials. This may serve as an illustration of the high complexity that structural mechanics equa-

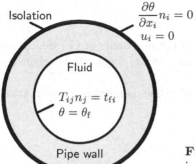

Isolation

$$\frac{\partial \theta}{\partial x_i} n_i = 0$$
$$u_i = 0$$

Fluid

$$T_{ij} n_j = t_{fi}$$
$$\theta = \theta_f$$

Pipe wall

Fig. 2.14. Example of a thermo-elastic problem in plane strain state

tions may take if non-linear effects appear. The practical relevance of hyperelasticity is due to the fact that it allows for a good description of large deformations of rubber-like materials. As an example, Fig. 2.15 shows the corresponding deformation of a hexahedral rubber block under compression.

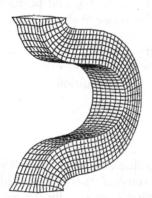

Fig. 2.15. Deformation of a rubber block under compression

A hyperelastic material is characterized by the fact that the stresses can be expressed as derivatives of a *strain energy density function* W with respect to the components $F_{ij} = \partial x_i / \partial a_j$ of the deformation gradient tensor $\mathbf{F}$:

$$T_{ij} = T_{ij}(\mathbf{F}) = \frac{\partial W}{\partial F_{ij}}(\mathbf{F}).$$

In this case one has a constitutive equation for the second Piola-Kirchhoff stress tensor of the form

$$P_{ij} = \rho_0(\gamma_1 \delta_{ij} + \gamma_2 G_{ij} + \gamma_3 G_{ik} G_{kj}) \tag{2.62}$$

with the Green-Lagrange strain tensor

$$G_{ij} = \frac{1}{2} \left(\frac{\partial u_i}{\partial a_j} + \frac{\partial u_j}{\partial a_i} + \frac{\partial u_k}{\partial a_i} \frac{\partial u_k}{\partial a_j} \right) . \tag{2.63}$$

The coefficients γ_1, γ_2, and γ_3 are functions of the invariants of $\mathbf{G}$ (see, for instance, [19]), i.e., they depend in a complex (non-linear) way on the derivatives of the displacements.

The relations (2.62) and (2.63), together with the momentum conservation equation in Lagrange formulation

$$\rho_0 \frac{D^2 x_i}{Dt^2} = \frac{\partial}{\partial a_j} \left(\frac{\partial x_i}{\partial a_k} P_{kj} \right) + \rho_0 f_i ,$$

give the system of differential equations for the unknown deformations x_i or the displacements $u_i = x_i - a_i$. The displacement boundary conditions are $u_i = u_{bi}$, as usual, and stress boundary conditions take the form

$$\frac{\partial x_i}{\partial a_k} P_{kj} n_j = t_{bi} .$$

For instance, for the problem illustrated in Fig. 2.15 at the top and bottom boundary the displacements are prescribed, while at the lateral (free) boundaries the stresses $t_{bi} = 0$ are given.

As can be seen from the above equations, in the case of hyperelasticity one is faced with a rather complex non-linear system of partial differential equations together with usually also non-linear boundary conditions.

2.5 Fluid Mechanical Problems

The general task in fluid mechanical problems is to characterize the flow behavior of liquids or gases, possibly with additional consideration of heat and species transport processes. For the description of fluid flows usually the Eulerian formulation is employed, because one is usually interested in the properties of the flow at certain locations in the flow domain.

We restrict ourselves to the case of linear viscous isotropic fluids known as *Newtonian fluids*, which are by far the most important ones for practical applications. Newtonian fluids are characterized by the following material law for the Cauchy stress tensor $\mathbf{T}$:

$$T_{ij} = \mu \left(\frac{\partial v_i}{\partial x_j} + \frac{\partial v_j}{\partial x_i} - \frac{2}{3} \frac{\partial v_k}{\partial x_k} \delta_{ij} \right) - p \delta_{ij} \tag{2.64}$$

with the *pressure* p and the *dynamic viscosity* μ. With this the conservation laws for mass, momentum, and energy take the form:

$$\frac{\partial \rho}{\partial t} + \frac{\partial (\rho v_i)}{\partial x_i} = 0, \tag{2.65}$$

$$\frac{\partial (\rho v_i)}{\partial t} + \frac{\partial (\rho v_i v_j)}{\partial x_j} = \frac{\partial}{\partial x_j} \left[\mu \left(\frac{\partial v_i}{\partial x_j} + \frac{\partial v_j}{\partial x_i} - \frac{2}{3} \frac{\partial v_k}{\partial x_k} \delta_{ij} \right) \right] - \frac{\partial p}{\partial x_i} + \rho f_i, \tag{2.66}$$

$$\frac{\partial (\rho e)}{\partial t} + \frac{\partial (\rho v_i e)}{\partial x_i} = \mu \left[\frac{\partial v_i}{\partial x_j} \left(\frac{\partial v_i}{\partial x_j} + \frac{\partial v_j}{\partial x_i} \right) - \frac{2}{3} \left(\frac{\partial v_i}{\partial x_i} \right)^2 \right] - p \frac{\partial v_i}{\partial x_i} \tag{2.67}$$

$$+ \frac{\partial}{\partial x_i} \left(\kappa \frac{\partial T}{\partial x_i} \right) + \rho q,$$

where in the energy balance (2.67) Fourier's law (2.20) is used again. Equation (2.66) is known as *(compressible) Navier-Stokes equation* (sometimes the full system (2.65)-(2.67) is also referred to as such). The unknowns are the velocity vector $\mathbf{v}$, the pressure p, the temperature T, the density ρ, and the internal energy e. Thus, one has 7 unknowns and only 5 equations. Therefore, the system has to be completed by two equations of state of the form

$$p = p(\rho, T) \quad \text{and} \quad e = e(\rho, T),$$

which define the thermodynamic properties of the fluid. These equations are called *thermal* and *caloric equation of states*, respectively. In many cases the fluid can be considered as an *ideal gas*. The thermal equation of state in this case reads:

$$p = \rho R T \tag{2.68}$$

with the *specific gas constant* R of the fluid. The internal energy in this case is only a function of the temperature, such that one has a caloric equation of state of the form $e = e(T)$. For a *caloric ideal gas*, for instance, one has

$$e = c_v T$$

with a constant specific heat capacity c_v (at constant volume).

Frequently, for flow problems it is not necessary to solve the equation system in the above most general form, i.e., it is possible to make additional assumptions to further simplify the system. The most relevant assumptions for practical applications are the *incompressibility* and the *inviscidity*, which, therefore, will be addressed in detail in the following.

2.5.1 Incompressible Flows

In many applications the fluid can be considered as approximatively incompressible. Due to the conservation of mass this is tantamount to a divergence-free velocity vector, i.e., $\partial v_i / \partial x_i = 0$ (see Sect. 2.2.1), and one speaks of an *incompressible flow.* For a criterion for the validity of this assumption the *Mach number*

$$\mathrm{Ma} = \frac{\bar{v}}{a}$$

is taken into account, where $\bar{v}$ is a characteristic flow velocity of the problem and a is the *speed of sound* in the corresponding fluid (at the corresponding temperature). Incompressibility usually is assumed if $\mathrm{Ma} < 0.3$. Flows of liquids in nearly all applications can be considered as incompressible, but this assumption is also valid for many flows of gases, which occur in practice.

For incompressible flows the divergence term in the material law (2.64) vanishes and the stress tensor becomes:

$$T_{ij} = \mu \left(\frac{\partial v_i}{\partial x_j} + \frac{\partial v_j}{\partial x_i} \right) - p\delta_{ij}. \tag{2.69}$$

The conservation equations for mass, momentum, and energy then read:

$$\frac{\partial v_i}{\partial x_i} = 0, \tag{2.70}$$

$$\frac{\partial (\rho v_i)}{\partial t} + \frac{\partial (\rho v_i v_j)}{\partial x_j} = \frac{\partial}{\partial x_j} \left[\mu \left(\frac{\partial v_i}{\partial x_j} + \frac{\partial v_j}{\partial x_i} \right) \right] - \frac{\partial p}{\partial x_i} + \rho f_i, \tag{2.71}$$

$$\frac{\partial (\rho e)}{\partial t} + \frac{\partial (\rho v_i e)}{\partial x_i} = \mu \frac{\partial v_i}{\partial x_j} \left(\frac{\partial v_i}{\partial x_j} + \frac{\partial v_j}{\partial x_i} \right) + \frac{\partial}{\partial x_i} \left(\kappa \frac{\partial T}{\partial x_i} \right) + \rho q. \tag{2.72}$$

One can observe that for isothermal processes in the incompressible case the energy equation does not need to be taken into account.

Neglecting the work performed by pressure and friction forces and assuming further that the specific heat is constant (which is valid in many cases), the energy conservation equation simplifies to a transport equation for the temperature:

$$\frac{\partial (\rho c_p T)}{\partial t} + \frac{\partial (\rho c_p v_i T)}{\partial x_i} = \frac{\partial}{\partial x_i} \left(\kappa \frac{\partial T}{\partial x_i} \right) + \rho q.$$

This again is the transport equation, with which we were already acquainted in Sect. 2.3.2 in the context of scalar heat transfer problems.

The equation system (2.70)-(2.72) has to be completed by boundary conditions and – in the unsteady case – by initial conditions. For the temperature the same conditions apply as for pure heat transfer problems as already discussed in Sect. 2.3.2. As boundary conditions for the velocity, for instance, the velocity components can be explicitly prescribed:

$$v_i = v_{\mathrm{b}i}.$$

Here, $\mathbf{v}_\mathrm{b}$ can be a known velocity profile at an inflow boundary or, in the case of an impermeable wall where a no-slip condition has to be fulfilled, a prescribed wall velocity ($v_i = 0$ for a fixed wall). Attention has to be paid to the fact that the velocities cannot be prescribed completely arbitrarily on the whole boundary Γ of the problem domain, since the equation system (2.70)-(2.72) only admits a solution, if the integral balance

$$\int_\Gamma v_{\mathrm{b}i} n_i \, \mathrm{d}\Gamma = 0$$

is fulfilled. This means that there flows as much mass into the problem domain as it flows out, which, of course, is physically evident for a "reasonably" formulated problem. At an outflow boundary, where usually the velocity is not known, a vanishing normal derivative for all velocity components can be prescribed.

If the velocity is prescribed at a part of the boundary, it is not possible to prescribe there additional conditions for the pressure. These are then intrinsicly determined already by the differential equations and the velocity boundary conditions.

In general, for incompressible flows, the pressure is uniquely determined only up to an additive constant (in the equations there appear only derivatives of the pressure). This can be fixed by *one* additional condition, e.g., by prescribing the pressure in a certain point of the problem domain or by an integral relation.

In Sect. 10.4 we will reconsider the different velocity boundary conditions in some more detail. Examples for incompressible flow problems are given in Sects. 2.6.2, 6.4, and 12.2.5.

2.5.2 Inviscid Flows

As one of the most important quantity in fluid mechanics the ratio of inertia and viscous forces in a flow is expressed by the *Reynolds number*

$$\mathrm{Re} = \frac{\bar{v} L \rho}{\mu} \,,$$

where $\bar{v}$ is again a characteristic flow velocity and L is a characteristic length of the problem (e.g., the pipe diameter for a pipe flow or the cross-sectional dimension of a body for the flow around it). The assumption of an inviscid flow, i.e., $\mu \approx 0$ can be made for "large" Reynolds numbers (e.g., $\mathrm{Re} > 10^7$). Away from solid surfaces this then yields a good approximation, since there the influence of the viscosity is low. Compressible flows at high Mach numbers (e.g., flows around airplanes or flows in turbomachines) are often treated as inviscid.

The neglection of the viscosity automatically entails the neglection of heat conduction (no molecular diffusivity). Also heat sources usually are neglected. Thus, in the inviscid case the conservation equations for mass, momentum, and energy read:

$$\frac{\partial \rho}{\partial t} + \frac{\partial(\rho v_i)}{\partial x_i} = 0\,, \tag{2.73}$$

$$\frac{\partial(\rho v_i)}{\partial t} + \frac{\partial(\rho v_i v_j)}{\partial x_j} = -\frac{\partial p}{\partial x_i} + \rho f_i\,, \tag{2.74}$$

$$\frac{\partial(\rho e)}{\partial t} + \frac{\partial(\rho v_i e)}{\partial x_i} = -p\frac{\partial v_i}{\partial x_i}\,. \tag{2.75}$$

This system of equations is called *Euler equations*. To complete the problem formulation one equation of state has to be added. For an ideal gas, for instance, one has:

$$p = R\rho e/c_v\,.$$

It should be noted that the neglection of the viscous terms results in a drastic change in the nature of the mathematical formulation, since all second derivatives in the equations disappear and, therefore, the equation system is of another type. This also causes changes in the admissible boundary conditions, since for a first-order system fewer conditions are necessary. For instance, at a wall only the normal component of the velocity can be prescribed and the tangential components are then determined automatically. For details concerning these aspects we refer to [12].

The further assumption of an irrotational flow in the inviscid case leads to the potential flow, which we have already discussed in Sect. 2.3.1.

2.6 Coupled Fluid-Solid Problems

For a variety of engineering applications it is not sufficient to consider phenomena of structural mechanics, fluid mechanics, or heat transfer individually, because there is a significant coupling of effects from two or three of these fields. Examples of such mechanically and/or thermally coupled fluid-solid problems can be found, for instance, in machine and plant building, engine manufacturing, turbomachinery, heat exchangers, offshore structures, chemical engineering processes, microsystem techniques, biology, or medicine, to mention only a few of them.

A schematic view of possible physical coupling mechanisms in such kind of problems is indicated in Fig. 2.16. The problems can be classified according to different couplings involved, yielding problem formulations of different complexity:

(1) *Flows acting on solids*: drag, lift, movement, and deformation of solids induced by pressure and friction fluid forces (e.g., aerodynamics of vehicles, wind load on buildings, water penetration of offshore structures, ...).
(2) *Solids acting on flows*: fluid flow induced by prescribed movement and/or deformation of solids (e.g., stirrers, turbomachines, nearby passing of two vehicles, flow processes in piston engines, ...).

(3) *Fluid-solid interactions*: movement and/or deformation of solids induced by fluid forces interacting with induced fluid flow (e.g., aeroelasticity of bridges or airplanes, injection systems, valves, pumps, airbags, ...).

(4) *Thermal couplings*: temperature-dependent material properties, buoyancy, convective heat transfer, thermal stresses and mechanical dissipation (e.g., heat exchangers, cooling of devices, thermal processes in combustion chambers, chemical reactors, ...).

In the following we will address some special modelling aspects which arise due to the coupling. Afterwards we will give a couple of exemplary applications for the above mentioned problem classes.

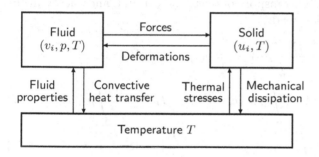

Fig. 2.16. Schematic view of mechanically and thermally coupled fluid-solid problems

2.6.1 Modeling

Let us consider a problem domain Ω consisting of a fluid part Ω_f and a solid part Ω_s, which regarding the shape as well as the location of fluid and solid parts can be quite arbitrary (see Fig. 2.17).

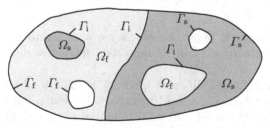

Fig. 2.17. Example of fluid-solid problem domain

The basis of the mathematical problem formulation of such coupled problems are the fundamental conservation laws for mass, momentum, moment of momentum, and energy (see Sect. 2.2), which are valid for any subvolume V (with surface S) of Ω either in fluid or solid parts of the problem domain. In the individual fluid and solid parts of the problem domain different material laws, as they are exemplarily given in Sects. 2.4 and 2.5, can be employed according to the corresponding needs within the subdomain.

If there is a movement of solid parts of the problem domain within fluid parts, it is convenient to consider the conservation equations for the fluid part in the so-called *arbitrary Lagrangian-Eulerian (ALE) formulation*. With this one can combine the advantages of the Lagrangian and Eulerian formulations as they usually are exploited for individual structural and fluid mechanics approaches, respectively. The principal idea of the ALE approach is that an observer is neither located at a fixed position in space nor moves with the material point, but can move "arbitrarily". Mathematically this can be expressed by employing a relative velocity in the convective terms of the conservation equations, i.e., for a (moving) control volume V with surface S the conservation equations governing transport of mass, momentum, and energy in the (integral) ALE fomulation read:

$$\frac{D}{Dt} \int_V \rho \, dV + \int_S \rho(v_j - v_j^g) n_j \, dS = 0, \qquad (2.76)$$

$$\frac{D}{Dt} \int_V \rho v_i \, dV + \int_S [\rho v_j(v_i - v_i^g) n_j - T_{ij}] \, dS = \int_V \rho f_i \, dV, \quad (2.77)$$

$$\frac{D}{Dt} \int_V \rho c_p T \, dV + \int_S \left[\rho c_p(v_i - v_i^g) T - \kappa \frac{\partial T}{\partial x_i} \right] n_i \, dS = \int_V \rho q \, dV, \quad (2.78)$$

where v_i^g is the velocity with which S may move, e.g., due to displacements of solid parts in the problem domain. In the context of a numerical scheme v_i^g is also called *grid velocity*. Note that with $v_i^g = 0$ and $v_i^g = v_i$ the pure Euler and Lagrange formulations are recovered, respectively. In the framework of a numerical scheme the so-called space (or geometric) conservation law

$$\frac{D}{Dt} \int_V dV + \int_S (v_j - v_j^g) n_j \, dS = 0, \qquad (2.79)$$

also plays an important role because it allows for an easy way to ensure the conservation properties for the moving control volumes. For details we refer, for instance, to [8].

Concerning the boundary conditions different kinds of boundaries have to be distinguished: a solid boundary Γ_s, a fluid boundary Γ_f, and a fluid-solid interface Γ_i (see Fig. 2.17). On solid and fluid boundaries Γ_s and Γ_f, standard conditions as for individual solid and fluid problems as discussed in the previous sections can be employed. On a fluid-solid interface Γ_i the velocities and the stresses have to fulfill the conditions

$$v_i = \frac{D u_i^b}{Dt} \quad \text{and} \quad \sigma_{ij} n_j = T_{ij} n_j, \qquad (2.80)$$

where u_i^b and Du_i^b/Dt are the displacement and velocity of the interface, respectively. In addition, if heat transfer is involved, the temperatures as well as the heat fluxes have to coincide on Γ_i. However, it is possible to treat the temperature globally over the full domain Ω, such that these thermal interface conditions do not have to be considered explicitly.

2.6.2 Examples of applications

In the following a variety of exemplary applications for coupled fluid-solid problems with different coupling mechanisms are given.

Flows Acting on Solids

This problem class involves problems where fluid flows exert pressure and friction forces on structures and inducing stresses there, but the corresponding solid deformations are so small that their influence on the flow can be neglected. Such problem situations typically occur, for instance, in aerodynamics applications (e.g., the flow around vehicles, airplanes, buildings, ...).

The problems can be solved by first solving the fluid flow problem yielding the fluid stresses $t_{fi} = T_{ij}n_j$, which can be used for the subsequent structural computation as (stress) boundary conditions $t_{bi} = t_{fi}$. Figure 2.18 illustrates a typical problem situation with the relevant boundary conditions.

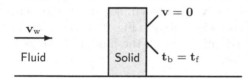

Fig. 2.18. Example for flow acting on solid with boundary conditions

Flow Induced by Prescribed Solid Movement

Problems with solid parts moving in a fluid, where the movement is not significantly influenced by the flow, often appear in technical industrial applications. For example, the mixing of fluids with rotating installations, the movement of turbines, or the passing-by of vehicles belong to these kinds of problems. In these cases there is usually no need to solve any equation for the solid part of the domain, but, due to the (prescribed) movement of the solid, the flow domain is time-dependent.

For such kind of problems it can be convenient to consider the conservation equations in a moving frame of reference related to the movement of the solid, such that relative to this system the boundary conditions can be handled in an easier way. In general, for a reference system rotating with an angular velocity

ω and translating with a velocity $\mathbf{c}$ (velocity of the origin) the momentum equation (2.10) transforms into (see, e.g., [23])

$$\frac{\partial(\rho\tilde{w}_i)}{\partial t} + \frac{\partial(\rho\tilde{w}_i\tilde{w}_j)}{\partial x_j} - \frac{\partial\tilde{T}_{ij}}{\partial x_j} = \rho\tilde{f}_i - \rho\tilde{b}_i \,, \tag{2.81}$$

where $\tilde{b}_i$ are the components of

$$\mathbf{b} = \left[\frac{D\mathbf{c}}{Dt} + 2\omega \times \mathbf{w} + \frac{D\omega}{Dt} \times \mathbf{x} + \omega \times (\omega \times \mathbf{x})\right], \tag{2.82}$$

and $\tilde{w}_i$ are the components of the velocity $\mathbf{w}$ relative to the moving system, which is related to $\mathbf{v}$ by

$$\mathbf{w} = \mathbf{v} - \omega \times \mathbf{x} - \mathbf{c} \,. \tag{2.83}$$

The tilde on the components indicates that these refer to the moving coordinate system. The additional terms in (2.81) are due to volume forces resulting from the movement of the frame of reference: the acceleration of the reference system $D\mathbf{c}/Dt$, the *Coriolis acceleration* $2\omega \times \mathbf{w}$, the *angular acceleration* $D\omega/Dt \times \mathbf{x}$, and the *centrifugal acceleration* $\omega \times (\omega \times \mathbf{x})$. With $\mathbf{c} = \mathbf{0}$ and $\omega = \mathbf{0}$ the original formulation (2.10) is recovered.

A simple example for this class of applications with a translatory movement is the passing-by of two objects (i.e., cars, trains, elevators, ...). The problem situation is illustrated schematically in Fig. 2.19 for rectangular objects moving with velocities $\mathbf{v}_1$ and $\mathbf{v}_2$. Along with the objects two moving coordinate systems with $\mathbf{c} = \mathbf{v}_1$ and $\mathbf{c} = \mathbf{v}_2$ can be employed together with the corresponding boundary conditions as indicated in Fig. 2.19. As an example, Fig. 2.20 shows the pressure distribution within the fluid at 4 points in time during the passing-by of the objects.

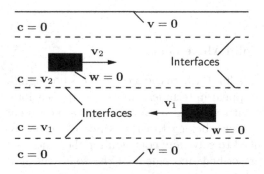

Fig. 2.19. Coordinate systems and boundary conditions for problems with translational solid movement

Stirrer configurations can be considered as typical applications for flows induced by a prescribed rotational solid movement. The problem situation is illustrated schematically in Fig. 2.21 for a stirrer device rotating with angular velocity ω_b. Along with the stirrer a rotating coordinate systems with $\omega = \omega_b$

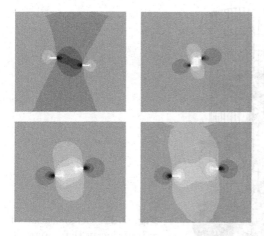

Fig. 2.20. Pressure distribution for passing-by problem at 4 points in time (top/left to bottom/right)

can be employed together with the corresponding boundary conditions as indicated in Fig. 2.21. Exemplarily, in Fig. 2.22 a helix stirrer configuration is shown where the corresponding flow is illustrated by some cross-sections of the axial velocity and a characteristic particle path.

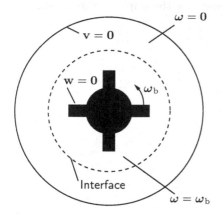

Fig. 2.21. Coordinate systems and boundary conditions for problems with rotational solid movement

Interaction of Flow and Solid Movement

Another important class of problems in technical applications is characterized by moving solid parts interacting with fluid flows. Here, the solid part of the problem can be described by a simple rigid body motion. As a simple representative example for this class of applications the interaction of a flow with a rotation of an impeller is displayed in Fig. 2.23. It shows the pressure distribution for four consecutive positions of the impeller illustrating the development

Fig. 2.22. Cross-section of axial velocity (left), particle path (right) for helix stirrer

of a sucking effect as in a pump. The flow, which enters at the upper channel, implies a turning moment to the impeller and the resulting impeller rotation interacts with the flow. The configuration is similar to that of a simple flow meter (measuring the flow rate by just counting the impeller rotations).

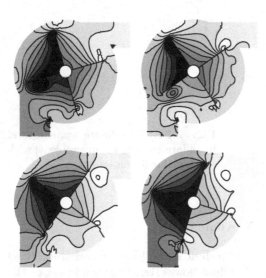

Fig. 2.23. Pressure distribution flow interacting with rotating impeller (dark area corresponds to low pressure).

Assuming that the impeller rotates without friction around the shaft and that the bending of the blades can be neglected, the impeller rotation can simply be modelled by

$$\frac{D^2\phi}{Dt^2} = \frac{3M_{\mathrm{f}}}{8md^2},\tag{2.84}$$

where $D^2\phi/Dt^2$, m, and d are the angular acceleration, the mass, and the diameter of the impeller, respectively. M_{f} are the moments supplied by the fluid, which are evaluated by integrating the pressure and friction forces over the surfaces of the blades.

Concerning the flow model again the technique involving fixed and moving frames of references as outlined in the previous section can be applied (see Fig. 2.21). The only difference is that now the angular velocity $\boldsymbol{\omega}$ of the rotating frame of reference is no longer constant, but varies with the impeller rotation by $\boldsymbol{\omega} = D\phi/Dt$.

Interaction of Flow and Solid Deformation

The most general problem class of mechanically coupled fluid-solid problems is the interaction of flows and solid deformations, which occur if flexible structures are involved (e.g., paper, fabrics, arteries, ...). As an example, Fig. 2.24 shows a corresponding problem situation with boundary conditions for the flow around an elastic cylinder in plane strain conditions, which is fixed at one point (right point).

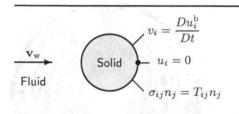

Fig. 2.24. Example for interaction of flow and solid deformation with boundary conditions

The deformation of the cylinder due to the pressure and shear stress forces of the flow can be seen in Fig. 2.25 showing an instantaneous distribution of the streamlines and the horizontal velocity. The deformation in turn influences the flow by inducing a moment via the movement of the cylinder boundary and by changing the flow geometry. In this way a strongly interacting dynamic process results.

Thermally coupled problem

As an example of a thermally coupled fluid-solid problem we consider a hollow thick walled massive pipe coaxially surrounded by another pipe conveying a fluid. The pipes are assumed to be infinitely long such that plane strain conditions for the solid part can be applied. The (two-dimensional) problem configuration with boundary conditions is sketched in Fig. 2.26 (left). The inner

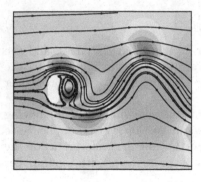

Fig. 2.25. Streamlines and horizontal fluid velocity for flow around cylinder with fluid-structure interaction

boundary of the solid pipe is fixed and imposed with a constant temperature $T = T_\mathrm{h}$. The outer boundary of the fluid pipe is at constant temperature $T = T_\mathrm{l}$ (with $T_\mathrm{l} < T_\mathrm{h}$). At the interface a fixed wall ($v_i = 0$) for the fluid can be applied, if the deformations are assumed to be small, while the flow forces define the (stress) boundary conditions $t_{\mathrm{b}i} = T_{ij}n_j$ for the solid. The characteristics of the problem solution can be seen in Fig. 2.26 (right) showing streamlines in the fluid part, displacements in the solid part, as well as the global temperature distribution.

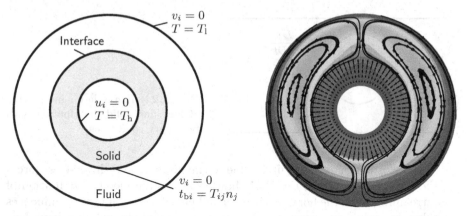

Fig. 2.26. Thermally coupled fluid-solid problem: configuration and boundary conditions (left), streamlines, displacements, and global temperature distribution (right)

Multi-Coupled Problem

Finally, as an example with multiple fluid-solid coupling mechanisms, we consider the multi-field problem determining the functionality of a complex antenna structure as they are employed, for instance, in space applications for tracking satellites and space probes (see Fig. 2.27).

Fig. 2.27. Deformation of antenna structure

Figure 2.28 schematically illustrates the problem situation and the interactions. The key parameters for the flow are the wind velocity and the site topography. The temperature distribution is influenced by the sun radiation, the ambient temperature, as well as by the flow. The structural deformation is affected by the antenna weight, the temperature gradients, and the flow forces. From the deformation of the structure, the pointing error describing the deviation of the axis of the reflector from the optical axis can be determined. This is the key parameter for the functionality of the antenna, since it has a significant influence on the power of the incoming electromagnetic signal. As an example the deformation resulting from a corresponding multi-field analysis is indicated in Fig. 2.27.

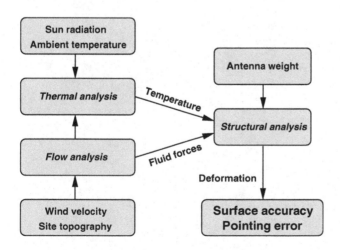

Fig. 2.28. Multi-field analysis of antennas

Exercises for Chap. 2

Exercise 2.1. Given is the deformation

$$\mathbf{x}(\mathbf{a}, t) = (a_1/4, e^t(a_2+a_3) + e^{-t}(a_2-a_3), e^t(a_2+a_3) - e^{-t}(a_2-a_3)) \,.$$

(i) Compute the Jacobi determinant of the mapping $\mathbf{x} = \mathbf{x}(\mathbf{a}, t)$ and formulate the reverse equation $\mathbf{a} = \mathbf{a}(\mathbf{x}, t)$. (ii) Determine the displacements and velocity components in Eulerian and Lagrangian descriptions. (iii) Compute the corresponding Green-Lagrange and the Green-Cauchy strain tensors.

Exercise 2.2. Show that balance of moment of momentum (2.12) can be expressed by the symmetry of the Cauchy stress tensor (see Sect. 2.2.3).

Exercise 2.3. Derive the differential form (2.14) of the energy conservation from the integral equation (2.13) (see Sect. 2.2.4).

Exercise 2.4. The velocity $v = v(x)$ of a (one-dimensional) pipe flow is described by the weak formulation: find $v = v(x)$ with $v(1) = 4$ such that

$$3 \int_1^3 v'\varphi \, dx - 2 \int_1^3 v'\varphi' \, dx = \int_1^3 \varphi \sin x \, dx + \varphi(3)$$

for all test functions φ. Derive the corresponding differential equation and boundary conditions.

3

Discretization of Problem Domain

Having fixed the mathematical model for the description of the underlying problem to be solved, the next step in the application of a numerical simulation method is to approximate the continuous problem domain (in space and time) by a discrete representation (i.e., nodes or subdomains), in which then the unknown variable values are determined. The discrete geometry representation usually is done in the form of a grid over the problem domain, such that the spatial discretization of the problem domain is also denoted as *grid generation*. In practice, where one often is faced with very complex geometries (e.g., airplanes, engine blocks, turbines, ...), this grid generation can be a very time-consuming task. Besides the question of the efficient generation of the grid, its properties with respect to its interaction with the accuracy and the efficiency of the subsequent computation in particular are of high practical interest. In this chapter we will address the most important aspects in these respects.

In order to keep the presentation simple, in the following considerations we will restrict ourselves mainly to the two-dimensional case. However, all methods – unless otherwise stated – usually are transferable without principal problems to the three-dimensional case (but mostly with a significantly increased "technical" effort).

3.1 Description of Problem Geometry

An important aspect when applying numerical methods to a concrete practical problem, which has to be taken into account before the actual discretization, is the interface of the numerical method to the problem geometry, i.e., how this is defined and how the geometric data are represented in the computer. For complex three-dimensional applications this is a non-trivial problem. Since it would be beyond the scope of this text to go deeper into this aspect, only the key ideas are briefly addressed. Detailed information can be found in the corresponding computer aided design (CAD) literature (e.g., [7]).

Nowadays in engineering practice the geometry information usually is available in a standardized data format (e.g., IGES, STEP, ...), which has been generated by means of a CAD system. These data provide the interface to the grid generation process. For the description of the geometry mainly the following techniques are employed:

- volume modeling,
- boundary modeling.

The volume modeling is based on the definition of a number of simple objects (e.g., cuboids, cylinders, spheres, ...), which can be combined by Boolean algebra operations. For instance, a square with a circular hole can be represented by the Boolean sum of the square and the negative of a circular disk (see Fig. 3.1).

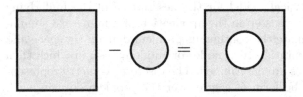

Fig. 3.1. Volume modeling of two-dimensional problem domain

The most common method to describe the geometry is the representation by boundary surfaces (in the three-dimensional case) or boundary curves (in the two-dimensional case). Such a description consists of a composition of (usually curvilinear) surfaces or curves with which the boundary of the problem domain is represented (see Fig. 3.2).

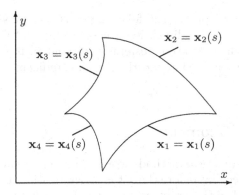

Fig. 3.2. Boundary modeling of two-dimensional problem domain

Curvilinear surfaces usually are described by *B-spline functions* or *Bezier curves*. For instance, a Bezier curve $\mathbf{x} = \mathbf{x}(s)$ of degree n over the parameter range $a \leq s \leq b$ is defined by

$$\mathbf{x}(s) = \sum_{i=0}^{n} \mathbf{b}_i B_i^n(s) \tag{3.1}$$

with the $n+1$ *control points* (or *Bezier points*) $\mathbf{b}_i$ and the so-called *Bernstein polynomials*

$$B_i^n(s) = \frac{1}{(b-a)^n} \binom{n}{i} (s-a)^i (b-s)^{n-i}.$$

In Fig. 3.3, as an example, a Bezier curve of degree 4 with the corresponding control points is illustrated.

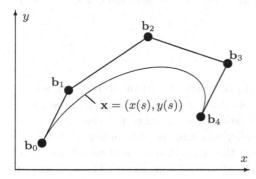

Fig. 3.3. Bezier curve of degree 4 with control points

The points on a Bezier curve can be determined in a numerically stable way from the prescribed control points by means of a relatively simple (recursive) procedure known as *de Casteljau algorithm*. In addition, such curve representations possess a number of very useful properties for the geometric data manipulation, which, however, we will not detail here (see, e.g., [7]).

By taking the corresponding tensor product, based on representations of the form (3.1) also *Bezier surfaces* $\mathbf{x} = \mathbf{x}(s,t)$ can be defined:

$$\mathbf{x}(s,t) = \sum_{i=0}^{n} \sum_{k=0}^{m} \mathbf{b}_{i,k} B_i^n(s) B_k^m(t) \tag{3.2}$$

with a two-dimensional parameter range $a_1 \leq s \leq b_1$, $a_2 \leq t \leq b_2$.

Since in practice this often causes problems, we mention that the boundary representations resulting from CAD systems often have gaps, discontinuities, or overlappings between neighboring surfaces or curves. Frequently, on the basis of such a representation it is not possible to generate a "reasonable" grid. For this it is necessary to suitably modify the input geometry (see Fig. 3.4). This usually is supported in commercial grid generators by special helping tools. Surface representations of the form (3.2) provide a good basis for such corrections because they allow in a relatively easy way, i.e., by simple conditions for the control points near the boundary, the achievement of smooth transitions between neighboring surface pieces (e.g., continuity of first and second derivatives).

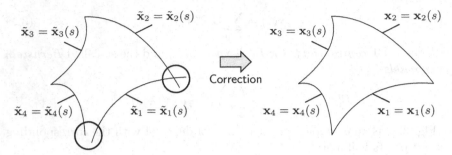

Fig. 3.4. Example for inaccurate boundary representations from CAD systems and corresponding corrections

3.2 Numerical Grids

All discretization methods, which will be addressed in the following chapters, first require a discretization of the spatial problem domain. This usually is done by the definition of a suitable grid structure covering the problem domain. A grid is defined by the grid cells, which in the two-dimensional case are formed by grid lines (see Fig. 3.5). In the three-dimensional case the cells are formed by grid surfaces which again are formed by grid lines. The intersections of the grid lines define the grid points.

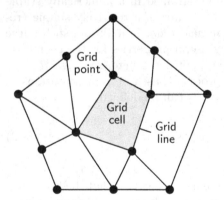

Fig. 3.5. Two-dimensional numerical grid with cells, lines, and points

In practice, often this grid generation for complex problem geometries is the most time consuming part of a numerical investigation. On the one hand, the grid should model the geometry as exactly as possible and, on the other hand, the grid should be "good" with respect to an efficient and accurate subsequent computation. Here, one has to take into account that there is a close interaction between the geometry discretization, the discretization of the equations, and the solution methods (for this also see the properties of the discretization methods addressed in Sect. 8.1).

In general the relation between the numerical solution method and the grid structure can be characterized as follows:

The more regular the grid, the more efficient the solution algorithms for the computation, but the more inflexible it is with respect to the modeling of complex geometries.

In practice, it is necessary to find here a reasonable compromise, where, in addition, the question of the actually required accuracy for the concrete application has to be taken into account. Also, the effort to create the grid (compared to the effort for the subsequent computation) should be within justifiable limits. In particular, this applies if due to a temporally varying problem domain the grid has to be modified or generated anew several times during the computation. In the next sections we will address the issues that are of most relevance for the aforementioned aspects.

3.2.1 Grid Types

The type of a grid is closely related to the discretization and solution techniques that are employed for the subsequent computation. There are a number of distinguishing features which are of importance in this respect.

A first classification feature is the form of the grid cells. While in the one-dimensional case subintervals are the only sensible choice, in the two-dimensional and three-dimensional cases triangles or quadrilaterals and tetrahedras, hexahedras, prisms, or pyramids, respectively, are common choices (rarely more general polygons and polyeders). In general, the cells also may have curvilinear boundaries.

Furthermore, the grids can be classified according to the following types:

- boundary-fitted grids,
- Cartesian grids,
- overlapping grids.

The characteristic features of corresponding grids are illustrated in Fig. 3.6. Boundary-fitted grids are characterized by the fact that all boundary parts of the problem domain are approximated by grid lines (boundary integrity). With Cartesian grids the problem domain is covered by a regular grid, such that at the boundary irregular grid cells may occur that require a special treatment. Using overlapping grids, which are also known as *Chimera grids*, different regions of the problem domain are discretized mostly independently from each other, with regular grids allowing for overlapping areas at the interfaces that also make a special treatment necessary.

Cartesian and overlapping grids are of interest only for very special applications (in particular in fluid mechanics), such that their application nowadays is not very common. Therefore, we will restrict our considerations in the following to the case of boundary-fitted grids.

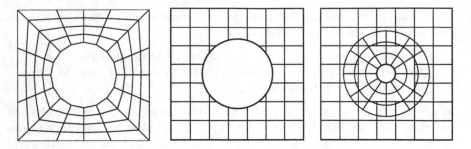

Fig. 3.6. Examples of boundary-fitted (left), Cartesian (middle), and overlapping (right) grids

3.2.2 Grid Structure

A practically very important distinguishing feature of numerical grids is the logical arrangement of the grid cells. In this respect, in general, the grids can be classified into two classes:

- structurerd grids,
- unstructured grids.

In Figs. 3.7 and 3.8 examples for both classes are given.

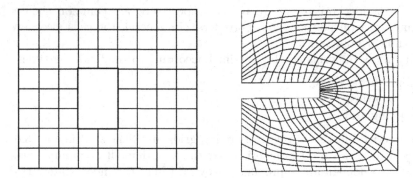

Fig. 3.7. Examples of structured grids

Structured grids are characterized by a regular arrangement of the grid cells. This means that there are directions along which the numbers of grid points is always the same, where, however, certain regions (obstacles) can be "masked out" (see left grid in Fig. 3.7). Thus, structured grids can be warped, but logically they are rectangular (or cuboidal in three dimensions). A consequence of the regular arrangement is that the neighboring relations betweeen the grid points follow a certain fixed pattern, which can be exploited in the discretization and solution schemes. Knowing the location of a grid

Fig. 3.8. Example of unstructured grid

point, the identity of the neighboring points is uniquely defined by the grid structure.

For unstructured grids there is no regularity in the arrangement of the grid points. The possibility of distributing the grid cells arbitrarily over the problem domain gives the best flexibility for the accurate modeling of the problem geometry. The grid cells can be adjusted optimally to the boundaries of the problem domain. Besides the locations of the grid points, for unstructured grids also the relations to neighboring grid cells have to be stored. Thus, a more involved data structure than for structured grids is necessary.

In Table 3.1 the most important advantages and disadvantages of structured and unstructured grids are summarized. For the practical application, a simple conclusion from this is that one should first try to create the grid as structured as possible. Deviations from this structure should be introduced only if this is necessary due to requirements of the problem geometry so that the grid quality is not too bad (see Sect. 8.3). The latter possibly also has to be viewed in connection with local grid refinement (see Sect. 12.1), which might be necessary for accuracy reasons and which can be realized relatively easily with unstructured grids. There are strong limitations in this respect with structured grids if one tries to maintain the structure.

Frequently, the grid structure is implicated with the shape of the grid cells, i.e., triangular or tetrahedral grids are referred to as unstructured, and quadrilateral or hexahedral grids as structured. However, the shape of the grid cells does not determine the structure of the grid since structured triangular grids and unstructured quadrilateral grids can be realized without problems. Also the discretization technique, which is employed for a given grid for the approximation of the equations, sometimes is implicated with the grid structure, i.e., finite-difference and finite-volume methods with structured grids and finite-element methods with unstructured grids. Also this is misleading. Each of the methods can be formulated for structured as well as unstructured grids, where, however, due to the specific properties of the discretization tech-

Table 3.1. Advantages and disadvantages of structured and unstructured grids (relative to each other)

Property	Structured	Unstructured
Modeling of complex geometries	−	+
Local (adaptive) grid refinement	−	+
Automatization of grid generation	−	+
Effort for grid generation	+	−
Programming effort	+	−
Data storage and management	+	−
Solution of algebraic equation systems	+	−
Parallelization and vectorization of solvers	+	−

niques the one or the other approach appears to be advantageous (see the corresponding discussions in Chaps. 4 and 5).

There are also mixed forms between structured and unstructured grids, whose utilization turns out to be useful for many applications because it is partially possible to combine the advantages of both approaches. Important variants of such grids are:

- block-structured grids,
- hierarchically structured grids.

Figures 3.9 and 3.10 show examples of these two grid types.

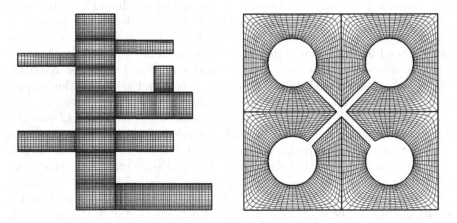

Fig. 3.9. Examples of block-structured grids

Block-structured grids are locally structured within each block, but globally unstructured (irregular block arrangement) and in this sense can be viewed as a compromise between the geometrically inflexible structured grids

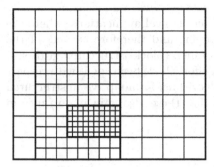

Fig. 3.10. Example of hierarchically structured grid

and the numerically costly unstructured grids. An adequate modeling of complex geometries is possible without restrictions and within the individual blocks efficient "structured" numerical techniques can be employed. However, special attention has to be paid to the treatment of the block interfaces. Local grid refinements can be done blockwise (see Fig. 3.11), and the so-called "hanging nodes" (i.e., discontinuous grid lines across block interfaces) that may occur require a special treatment in the solution method. Furthermore, a block structure of the grid provides a natural basis for the parallelization of the numerical scheme (this aspect will be detailed in Sect. 12.3).

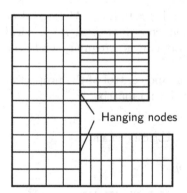

Fig. 3.11. Example of blockwise locally refined block-structured grid

In hierarchically structured grids, starting from a (block-)structured grid, certain regions of the problem domain are locally refined again in a structured way. With such grids there is much freedom for local adaptive grid refinement and within the individual subregions still efficient "structured" solvers can be used. Also in this case the interfaces between differently discretized regions need special attention. Presently, hierarchically structured grids are not very common in practice because there are hardly any numerical codes in which such an approach is consistently realized.

For finite-element computations in the field of structural mechanics usually unstructured grids are employed. In fluid mechanical applications, where the number of grid points usually is much larger (and therefore aspects of the efficiency of the solver become more important), (block-)structured grids still dominate. However, in recent years a lot of efforts have been spent with respect to the development of efficient solution algorithms for adaptive unstructured and hierarchically structured grids, such that these also gain importance in the area of fluid mechanics.

3.3 Generation of Structured Grids

In the following sections we outline the two most common methods for the generation of structured grids: the *algebraic grid generation* and the *elliptic grid generation*. The methods can be used as well for the blockwise generation of block-structured grids, where additionally the difficulty of discontinuous block interfaces has to be taken into account (which, however, we will not discuss further).

In general, the task of generating a structured grid consists in finding a unique mapping

$$(x,y) = (x(\xi,\eta), y(\xi,\eta)) \quad \text{or} \quad (\xi,\eta) = (\xi(x,y), \eta(x,y)) \qquad (3.3)$$

between given discrete values $\xi = 0, 1, \ldots, N$ and $\eta = 0, 1, \ldots, M$ (logical or computational domain) and the physical problem coordinates (x,y) (physical or problem domain). Generally, the physical domain is irregular, while the logical domain is regular (see Fig. 3.12).

Important quantities for the characterization and control of the properties of a structured grid are the components of the Jacobi matrix

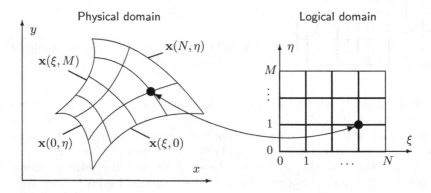

Fig. 3.12. Relation between coordinates and grid points in physical and logical domains

$$\mathbf{J} = \begin{bmatrix} \dfrac{\partial \xi}{\partial x} & \dfrac{\partial \xi}{\partial y} \\[2ex] \dfrac{\partial \eta}{\partial x} & \dfrac{\partial \eta}{\partial y} \end{bmatrix}$$

of the mapping defined by the relations (3.3). To abbreviate the notation we denote the derivatives of the physical with respect to the logical coordinates (or vice versa) with a corresponding index, e.g., $x_\xi = \partial x/\partial \xi$ or $\eta_x = \partial \eta/\partial x$. These quantities also are called *metrics* of the grid.

For the required uniqueness of the relation between physical and logical coordinates the determinant of the Jacobi matrix $\mathbf{J}$ may not vanish:

$$\det(\mathbf{J}) = \xi_x \eta_y - \xi_y \eta_x \neq 0.$$

In the next two sections we will see how corresponding coordinate transformations can be obtained either by algebraic relations or by the solution of differential equations.

3.3.1 Algebraic Grid Generation

The starting point for an algebraic grid generation is the prescription of grid points at the boundary of the problem domain (advantageously in physical coordinates):

$$\mathbf{x}(\xi,0) = \mathbf{x}_\mathrm{s}(\xi), \quad \mathbf{x}(\xi,M) = \mathbf{x}_\mathrm{n}(\xi) \quad \text{for} \quad \xi = 0,\dots,N,$$

$$\mathbf{x}(0,\eta) = \mathbf{x}_\mathrm{w}(\eta), \quad \mathbf{x}(N,\eta) = \mathbf{x}_\mathrm{e}(\eta) \quad \text{for} \quad \eta = 0,\dots,M.$$

For boundary-fitted grids the prescribed grid points are located on the corresponding boundary curves $\mathbf{x}_1,\dots,\mathbf{x}_4$. For the corner points the compatibility conditions

$$\mathbf{x}_\mathrm{s}(0) = \mathbf{x}_\mathrm{w}(0), \quad \mathbf{x}_\mathrm{s}(N) = \mathbf{x}_\mathrm{e}(0), \quad \mathbf{x}_\mathrm{n}(0) = \mathbf{x}_\mathrm{w}(M), \quad \mathbf{x}_\mathrm{n}(N) = \mathbf{x}_\mathrm{e}(M)$$

have to be fulfilled (see Fig. 3.13).

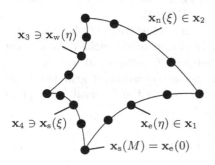

Fig. 3.13. Prescription of boundary grid points for algebraic generation of boundary-fitted grids

By algebraic grid generation the points in the interior of the domain are determined from the boundary grid points by an interpolation rule. Using a simple linear interpolation results, for instance, in the following relation

$$\mathbf{x}(\xi,\eta) = (1 - \frac{\eta}{M})\mathbf{x}_s(\xi) + \frac{\eta}{M}\mathbf{x}_n(\xi) + (1 - \frac{\xi}{N})\mathbf{x}_w(\eta) + \frac{\xi}{N}\mathbf{x}_e(\eta)$$
$$- \frac{\xi}{N}\left[\frac{\eta}{M}\mathbf{x}_n(M) + (1 - \frac{\eta}{M})\mathbf{x}_s(M)\right] \qquad (3.4)$$
$$-(1 - \frac{\xi}{N})\left[\frac{\eta}{M}\mathbf{x}_n(0) + (1 - \frac{\eta}{M})\mathbf{x}_s(0)\right],$$

which is known as *transfinite interpolation*. With formula (3.4) the coordinates of all interior grid points (for $\xi = 1, \ldots, N-1$ and $\eta = 1, \ldots, M-1$) are defined in terms of the given boundary grid points. For the problem domain shown in Fig. 3.13 with the given distribution of boundary grid points from (3.4) the grid shown in Fig. 3.14 results. Another grid generated with this method is shown in Fig. 3.16 (left). Generalizations of the transfinite interpolation, which are possible in different directions, result, for instance, by a prior partitioning of the problem domain into different subdomains or by the usage of higher-order interpolation rules (see, e.g., [13]).

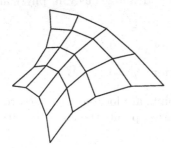

Fig. 3.14. Grid generated by transfinite interpolation

Frequently, it is desirable to cluster grid lines in certain areas of the problem domain in order to achieve there locally a higher discretization accuracy. Usually these are regions in which high gradients of the problem variables occur (e.g., in the vicinity of walls for flow problems).

By employing the transfinite interpolation a clustering of the grid lines can be achieved by a corresponding concentration of the grid points along the boundary curves. For this so-called *stretching functions* can be employed. A simple stretching function allowing for a concentration of grid points x_i ($i = 1, \ldots, N - 1$), for instance at the right end of the interval $[x_0, x_N]$, is given by

$$x_i = x_0 + \frac{\alpha^i - 1}{\alpha^N - 1}(x_N - x_0) \quad \text{for all } i = 0, \ldots, N, \qquad (3.5)$$

where the parameter $0 < \alpha < 1$ (expansion factor) serves to control the desired clustering of the grid points. The closer α is to zero, the denser the grid points are near x_N. In Fig. 3.15 the distributions resulting from different values for α are indicated.

Equation (3.5) is based on the well-known totals formula for geometric series. The grids generated this way have the property that the ratio of neighboring grid point distances is always constant, i.e.,

$$\frac{x_{i+1} - x_i}{x_i - x_{i-1}} = \alpha \quad \text{for all} \ \ i = 1, \ldots, N - 1.$$

By applying formula (3.5) to certain subregions a concentration of grid points can be achieved at arbitrary locations. Two- or three-dimensional concentrations of grid points can be obtained by a corresponding application of (3.5) in each spatial direction.

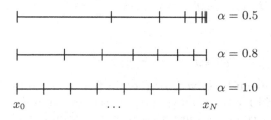

$\alpha = 0.5$

$\alpha = 0.8$

$\alpha = 1.0$

x_0 ... x_N

Fig. 3.15. Grid point clustering for $N = 7$ with different expansion factors α

From the above procedure already the general assets and drawbacks of algebraic grid generation become apparent. They are very easy to implement (including the clustering of grid lines), require little computational effort, and for simpler geometries possess enough flexibility for a quick generation of "reasonable" grids. All geometric quantities (metrics) needed for the subsequent computation can be computed analytically, such that no additional numerical errors arise. However, for more complex problem domains such methods are less suitable, because irregularities (e.g., kinks) in the boundary of the problem domain propagate into the interior and the proper control of the "smoothness" and the "distortion" of the grids is relatively difficult.

3.3.2 Elliptic Grid Generation

An alternative approach for the generation of structured grids is provided by techniques based on the solution of suitable partial differential equations. In this context one distinguishes between *hyperbolic*, *parabolic*, and *elliptic* methods, corresponding to the type of the underlying differential equation. We will address here only the elliptic method, which is the most widespread in practice.

With elliptic grid generation one creates the grid, for instance via a system of differential equations of the form

$$\xi_{xx} + \xi_{yy} = 0 \, ,$$
$$\eta_{xx} + \eta_{yy} = 0 \, , \tag{3.6}$$

which is solved in the problem domain with prescribed boundary grid points as boundary conditions. Here, one exploits the fact that elliptic differential equations, as for instance the above Laplace equation (3.6), fulfill a maximum principle stating that extremal values always are taken at the boundary. This ensures that a monotonic prescription of the boundary grid points always results in a valid grid, i.e., crossovers of grid lines do not occur.

For the actual determination of the grid coordinates the equations (3.6) have to be solved in the logical problem domain, i.e., dependent und independent variables have to be interchanged. In this way (by applying the chain rule) the following equations result:

$$c_1 x_{\xi\xi} - 2c_2 x_{\xi\eta} + c_3 x_{\eta\eta} = 0 \, ,$$
$$c_1 y_{\xi\xi} - 2c_2 y_{\xi\eta} + c_3 y_{\eta\eta} = 0 \tag{3.7}$$

with

$$c_1 = x_\eta^2 + y_\eta^2 \, , \quad c_2 = x_\xi x_\eta + y_\xi y_\eta \, , \quad c_3 = x_\xi^2 + y_\xi^2 \, .$$

This differential equation system can be solved for the physical grid coordinates x and y. As numerical grid for the solution process the logical grid is taken and the required boundary conditions are defined by the given boundary curves (in the physical domain). Note that the system (3.7) is non-linear, such that an iterative solution process is required (see Sect. 7.2). This can be started, for instance, with a grid that is generated algebraically beforehand.

By adding source terms to (3.6) a clustering of grid points in certain regions of the problem domain can be controlled:

$$\xi_{xx} + \xi_{yy} = P(\xi, \eta),$$
$$\eta_{xx} + \eta_{yy} = Q(\xi, \eta).$$

Regarding the (non-trivial) question of how to choose the functions P and Q in order to achieve a certain grid point distribution, we refer to the corresponding literature (e.g., [15]).

To illustrate the characteristic properties of algebraically and elliptically generated grids, Fig. 3.16 shows corresponding grids obtained for the same problem geometry with typical representatives of both approaches. One can observe the considerably "smoother" grid lines resulting from the elliptic method.

Concerning the advantages and disadvantages of the elliptic grid generation one can state that due to the necessity of solving (relatively simple) partial differential equations these methods are computationally more costly than algebraic methods. An advantage is that also in the case of boundary irregularities smooth grids in the interior result. The metric of the grid usually

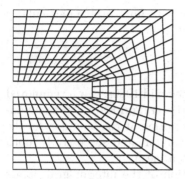 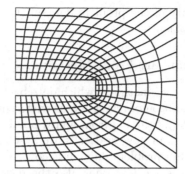

Fig. 3.16. Example for different properties of algebraically (left) and elliptically (right) generated grids

has to be determined numerically. However, this usually is not critical due to the smoothness of the grid. The concentration of grid points in certain regions of the problem domain is basically simple, but the "proper" choice of the corresponding functions P and Q in the concrete case might be problematic.

3.4 Generation of Unstructured Grids

A vital motivation for the usage of unstructured grids is the desire to automatize the grid generation process as far as possible. The ideal case would be to start form the description of the problem geometry by boundary curves or surfaces, so that a "feasible" grid is generated without any further intervention of the user. Since one comes closest to this ideal case with triangles or tetrahedras, usually these cell types are employed for unstructured grids. Recently, also interesting approaches for the automatic generation of quadrilateral or hexahedral grids for arbitrary geometries have been developed. The so-called *paving method* appears to be particularly promising (see, e.g., [5]). However, we will concentrate here on triangular grids only.

In practice, the most common techniques for the generation of unstructured triangular grids are *advancing-front methods* and *Delaunay triangulations*, which exist in a variety of variants. We will concentrate on identifying the basic ideas of these two approaches. Also combinations of both techniques, which try to exploit respective advantages (or avoid the disadvantages) are employed.

Other techniques for the generation of unstructured grids are *quadtree methods* (*octree methods* in three dimensions), which are based on recursive subdivisions of the problem domain. These methods are easy to implement, require little computational effort, and usually produce quite "good" grids in the interior of the problem domain. However, an essential disadvantage is that the corresponding grids often have a very irregular structure in the vicinity of the boundary of the problem domain (which is very unfavorable particularly for flow problems).

3.4.1 Advancing Front Methods

Advancing front methods, which date back to the mid 1980s, can be employed for the generation of triangular as well as quadrilateral grids (or also combinations of both). Starting from a grid point distribution at the boundary of the problem domain new grid cells are systematically created successively, until finally the full problem domain is covered with a mesh.

Let us assume first that also the distribution of the interior grid points already is prescribed. In this case the advancing front methods for generating triangular grids is as follows:

(i)   All edges along the inner boundary of the problem domain are successively numbered clockwise (not applicable if there are no inner boundaries). The edges along the outer boundary are successively numbered counterclockwise. The full numbering is consecutively stored in a vector **k** defining the advancing front.

(ii) For the last edge in the vector **k** all grid points are searched which are located on or within the advancing front. From these (admissible) grid points one is selected according to a certain criterion, e.g., the one for which the sum of the distances to the two grid points of the last edge is smallest. With the selected and the two points of the last edge a new triangle is formed.

(iii) The edges of the new triangle, which are contained in **k**, are deleted and the numbering of the remaining edges in **k** is adjusted (compression). The edges of the new triangle, which are not contained in **k**, are added at the end of **k**.

(iv) Steps (ii) and (iii) are repeated until all edges in **k** are deleted.

The procedure is illustrated in Fig. 3.18 for a simple example (without inner boundary).

The advancing front method also can be carried out without prescribing interior grid points, which, of course, only makes the algorithm interesting with respect to a mostly automatized grid generation. Here, at the beginning of step (ii), following a certain rule first a new interior grid point is created, e.g., such that this forms with the two points of the actual edge an equilateral triangle. However, one should check if the point created this way is admissible. In other words, it must be located such that the generated triangle does not intersect with an already created triangle and it should not be too close to an already existing point because this would lead (maybe only at a later stage of the algorithm) to a strongly degenerated triangle. If the created point is not admissible in this sense, another point can be created instead or a point already existing on the advancing front can be selected following the same criteria as above.

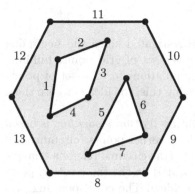

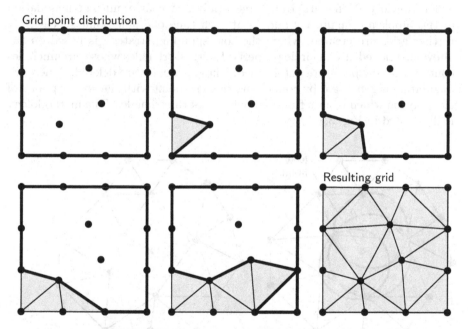

Fig. 3.17. Numbering of edges at inner and outer boundaries for advancing front method

Grid point distribution

Resulting grid

Fig. 3.18. Example for generation of unstructured grid with advancing front method (the thick lines represent the respective actual advancing fronts)

An advantage of the advancing front method is the simple possibility of the automatic generation of the interior grid points (with "good" quality of the triangles). In addition, it is always ensured – also in non-convex cases – that the boundary of the problem domain is represented by grid lines because the boundary discretization defines the starting point of the method and is not modified during the process. A disadvantage of the method must be seen in the relatively high computational effort, which, in particular, is necessary for checking if points are admissible and have tolerable distances.

3.4.2 Delaunay Triangulations

Delaunay triangulations, which have been investigated since the beginning of the 1980s, are unique triangulations of a given set of grid points fulfilling certain properties, i.e., out of all possible triangulations for this set of points these properties determine a unique one. Knowing this, the idea is to use these properties to create such triangulations.

One such approach, which is known as *Bowyer-Watson algorithm*, is based on the property that the circle through the three corner points (circumcircle) of an arbitrary triangle contains no other points (the circumcircle of a triangle is uniquely determined and its midpoint is located at the intersection point of the perpendicular bisectors of the triangle sides). The corresponding grid generation algorithm starts from a (generally very coarse) initial triangulation of the problem domain. By adding at each time one new point successively further grids are created, where the corresponding strategy is based on the above mentioned circumcircle property. First, all triangles whose circumcircles contain the new point are determined. These triangles are deleted. A new triangulation is generated by connecting the new point with the corner points of the polygon which results from the deletion of the triangles. The methodology is illustrated in Fig. 3.19.

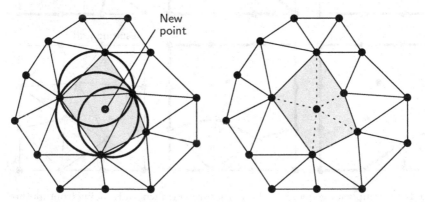

Fig. 3.19. Insertion of new grid point in a Delaunay triangulation with the Bowyer-Watson algorithm

The above triangulation procedure can be started "from scratch" by defining a "supertriangle" fully containing the problem domain. The nodes of the supertriangle are temporarily added to the list of nodes and the point insertion procedure is carried out as described above. Finally, after having inserted all points, all triangles which contain one or more vertices of the supertriangle are removed. As an example, the procedure is illustrated in Fig. 3.20 for the Delaunay triangulation of an ellipse with one internal grid point. The points

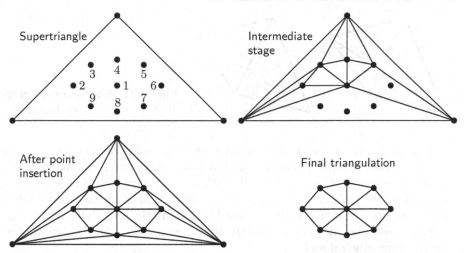

Fig. 3.20. Supertriangle and point insertion process for Delaunay triangulation of an ellipse

are inserted in the order of the indicated numbering. The intermediate stage shows the situation after the insertion of point 5.

If the locations at which the grid points should be added are prescribed, the above procedure simply is carried out as long as all points are inserted. However, as for the advancing-front method, also the Bowyer-Watson algorithm can be combined with a strategy for an automatic generation of grid points. A relatively simple approach for this can be realized, for instance, in connection with a priority list on the basis of a certain property of the triangles (e.g., the diameter of the circumcircle). The triangle with the highest priority is checked with respect to a certain criterion, e.g., the diameter of the circumcircle should be larger than a prescribed value. If this criterion is fulfilled, a new grid point is generated (e.g., the midpoint of the circumcircle) for this triangle. According to this procedure new triangles are inserted into the grid and added to the priority list. The algorithm stops if there is no more triangle fulfilling the given criterion.

Other methods for the generation of a Delaunay triangulation are based on the so called *edge-swapping technique*. Here, each time pairs of neighboring triangles possessing a common edge are considered. By the "swapping" of the common edge two other triangles are created, and based on a certain criterion one decides which of the two configurations is selected (see Fig. 3.21). Such a criterion can be, for instance, that the smallest angle occurring in the two triangles is maximal. For the example shown in Fig. 3.21 in this case the right configuration would be selected.

Methods for the generation of a Delaunay triangulations usually are less computationally intensive than advancing front methods because there are no elaborate checking routines with respect to intersections and minimal dis-

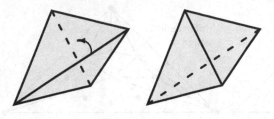

Fig. 3.21. Edge-swapping technique for neighboring triangles

tances necessary. A major disadvantage of these methods, however, is that for non-convex problem domains no boundary integrity is guaranteed, i.e., triangles can arise that are located (at least partially) outside the problem domain. In this case the grid has to be suitably modified in the corresponding region, where it is then not always possible to fulfill the underlying basic property. In particular, this can cause difficulties for corresponding generalizations for the three-dmensional case.

Exercises for Chap. 3

Exercise 3.1. For a two-dimensional problem domain the boundary curves $\mathbf{x}_1 = (s,0)$, $\mathbf{x}_2 = (1+2s-2s^2, s)$, $\mathbf{x}_3 = (s, 1-3s+3s^2)$, and $\mathbf{x}_4 = (0,s)$ with $0 \le s \le 1$ are given. (i) Determine the coordinate transformation resulting from the application of the transfinite interpolation (3.4) when the prescribed boundary points are the ones defined by $s = i/4$ with $i = 0, \dots, 4$. (ii) Discuss the uniqueness of the mapping. (iii) Use the stretching function (3.5) with $\alpha = 2/3$ for the clustering of grid points along the boundary curve $\mathbf{x}_4$. (iv) Transform the membrane equation (2.17) to the coordinates (ξ, η).

Exercise 3.2. For the problem geometry shown in Fig. 3.22 generate a triangular grid with the advancing front method with the black grid points. Afterwards insert the white grid points into the grid by means of the Bowyer-Watson algorithm.

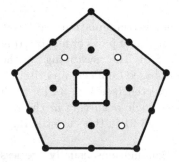

Fig. 3.22. Problem domain and grid point distribution for Exercise 3.2

4

Finite-Volume Methods

Finite-volume methods (FVM) – sometimes also called box methods – are mainly employed for the numerical solution of problems in fluid mechanics, where they were introduced in the 1970s by McDonald, MacCormack, and Paullay. However, the application of the FVM is not limited to flow problems. An important property of finite-volume methods is that the balance principles, which are the basis for the mathematical modelling of continuum mechanical problems, per definition, also are fulfilled for the discrete equations (conservativity). In this chapter we will discuss the most important basics of finite-volume discretizations applied to continuum mechanical problems. For clarity in the presentation of the essential principles we will restrict ourselves mainly to the two-dimensional case.

4.1 General Methodology

In general, the FVM involves the following steps:

(1) Decomposition of the problem domain into control volumes.
(2) Formulation of integral balance equations for each control volume.
(3) Approximation of integrals by numerical integration.
(4) Approximation of function values and derivatives by interpolation with nodal values.
(5) Assembling and solution of discrete algebraic system.

In the following we will outline in detail the individual steps (the solution of algebraic systems will be the topic of Chap. 7). We will do this by example for the general stationary transport equation (see Sect. 2.3.2)

$$\frac{\partial}{\partial x_i} \left(\rho v_i \phi - \alpha \frac{\partial \phi}{\partial x_i} \right) = f \tag{4.1}$$

for some problem domain Ω. We remark that a generalization of the FVM to other types of equations as given in Chap. 2 is straightforward (in Chap. 10 this will be done for the Navier-Stokes equations).

The starting point for a finite-volume discretization is a decomposition of the problem domain Ω into a finite number of subdomains V_i $(i = 1, \ldots, N)$, called *control volumes* (CVs), and related nodes where the unknown variables are to be computed. The union of all CVs should cover the whole problem domain. In general, the CVs also may overlap, but since this results in unnecessary complications we consider here the non-overlapping case only. Since finally each CV gives one equation for computing the nodal values, their final number (i.e., after the incorporation of boundary conditions) should be equal to the number of CVs. Usually, the CVs and the nodes are defined on the basis of a numerical grid, which, for instance, is generated with one of the techniques described in Chap. 3. In order to keep the usual terminology of the FVM, we always talk of volumes (and their surfaces), although strictly speaking this is only correct for the three-dimensional case.

For one-dimensional problems the CVs are subintervals of the problem interval and the nodes can be the midpoints or the edges of the subintervals (see Fig. 4.1).

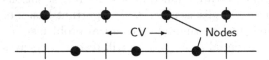

Fig. 4.1. Definitions of CVs and edge (top) and cell-oriented (bottom) arrangement of nodes for one-dimensional grids

In the two-dimensional case, in principle, the CVs can be arbitrary polygons. For quadrilateral grids the CVs usually are chosen identically with the grid cells. The nodes can be defined as the vertices or the centers of the CVs (see Fig. 4.2), often called edge or cell-centered approaches, respectively. For triangular grids, in principle, one could do it similarly, i.e., the triangles define the CVs and the nodes can be the vertices or the centers of the triangles. However, in this case other CV definitions are usually employed. One approach is closely related to the Delaunay triangulation discussed in Sect. 3.4.2. Here, the nodes are chosen as the vertices of the triangles and the CVs are defined as the polygons formed by the perpendicular bisectors of the sides of the surrounding triangles (see Fig. 4.3). These polygons are known as *Voronoi polygons* and in the case of convex problems domains and non-obtuse triangles there is a one-to-one correspondance to a Delaunay triangulation with its "nice" properties. However, this approach may fail for arbitrary triangulations. Another more general approach is to define a polygonal CV by joining the centroids and the midpoints of the edges of the triangles surrounding a node leading to the so-called *Donald polygons* (see Fig. 4.4).

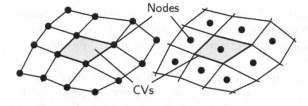

Fig. 4.2. Edge-oriented (left) and cell-oriented (right) arrangements of nodes for quadrilateral grids

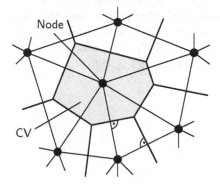

Fig. 4.3. Definition of CVs and nodes for triangular grids with Voronoi polygons

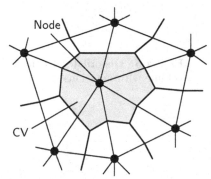

Fig. 4.4. Definition of CVs and nodes for triangular grids with Donald polygons

For three-dimensional problems on the basis of hexahedral or tetrahedral grids similar techniques as in the two-dimensional case can be applied (see, e.g., [26]).

After having defined the CVs, the balance equations describing the problem are formulated in integral form for each CV. Normally, these equations are directly available from the corresponding continuum mechanical conservation laws (applied to a CV), but they can also be derived by integration from the corresponding differential equations. By integration of (4.1) over an arbitrary control volume V and application of the Gauß integral theorem, one obtains:

$$\int\limits_S \left(\rho v_i \phi - \alpha \frac{\partial \phi}{\partial x_i} \right) n_i \, \mathrm{d}S = \int\limits_V f \, \mathrm{d}V \, , \qquad (4.2)$$

where S is the surface of the CV and n_i are the components of the unit normal vector to the surface. The integral balance equation (4.2) constitutes the starting point for the further discretization of the considered problem with an FVM.

As an example we consider quadrilateral CVs with a cell-oriented arrangement of nodes (a generalization to arbitrary polygons poses no principal difficulties). For a general quadrilateral CV we use the notations of the distinguished points (midpoint, midpoints of faces, and edge points) and the unit normal vectors according to the so-called compass notation as indicated in Fig. 4.5. The midpoints of the directly neighboring CVs we denote – again in compass notation – with capital letters S, SE, etc. (see Fig. 4.6).

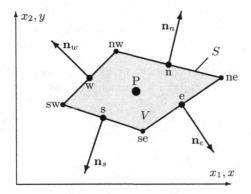

Fig. 4.5. Quadrilateral control volume with notations

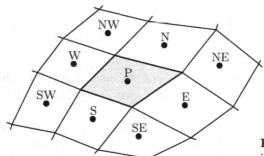

Fig. 4.6. Notations for neighboring control volumes

The surface integral in (4.2) can be split into the sum of the four surface integrals over the cell faces S_c ($c = e, w, n, s$) of the CV, such that the balance equation (4.2) can be written equivalently in the form

$$\sum_c \int_{S_c} \left(\rho v_i \phi - \alpha \frac{\partial \phi}{\partial x_i} \right) n_{ci} \, dS_c = \int_V f \, dV. \tag{4.3}$$

The expression (4.3) represents a balance equation for the convective and diffusive fluxes F_c^C and F_c^D through the CV faces, respectively, with

$$F_c^C = \int_{S_c} (\rho v_i \phi)\, n_{ci}\, \mathrm{d}S_c \quad \text{and} \quad F_c^D = -\int_{S_c} \left(\alpha \frac{\partial \phi}{\partial x_i} \right) n_{ci}\, \mathrm{d}S_c\,.$$

For the face S_e, for instance, the unit normal vector $\mathbf{n_e} = (n_{e1}, n_{e2})$ is defined by the following (geometric) conditions:

$$(\mathbf{x_{ne}} - \mathbf{x_{se}}) \cdot \mathbf{n_e} = 0 \quad \text{und} \quad |\mathbf{n_e}| = \sqrt{n_{e1}^2 + n_{e2}^2} = 1\,.$$

From this one obtains the representation

$$\mathbf{n_e} = \frac{(y_{ne} - y_{se})}{\delta S_e} \mathbf{e_1} - \frac{(x_{ne} - x_{se})}{\delta S_e} \mathbf{e_2}\,, \tag{4.4}$$

where

$$\delta S_e = |\mathbf{x_{ne}} - \mathbf{x_{se}}| = \sqrt{(x_{ne} - x_{se})^2 + (y_{ne} - y_{se})^2}$$

denotes the length of the face S_e. Analogous relations result for the other CV faces.

For neighboring CVs with a common face the absolute value of the total flux $F_c = F_c^C + F_c^D$ through this face is identical, but the sign differs. For instance, for the CV around point P the flux F_e is equal to the flux $-F_w$ for the CV around point E (since $(\mathbf{n_e})_P = -(\mathbf{n_w})_E$). This is exploited for the implementation of the method in order to avoid on the one hand a double computation for the fluxes and on the other hand to ensure that the corresponding absolute fluxes really are equal (important for conservativity, see Sect. 8.1.4). In the case of quadrilateral CVs the computation can be organized in such a way that, starting from a CV face at the boundary of the problem domain, for instance, only F_e und F_n have to be computed.

It should be noted that up to this point we haven't introduced any approximation, i.e., the flux balance (4.3) is still exact. The actual discretization now mainly consists in the approximation of the surface integrals and the volume integral in (4.3) by suitable averages of the corresponding integrands at the CV faces. Afterwards, these have to be put into proper relation to the unknown function values in the nodes.

4.2 Approximation of Surface and Volume Integrals

We start with the approximation of the surface integrals in (4.3), which for a cell-centered variable arrangement suitably is carried out in two steps:

(1) Approximation of the surface integrals (fluxes) by values on the CV faces.

(2) Approximation of the variable values at the CV faces by node values.

As an example let us consider the approximation of the surface integral

$$\int_{S_e} w_i n_{ei} \, \mathrm{d}S_e$$

over the face S_e of a CV for a general integrand function $\mathbf{w} = (w_1(\mathbf{x}), w_2(\mathbf{x}))$ (the other faces can be treated in a completely analogous way).

The integral can be approximated in different ways by involving more or less values of the integrand at the CV face. The simplest possibility is an approximation by just using the midpoint of the face:

$$\int_{S_e} w_i n_{ei} \, \mathrm{d}S_e \approx g_e \, \delta S_e \,, \tag{4.5}$$

where we denote with $g_e = w_{ei} n_{ei}$ the normal component of $\mathbf{w}$ at the location e. With this, one obtains an approximation of 2nd order (with respect to the face length δS_e) for the surface integral, which can be checked by means of a Taylor series expansion (Exercise 4.1). The integration formula (4.5) corresponds to the *midpoint rule* known from numerical integration.

Other common integration formulas, that can be employed for such approximations are, for instance, the *trapezoidal rule* and the *Simpson rule*. The corresponding formulas are summarized in Table 4.1 with their respective orders (with respect to δS_e).

Table 4.1. Approximations for surface integrals over the face S_e

Name	Formula	Order
Midpoint rule	$\delta S_e g_e$	2
Trapezoidal rule	$\delta S_e (g_{ne} + g_{se})/2$	2
Simpson rule	$\delta S_e (g_{ne} + 4g_e + g_{se})/6$	4

For instance, by applying the midpoint rule for the approximation of the convective and diffusive fluxes through the CV faces in (4.3), we obtain the approximations:

$$F_c^C \approx \underbrace{\rho v_i n_{ci} \delta S_c}_{\dot{m}_c} \phi_c \quad \text{and} \quad F_c^D \approx -\alpha n_{ci} \delta S_c \left(\frac{\partial \phi}{\partial x_i} \right)_c ,$$

where, for simplicity, we have assumed that v_i, ρ, and α are constant across the CV. $\dot{m}_c$ denotes the mass flux through the face S_c. Inserting the definition of the normal vector, we obtain, for instance, for the convective flux through the face S_e, the approximation

$$F_e^C \approx \dot{m}_e \phi_e = \rho[v_1(y_{ne} - y_{se}) - v_2(x_{ne} - x_{se})].$$

Before we turn to the further discretization of the fluxes, we first deal with the approximation of the volume integral in (4.3), which normally also is carried out by means of numerical integration. The assumption that the value f_P of f in the CV center represents an average value over the CV leads to the two-dimensional midpoint rule:

$$\int_V f \, dV \approx f_P \, \delta V,$$

where δV denotes the volume of the CV, which for a quadrilateral CV is given by

$$\delta V = \frac{1}{2}|(x_{se} - x_{nw})(y_{ne} - y_{sw}) - (x_{ne} - x_{sw})(y_{se} - y_{nw})|.$$

An overview of the most common two-dimensional integration formulas for Cartesian CVs with the corresponding error order (with respect to δV) is given in Fig. 4.7 showing a schematical representation with the corresponding location of integration points and weighting factors. As a formula this means, e.g., in the case of the Simpson rule, an approximation of the form:

$$\int_V f \, dV \approx \frac{\delta V}{36}(16 f_P + 4 f_e + 4 f_w + 4 f_n + 4 f_s + f_{ne} + f_{se} + f_{ne} + f_{se}).$$

It should be noted that the formulas for the two-dimensional numerical integration can be used to approximate the surface integrals occurring in three-dimensional applications. For three-dimensional volume integrals analogous integration formulas as for the two-dimensional case are available.

In summary, by applying the midpoint rule (to which we will retrict ourselves) we now have the following approximation for the balance equation (4.3):

$$\underbrace{\sum_c \dot{m}_c \phi_c}_{\text{conv. fluxes}} - \underbrace{\sum_c \alpha n_{ci} \, \delta S_c \left(\frac{\partial \phi}{\partial x_i}\right)_c}_{\text{diff. fluxes}} = \underbrace{f_P \, \delta V.}_{\text{source}} \qquad (4.6)$$

In the next step it is necessary to approximate the function values and derivatives of ϕ at the CV faces occurring in the convective and diffusive flux expressions, respectively, by variable values in the nodes (here the CV centers). In order to clearly outline the essential principles, we will first explain the corresponding approaches for a two-dimensional Cartesian CV as indicated in Fig. 4.8. In this case the unit normal vectors n_c along the CV faces are given by

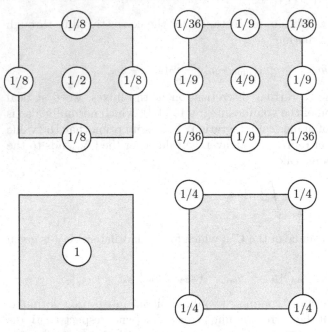

Fig. 4.7. Schematic representation of numerical integration formulas for two-dimensional volume integrals over a Cartesian CV

$$\mathbf{n}_e = \mathbf{e}_1\,, \quad \mathbf{n}_w = -\mathbf{e}_1\,, \quad \mathbf{n}_n = \mathbf{e}_2\,, \quad \mathbf{n}_s = -\mathbf{e}_2$$

and the expressions for the mass fluxes through the CV faces simplify to

$$\dot{m}_e = \rho v_1(y_n - y_s)\,, \quad \dot{m}_n = \rho v_2(x_c - x_w)\,,$$
$$\dot{m}_w = \rho v_1(y_s - y_n)\,, \quad \dot{m}_s = \rho v_2(x_w - x_e)\,.$$

Particularities that arise due to non-Cartesian grids will be considered in Sect. 4.5.

4.3 Discretization of Convective Fluxes

For the further approximation of the convective fluxes F_c^C, it is necessary to approximate ϕ_c by variable values in the CV centers. In general, this involves using neighboring nodal values ϕ_E, ϕ_P, ... of ϕ_c. The methods most frequently employed in practice for the approximation will be explained in the following, where we can restrict ourselves to one-dimensional considerations for the face S_e, since the other faces and the second (or third) spatial dimension can be treated in a fully analogous way. Traditionally, the corresponding approximations are called differencing techniques, since they result

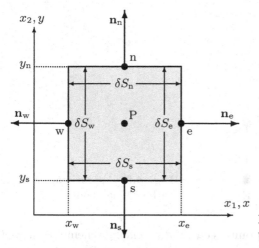

Fig. 4.8. Cartesian control volume with notations

in formulas analogous to finite-difference methods. Strictly speaking, these are interpolation techniques.

4.3.1 Central Differences

For the *central differencing scheme (CDS)* ϕ_e is approximated by linear interpolation with the values in the neighboring nodes P und E (see Fig. 4.9):

$$\phi_e \approx \gamma_e \phi_E + (1 - \gamma_e)\phi_P . \tag{4.7}$$

The interpolation factor γ_e is defined by

$$\gamma_e = \frac{x_e - x_P}{x_E - x_P} .$$

The approximation (4.7) has, for an equidistant grid as well as for a non-equidistant grid, an interpolation error of 2nd order. This can be seen from a Taylor series expansion of ϕ around the point x_P:

$$\phi(x) = \phi_P + (x - x_P)\left(\frac{\partial \phi}{\partial x}\right)_P + \frac{(x - x_P)^2}{2}\left(\frac{\partial^2 \phi}{\partial x^2}\right)_P + T_H ,$$

where T_H denotes the terms of higher order. Evaluating this series at the locations x_e and x_E and taking the difference leads to the relation

$$\phi_e = \gamma_e \phi_E + (1 - \gamma_e)\phi_P - \frac{(x_e - x_P)(x_E - x_e)}{2}\left(\frac{\partial^2 \phi}{\partial x^2}\right)_P + T_H ,$$

which shows that the leading error term depends quadratically on the grid spacing.

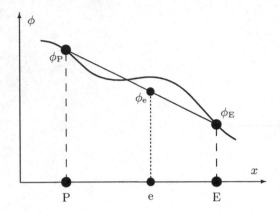

Fig. 4.9. Approximation of ϕ_e with CDS method

By involving additional grid points, central differencing schemes of higher order can be defined. For instance, an approximation of 4th order for an equidistant grid is given by

$$\phi_e = \frac{1}{48}(-3\phi_{EE} + 27\phi_E + 27\phi_P - 3\phi_W),$$

where EE denotes the "east" neighboring point of E (see Fig. 4.11). Note that an application of this formula only makes sense if it is used together with an integration formula of 4th order, e.g., the Simpson rule. Only in this case is the total approximation of the convective flux also of 4th order.

When using central differencing approximations unphysical oscillations may appear in the numerical solution (the reasons for this problem will be discussed in detail in Sect. 8.1). Therefore, one often uses so-called *upwind approximations*, which are not sensitive or less sensitive to this problem. The principal idea of these methods is to make the interpolation dependent on the direction of the velocity vector. Doing so, one exploits the transport property of convection processes, which means that the convective transport of ϕ only takes place "downstream". In the following we will discuss two of the most important upwind techniques.

4.3.2 Upwind Techniques

The simplest upwind method results if ϕ is approximated by a step function. Here, ϕ_e is determined depending on the direction of the mass flux as follows (see Fig. 4.10):

$$\phi_e = \phi_P, \quad \text{if } \dot{m}_e > 0,$$
$$\phi_e = \phi_E, \quad \text{if } \dot{m}_e < 0.$$

This method is called *upwind differencing scheme (UDS)*. A Taylor series expansion of ϕ around the point x_P, evaluated at the point x_e, gives:

$$\phi_e = \phi_P + (x_e - x_P)\left(\frac{\partial\phi}{\partial x}\right)_P + \frac{(x_e - x_P)^2}{2}\left(\frac{\partial^2\phi}{\partial x^2}\right)_P + T_H .$$

This shows that the UDS method (independent of the grid) has an interpolation error of 1st order. The leading error term in the resulting approximation of the convective flux F_e^C becomes

$$\underbrace{\dot{m}_e(x_e - x_P)}_{\alpha_{num}}\left(\frac{\partial\phi}{\partial x}\right)_P .$$

The error caused by this is called *artificial* or *numerical diffusion*, since the error term can be interpreted as a diffusive flux. The coefficient α_{num} is a measure for the amount of the numerical diffusion. If the transport direction is nearly perpendicular to the CV face, the approximation of the convective fluxes resulting with the UDS method is comparably good (the derivative $(\partial\phi/\partial x)_P$ is then small). Otherwise the approximation can be quite inaccurate and for large mass fluxes (i.e., large velocities) it can then be necessary to employ very fine grids (i.e., $x_e - x_P$ very small) for the computation in order to achieve a solution with an adequate accuracy. The disadvantage of the relatively poor accuracy is confronted by the advantage that the UDS method leads to an unconditionally bounded solution algorithm. We will discuss this aspect in more detail in Sect. 8.1.5.

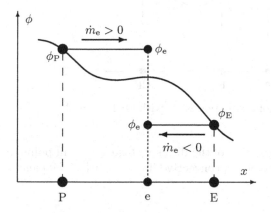

Fig. 4.10. Mass flux dependent approximation of ϕ_e with UDS method

An upwind approximation frequently employed in practice is the quadratic upwind interpolation, which in the literature is known as the QUICK method (Quadratic Upwind Interpolation for Convective Kinematics). Here, a quadratic polynomial is fitted through the two neighboring points P and E, and a third point, which is located upstream (W or EE depending on the flow direction). Evaluating this polynomial at point e one obtains the approximation (see also Fig. 4.11):

$$\phi_e = a_1\phi_E - a_2\phi_W + (1 - a_1 + a_2)\phi_P , \quad \text{if } \dot{m}_e > 0 ,$$
$$\phi_e = b_1\phi_P - b_2\phi_{EE} + (1 - b_1 + b_2)\phi_E , \quad \text{if } \dot{m}_e < 0 ,$$

where

$$a_1 = \frac{(2 - \gamma_w)\gamma_e^2}{1 + \gamma_e - \gamma_w} , \qquad a_2 = \frac{(1 - \gamma_e)(1 - \gamma_w)^2}{1 + \gamma_e - \gamma_w} ,$$
$$b_1 = \frac{(1 + \gamma_w)(1 - \gamma_e)^2}{1 + \gamma_{ee} - \gamma_e} , \qquad b_2 = \frac{\gamma_{ee}^2\gamma_e}{1 + \gamma_{ee} - \gamma_e} .$$

For an equidistant grid one has:

$$a_1 = \frac{3}{8} , \quad a_2 = \frac{1}{8} , \quad b_1 = \frac{3}{8} , \quad b_2 = \frac{1}{8} .$$

In this case the QUICK method possesses an interpolation error of 3rd order. However, if it is used together with numerical integration of only 2nd order the overall flux approximation also is only of 2nd order, but it is somewhat more accurate than with the CDS method.

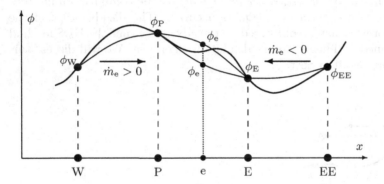

Fig. 4.11. Mass flux dependent approximation of ϕ_e with QUICK method

Before we turn to the discretization of the diffusive fluxes, we will point to a special technique for the treatment of convective fluxes, which is frequently employed for transport equations.

4.3.3 Flux-Blending Technique

The principal idea of *flux-blending*, which goes back to Khosla und Rubin (1974), is to mix different approximations for the convective flux. In this way one attempts to combine the advantages of an accurate approximation of a higher order scheme with the better robustness and boundedness properties of a lower order scheme (mostly the UDS method).

To explain the method we again consider exemplarily the face S_e of a CV. The corresponding approximations for ϕ_e in the convective flux F_e^C for the two methods to be combined are denoted by ϕ_e^{ML} and ϕ_e^{MH}, where ML and MH are the lower and higher order methods, respectively. The approximation for the combined method reads:

$$\phi_e \approx (1 - \beta)\phi_e^{\text{ML}} + \beta\phi_e^{\text{MH}} = \phi_e^{\text{ML}} + \underbrace{\beta(\phi_e^{\text{MH}} - \phi_e^{\text{ML}})}_{b_\beta^{\phi,e}} . \qquad (4.8)$$

From (4.8) for $\beta=0$ and $\beta=1$ the methods ML and MH, respectively, result. However, it is possible to choose for β any other value between 0 and 1, allowing to control the portions of the corresponding methods according to the needs of the underlying problem. However, due to the loss in accuracy, values $\beta < 1$ should be selected only if with $\beta = 1$ on the given grid no "reasonable" solution can be obtained (see Sect. 8.1.5) and a finer grid is not possible due to limitations in memory or computing time.

Also, if $\beta = 1$ (i.e., the higher order method) is employed, it can be beneficial to use the splitting according to (4.8) in order to treat the term $b_\beta^{\phi,e}$ "explicitly" in combination with an iterative solver. This means that this term is computed with (known) values of ϕ from the preceding iteration and added to the source term. This may lead to a more stable iterative solution procedure, since this (probably critical) term then makes no contribution to the system matrix, which becomes more diagonally dominant. It should be pointed out that this modification has no influence on the converged solution, which is identical to that obtained with the higher order method MH alone. We will discuss this approach in some more detail at the end of Sect. 7.1.4.

4.4 Discretization of Diffusive Fluxes

For the approximation of diffusive fluxes it is necessary to approximate the values of the normal derivative of ϕ at the CV faces by nodal values in the CV centers. For the east face S_e of the CV, which we will again consider exemplarily, one has to approximate (in the Cartesian case) the derivative $(\partial\phi/\partial x)_e$. For this, difference formulas as they are common in the framework of the finite-difference method can be used (see, e.g., [9]).

The simplest approximation one obtains when using a central differencing formula

$$\left(\frac{\partial\phi}{\partial x}\right)_e \approx \frac{\phi_E - \phi_P}{x_E - x_P} , \qquad (4.9)$$

which is equivalent to the assumption that ϕ is a linear function between the points x_P and x_E (see Fig. 4.12). For the discussion of the error of this approximation, we consider the difference of the Taylor series expansion around x_e at the locations x_P and x_E:

$$\left(\frac{\partial\phi}{\partial x}\right)_e = \frac{\phi_E - \phi_P}{x_E - x_P} + \frac{(x_e - x_P)^2 - (x_E - x_e)^2}{2(x_E - x_P)}\left(\frac{\partial^2\phi}{\partial x^2}\right)_e$$
$$- \frac{(x_e - x_P)^3 + (x_E - x_e)^3}{6(x_E - x_P)}\left(\frac{\partial^3\phi}{\partial x^3}\right)_e + T_H.$$

One can observe that for an equidistant grid an error of 2nd order results, since in this case the coefficient in front of the second derivative is zero. In the case of non-equidistant grids, one obtains by a simple algebraic rearrangement that this leading error term is proportional to the grid spacing and the expansion rate ξ_e of neighboring grid spacings:

$$\frac{(1 - \xi_e)(x_e - x_P)}{2}\left(\frac{\partial^2\phi}{\partial x^2}\right)_e \quad \text{with} \quad \xi_e = \frac{x_E - x_e}{x_e - x_P}.$$

This means that the portion of the 1st order error term gets larger the more the expansion rate deviates from 1. This aspect should be taken into account in the grid generation such that neighboring CVs do not differ that much in the corresponding dimensions (see also Sect. 8.3).

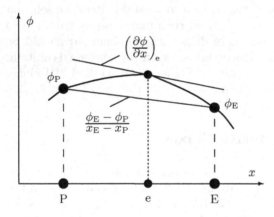

Fig. 4.12. Central differencing formula for approximation of 1st derivative at CV face

One obtains a 4th order approximation of the derivative at the CV face for an equidistant grid by

$$\left(\frac{\partial\phi}{\partial x}\right)_e \approx \frac{1}{24\Delta x}(\phi_W - 27\phi_P + 27\phi_E - \phi_{EE}), \tag{4.10}$$

which, for instance, can be used together with the Simpson rule to obtain an overall approximation for the diffusive flux of 4th order.

Although principally there are also other possibilities for approximating the derivatives (e.g., forward or backward differencing formulas), in practice almost only central differencing formulas are employed, which possess the best accuracy for a given number of grid points involved in the discretization. Problems with boundedness, as for the convective fluxes, do not exist. Thus, there is no reason to use less accurate approximations. For CVs located at the boundary of the problem domain, it might be necessary to employ forward or backward differencing formulas because there are no grid points beyond the boundary (see Sect. 4.7).

4.5 Non-Cartesian Grids

The previous considerations with respect to the discretization of the convective and diffusive fluxes were confined to the case of Cartesian grids. In this section we will discuss necessary modifications for general (quadrilateral) CVs.

For the convective fluxes, simple generalizations of the schemes introduced in Sect. 4.3 (e.g., UDS, CDS, QUICK, ...) can be employed for the approximation of ϕ_c. For instance, a corresponding CDS approximation for ϕ_e reads:

$$\phi_e \approx \frac{|\mathbf{x}_{\tilde{e}} - \mathbf{x}_P|}{|\mathbf{x}_E - \mathbf{x}_P|} \phi_E + \frac{|\mathbf{x}_E - \mathbf{x}_{\tilde{e}}|}{|\mathbf{x}_E - \mathbf{x}_P|} \phi_P , \qquad (4.11)$$

where $\mathbf{x}_{\tilde{e}}$ is the intersection of the connnecting line of the points P and E with the (probably extended) CV face S_e (see Fig. 4.13). For the convective flux through S_e this results in the following approximation:

$$F_e^C \approx \frac{\dot{m}_e}{|\mathbf{x}_E - \mathbf{x}_P|} \left(|\mathbf{x}_{\tilde{e}} - \mathbf{x}_P| \phi_E + |\mathbf{x}_E - \mathbf{x}_{\tilde{e}}| \phi_P \right) .$$

When the grid at the corresponding face has a "kink", an additional error results because the points $\mathbf{x}_{\tilde{e}}$ and $\mathbf{x}_e$ do not coincide (see Fig. 4.13). This aspect should be taken into account for the grid generation (see also Sect. 8.3).

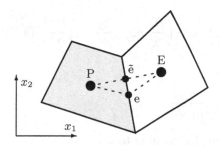

Fig. 4.13. Central difference approximation of convective fluxes for non-Cartesian control volumes

Let us turn to the approximation of the diffusive fluxes, for which farther reaching distinctions to the Cartesian case arise as for the convective fluxes. Here, for the required approximation of the normal derivative of ϕ in the center of the CV face there are a variety of different possibilities, depending on the directions in which the derivative is approximated, the locations where the appearing derivatives are evaluated, and the node values which are used for the interpolation. As an example we will give here one variant and consider only the CV face S_e.

Since along the normal direction in general there are no nodal points, the normal derivative has to be expressed by derivatives along other suitable directions. For this we use here the coordinates $\tilde{\xi}$ and $\tilde{\eta}$ defined according to Fig. 4.14. The direction $\tilde{\xi}$ is determined by the connecting line between points P and E, and the direction $\tilde{\eta}$ is determined by the direction of the CV face. Note that $\tilde{\xi}$ and $\tilde{\eta}$, because of a distortion of the grid, can deviate from the directions ξ und η, which are defined by the connecting lines of P with the CV face centers e and n. The larger these deviations are, the larger the discretization error becomes. This is another aspect that has to be taken into account when generating the grid (see also Sect. 8.3).

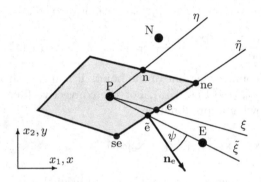

Fig. 4.14. Approximation of diffusive fluxes for non-Cartesian control volumes

A coordinate transformation $(x, y) \to (\tilde{\xi}, \tilde{\eta})$ results for the normal derivative in the following representation:

$$\frac{\partial \phi}{\partial x} n_{e1} + \frac{\partial \phi}{\partial y} n_{e2} = \frac{1}{J}\left[\left(\frac{\partial y}{\partial \tilde{\eta}} n_{e1} - \frac{\partial x}{\partial \tilde{\eta}} n_{e2}\right)\frac{\partial \phi}{\partial \tilde{\xi}} + \left(\frac{\partial x}{\partial \tilde{\xi}} n_{e2} - \frac{\partial y}{\partial \tilde{\xi}} n_{e1}\right)\frac{\partial \phi}{\partial \tilde{\eta}}\right] \quad (4.12)$$

with the Jacobi determinant

$$J = \frac{\partial x}{\partial \tilde{\xi}}\frac{\partial y}{\partial \tilde{\eta}} - \frac{\partial y}{\partial \tilde{\xi}}\frac{\partial x}{\partial \tilde{\eta}}.$$

The metric quantities can be approximated according to

$$\frac{\partial \mathbf{x}}{\partial \tilde{\xi}} \approx \frac{\mathbf{x}_E - \mathbf{x}_P}{|\mathbf{x}_E - \mathbf{x}_P|} \quad \text{and} \quad \frac{\partial \mathbf{x}}{\partial \tilde{\eta}} \approx \frac{\mathbf{x}_{ne} - \mathbf{x}_{se}}{\delta S_e}, \quad (4.13)$$

which results for the Jacobi determinant in the approximation

$$J_e \approx \frac{(x_E - x_P)(y_{ne} - y_{se}) - (y_E - y_P)(x_{ne} - x_{se})}{|\mathbf{x}_E - \mathbf{x}_P|\delta S_e} = \cos \psi,$$

where ψ denotes the angle between the direction $\tilde{\xi}$ and $\mathbf{n}_e$ (see Fig. 4.14). ψ is a measure for the deviation of the grid from orthogonality ($\psi = 0$ for an orthogonal grid).

The derivatives with respect to $\tilde{\xi}$ and $\tilde{\eta}$ in (4.12) can be approximated in the usual way with a finite-difference formula. For example, the use of a central difference of 2nd order gives:

$$\frac{\partial \phi}{\partial \tilde{\xi}} \approx \frac{\phi_E - \phi_P}{|\mathbf{x}_E - \mathbf{x}_P|} \quad \text{and} \quad \frac{\partial \phi}{\partial \tilde{\eta}} \approx \frac{\phi_{ne} - \phi_{se}}{\delta S_e}. \tag{4.14}$$

Inserting the approximations (4.13) and (4.14) into (4.12) and using the component representation (4.4) of the unit normal vector $\mathbf{n}_e$ we finally obtain the following approximation for the diffusive flux through the CV face S_e:

$$F_e^D \approx D_e(\phi_E - \phi_P) + N_e(\phi_{ne} - \phi_{se}) \tag{4.15}$$

with

$$D_e = \frac{\alpha \left[(y_{ne} - y_{se})^2 + (x_{ne} - x_{se})^2\right]}{(x_{ne} - x_{se})(y_E - y_P) - (y_{ne} - y_{se})(x_E - x_P)}, \tag{4.16}$$

$$N_e = \frac{\alpha \left[(y_{ne} - y_{se})(y_E - y_P) + (x_{ne} - x_{se})(x_E - x_P)\right]}{(y_{ne} - y_{se})(x_E - x_P) - (x_{ne} - x_{se})(y_E - y_P)}. \tag{4.17}$$

The coefficient N_e represents the portion that arise due to the non-orthogonality of the grid. If the grid is orthogonal, $\mathbf{n}_e$ and $\mathbf{x}_E - \mathbf{x}_P$ have the same direction such that $N_e = 0$. The coefficient N_e (and the corresponding values for the other CV faces) should be kept as small as possible (see als Sect. 8.3).

The values for ϕ_{ne} and ϕ_{se} in (4.15) can be approximated, for instance, by linear interpolation of four neighboring nodal values:

$$\phi_{ne} = \frac{\gamma_P \phi_P + \gamma_E \phi_E + \gamma_N \phi_N + \gamma_{NE} \phi_{NE}}{\gamma_P + \gamma_E + \gamma_N + \gamma_{NE}}$$

with suitable interpolation factors γ_P, γ_E, γ_N, and γ_{NE} (see Fig. 4.15).

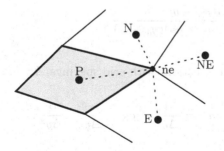

Fig. 4.15. Interpolation of values in CV edges for discretization of diffusive fluxes for non-Cartesian CV

4.6 Discrete Transport Equation

Let us now return to our example of the general two-dimensional transport equation (4.3) and apply the approximation techniques introduced in the preceding sections to it.

We employ exemplarily the midpoint rule for the integral approximations, the UDS method for the convective flux, and the CDS method for the diffusive flux. Additionally, we assume that we have velocity components $v_1, v_2 > 0$ and that the grid is a Cartesian one. With these assumptions one obtains the following approximation of the balance equation (4.3):

$$
\left(\rho v_1 \phi_P - \alpha \frac{\phi_E - \phi_P}{x_E - x_P} \right) (y_n - y_s)
$$

$$
- \left(\rho v_1 \phi_W - \alpha \frac{\phi_P - \phi_W}{x_P - x_W} \right) (y_n - y_s)
$$

$$
+ \left(\rho v_2 \phi_P - \alpha \frac{\phi_N - \phi_P}{y_N - y_P} \right) (x_e - x_w)
$$

$$
- \left(\rho v_2 \phi_S - \alpha \frac{\phi_P - \phi_S}{y_P - y_s} \right) (x_e - x_w) = f_P (y_n - y_s)(x_e - x_w) .
$$

A simple rearrangement gives a relation of the form

$$
a_P \phi_P = a_E \phi_E + a_W \phi_W + a_N \phi_N + a_S \phi_S + b_P \tag{4.18}
$$

with the coefficients

$$
a_E = \frac{\alpha}{(x_E - x_P)(x_e - x_w)} ,
$$

$$
a_W = \frac{\rho v_1}{x_e - x_w} + \frac{\alpha}{(x_P - x_W)(x_e - x_w)} ,
$$

$$
a_N = \frac{\alpha}{(y_N - y_P)(y_n - y_s)} ,
$$

$$
a_S = \frac{\rho v_2}{y_n - y_s} + \frac{\alpha}{(y_P - y_s)(y_n - y_s)} ,
$$

$$
a_P = \frac{\rho v_1}{x_e - x_w} + \frac{\alpha(x_E - x_W)}{(x_P - x_W)(x_E - x_P)(x_e - x_w)} +
$$

$$
\frac{\rho v_2}{y_n - y_s} + \frac{\alpha(y_N - y_s)}{(y_P - y_s)(y_N - y_P)(y_n - y_s)} ,
$$

$$
b_P = f_P .
$$

If the grid is equidistant in each spatial direction (with grid spacings Δx and Δy), the coefficients become:

$$
a_E = \frac{\alpha}{\Delta x^2}, \quad a_W = \frac{\rho v_1}{\Delta x} + \frac{\alpha}{\Delta x^2}, \quad a_N = \frac{\alpha}{\Delta y^2}, \quad a_S = \frac{\rho v_2}{\Delta y} + \frac{\alpha}{\Delta y^2},
$$

$$
a_P = \frac{\rho v_1}{\Delta x} + \frac{2\alpha}{\Delta x^2} + \frac{\rho v_2}{\Delta y} + \frac{2\alpha}{\Delta y^2}, \quad b_P = f_P .
$$

In this particular case (4.18) coincides with a discretization that would result from a corresponding finite-difference method (for general grids this normally is not the case).

It can be seen that – independent from the grid employed – one has for the coefficients in (4.18) the relation

$$a_P = a_E + a_W + a_N + a_S \, .$$

This is characteristic for finite-volume discretizations and expresses the conservativity of the method. We will return to this important property in Sect. 8.1.4.

Equation (4.18) is valid in this form for all CVs, which are not located at the boundary of the problem domain. For boundary CVs the approximation (4.18) includes nodal values outside the problem domain, such that they require a special treatment depending on the given type of boundary condition.

4.7 Treatment of Boundary Conditions

We consider the three boundary condition types that most frequently occur for the considered type of problems (see Chap. 2): a prescibed variable value, a prescibed flux, and a symmetry boundary. For an explanation of the implementation of such conditions into a finite-volume method, we consider as an example a Cartesian CV at the west boundary (see Fig. 4.16) for the transport equation (4.3). Correspondingly modified approaches for the non-Cartesian case or for other types of equations can be formulated analogously (for this see also Sect. 10.4).

Let us start with the case of a prescribed boundary value $\phi_w = \phi^0$. For the convective flux at the boundary one has the approximation:

$$F_w^C \approx \dot{m}_w \phi_w = \dot{m}_w \phi^0 \, .$$

With this the approximation of F_w^C is known (the mass flux $\dot{m}_w$ at the boundary is also known) and can simply be introduced in the balance equation (4.6). This results in an additional contribution to the source term b_P.

The diffusive flux through the boundary is determined with the same approach as in the interior of the domain (see (4.18)). Analogously to (4.9) the derivative at the boundary can be approximated as follows:

$$\left(\frac{\partial \phi}{\partial x} \right)_w \approx \frac{\phi_P - \phi_w}{x_P - x_w} = \frac{\phi_P - \phi^0}{x_P - x_w} \, . \tag{4.19}$$

This corresponds to a forward difference formula of 1st order. Of course, it is also possible to apply more elaborate formulas of higher order. However, since the distance between the boundary point w and the point P is smaller than the distance between two inner points (half as much for an equidistant grid, see Fig. 4.16), a lower order approximation at the boundary usually does not influence the overall accuracy that much.

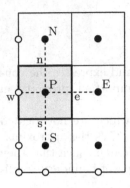

Fig. 4.16. Cartesian boundary CV at west boundary with notations

In summary, one has for the considered boundary CV a relation of the form (4.18) with the modified coefficients:

$$a_{\mathrm{W}} = 0\,,$$

$$a_{\mathrm{P}} = \frac{\rho v_1}{x_{\mathrm{e}} - x_{\mathrm{w}}} + \frac{\alpha(x_{\mathrm{E}} - x_{\mathrm{w}})}{(x_{\mathrm{P}} - x_{\mathrm{w}})(x_{\mathrm{E}} - x_{\mathrm{P}})(x_{\mathrm{e}} - x_{\mathrm{w}})} +$$

$$\frac{\rho v_2}{y_{\mathrm{n}} - y_{\mathrm{s}}} + \frac{\alpha(y_{\mathrm{N}} - y_{\mathrm{s}})}{(y_{\mathrm{P}} - y_{\mathrm{s}})(y_{\mathrm{N}} - y_{\mathrm{P}})(y_{\mathrm{n}} - y_{\mathrm{s}})}\,,$$

$$b_{\mathrm{P}} = f_{\mathrm{P}} + \left[\frac{\rho v_1}{x_{\mathrm{e}} - x_{\mathrm{w}}} + \frac{\alpha}{(x_{\mathrm{P}} - x_{\mathrm{w}})(x_{\mathrm{e}} - x_{\mathrm{w}})}\right]\phi^0\,.$$

All other coefficients are computed as for a CV in the interior of the problem domain.

Let us now consider the case where the flux $F_{\mathrm{w}} = F^0$ is prescribed at the west boundary. The flux through the CV face is obtained by dividing F^0 through the length of the face $x_{\mathrm{e}} - x_{\mathrm{w}}$. The resulting value is introduced in (4.6) as total flux and the modified coefficients for the boundary CV become:

$$a_{\mathrm{W}} = 0\,,$$

$$a_{\mathrm{P}} = \frac{\rho v_1}{x_{\mathrm{e}} - x_{\mathrm{w}}} + \frac{\alpha}{(x_{\mathrm{E}} - x_{\mathrm{P}})(x_{\mathrm{e}} - x_{\mathrm{w}})} +$$

$$\frac{\rho v_2}{y_{\mathrm{n}} - y_{\mathrm{s}}} + \frac{\alpha(y_{\mathrm{N}} - y_{\mathrm{s}})}{(y_{\mathrm{P}} - y_{\mathrm{s}})(y_{\mathrm{N}} - y_{\mathrm{P}})(y_{\mathrm{n}} - y_{\mathrm{s}})}\,,$$

$$b_{\mathrm{P}} = f_{\mathrm{P}} + \frac{F^0}{x_{\mathrm{e}} - x_{\mathrm{w}}}\,.$$

All other coefficients remain unchanged.

Sometimes it is possible to exploit symmetries of a problem in order to downsize the problem domain to save computing time or get a higher accuracy (with a finer grid) with the same computational effort. In such cases one has to consider symmetry planes or symmetry lines at the corresponding problem boundary. In this case one has the boundary condition:

$$\frac{\partial \phi}{\partial x_i} n_i = 0. \tag{4.20}$$

From this condition it follows that the diffusive flux through the symmetry boundary is zero (see (4.18)). Since also the normal component of the velocity vector has to be zero at a symmetry boundary (i.e., $v_i n_i = 0$), the mass flux and, therefore, the convective flux through the boundary is zero. Thus, in the balance equation (4.6) the total flux through the corresponding CV face can be set to zero. For the boundary CV in Fig. 4.16 this results in the following modified coefficients:

$$a_\mathrm{W} = 0,$$
$$a_\mathrm{P} = \frac{\rho v_1}{x_\mathrm{e} - x_\mathrm{w}} + \frac{\alpha}{(x_\mathrm{E} - x_\mathrm{P})(x_\mathrm{e} - x_\mathrm{w})} +$$
$$\frac{\rho v_2}{y_\mathrm{n} - y_\mathrm{s}} + \frac{\alpha(y_\mathrm{N} - y_\mathrm{S})}{(y_\mathrm{P} - y_\mathrm{S})(y_\mathrm{N} - y_\mathrm{P})(y_\mathrm{n} - y_\mathrm{s})} .$$

If required, the (unknown) variable value at the boundary can be determined by a finite-difference approximation of the boundary condition (4.20). In the considered case, for instance, with a forward difference formula (cp. (4.19)) one simply obtains $\phi_\mathrm{w} = \phi_\mathrm{P}$.

As with all other discretization techniques, the algebraic system of equations resulting from a finite-volume discretization has a unique solution only if the boundary conditions at all boundaries of the problem domain are taken into account (e.g., as outlined above). Otherwise there would be more unknowns than equations.

4.8 Algebraic System of Equations

As exemplarily outlined in Sect. 4.6 for the general scalar transport equation, a finite-volume discretization for each CV results in an algebraic equation of the form:

$$a_\mathrm{P} \phi_\mathrm{P} - \sum_c a_c \phi_c = b_\mathrm{P} ,$$

where the index c runs over all neighboring points that are involved in the approximation as a result of the discretization scheme employed. Globally, i.e., for all control volumes V_i ($i = 1, \ldots, N$) of the problem domain, this gives a linear system of N equations

$$a_{\mathrm{P}}^{i}\phi_{\mathrm{P}}^{i} - \sum_{c} a_{c}^{i}\phi_{c}^{i} = b_{\mathrm{P}}^{i} \quad \text{for all} \quad i = 1, \ldots, N \qquad (4.21)$$

for the N unknown nodal values ϕ_{P}^{i} in the CV centers.

After introducing a corresponding numbering of the CVs (or nodal values), in the case of a Cartesian grid the system (4.21) has a fully analogous structure that also would result from a finite-difference approximation. To illustrate this, we consider first the one-dimensional case. Let the problem domain be the interval $[0, L]$, which we divide into N not necessarily equidistant CVs (subintervals) (see Fig 4.17).

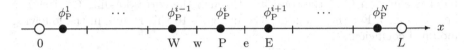

Fig. 4.17. Arrangement of CVs and nodes for 1-D transport problem

Using the second-order central differencing scheme, the discrete equations have the form:

$$a_{\mathrm{P}}^{i}\phi_{\mathrm{P}}^{i} - a_{\mathrm{E}}^{i}\phi_{\mathrm{E}}^{i} - a_{\mathrm{W}}^{i}\phi_{\mathrm{W}}^{i} = b_{\mathrm{P}}^{i}. \qquad (4.22)$$

With the usual lexicographical numbering of the nodal values as given in Fig. 4.17 one has:

$$\phi_{\mathrm{W}}^{i} = \phi_{\mathrm{P}}^{i-1} \quad \text{for all} \quad i = 2, \ldots, N,$$
$$\phi_{\mathrm{E}}^{i} = \phi_{\mathrm{P}}^{i+1} \quad \text{for all} \quad i = 1, \ldots, N-1.$$

Thus, the result is a linear system of equations which can be represented in matrix form as follows:

$$\underbrace{\begin{bmatrix} a_{\mathrm{P}}^{1} & -a_{\mathrm{E}}^{1} & & & & \\ -a_{\mathrm{W}}^{2} & a_{\mathrm{P}}^{2} & -a_{\mathrm{E}}^{2} & & 0 & \\ & \cdot & \cdot & \cdot & & \\ & & -a_{\mathrm{W}}^{i} & a_{\mathrm{P}}^{i} & -a_{\mathrm{E}}^{i} & \\ & & & \cdot & \cdot & \cdot \\ & 0 & & & \cdot & \cdot & -a_{\mathrm{E}}^{N-1} \\ & & & & -a_{\mathrm{W}}^{N} & a_{\mathrm{P}}^{N} \end{bmatrix}}_{\mathbf{A}} \underbrace{\begin{bmatrix} \phi_{\mathrm{P}}^{1} \\ \cdot \\ \phi_{\mathrm{P}}^{i-1} \\ \phi_{\mathrm{P}}^{i} \\ \phi_{\mathrm{P}}^{i+1} \\ \cdot \\ \phi_{\mathrm{P}}^{N} \end{bmatrix}}_{\boldsymbol{\phi}} = \underbrace{\begin{bmatrix} b_{\mathrm{P}}^{1} \\ b_{\mathrm{P}}^{2} \\ \cdot \\ b_{\mathrm{P}}^{i} \\ \cdot \\ \cdot \\ b_{\mathrm{P}}^{N} \end{bmatrix}}_{\mathbf{b}}.$$

When using a QUICK discretization or a central differencing scheme of 4th order, there are also coefficients for the farther points EE and WW (see Fig. 4.18):

$$a_P \phi_P - a_{EE} \phi_{EE} - a_E \phi_E - a_W \phi_W - a_{WW} \phi_{WW} = b_P, \qquad (4.23)$$

i.e., in the corresponding coefficient matrix $\mathbf{A}$ two additional non-zero diagonals appear:

$$\mathbf{A} = \begin{bmatrix}
a_P^1 & -a_E^1 & -a_{EE}^1 & & & & & \\
-a_W^2 & a_P^2 & -a_E^2 & -a_{EE}^2 & & & 0 & \\
-a_{WW}^3 & -a_W^3 & a_P^3 & -a_E^3 & -a_{EE}^3 & & & \\
& \cdot & \cdot & \cdot & \cdot & \cdot & & \\
& & -a_{WW}^i & -a_W^i & a_P^i & -a_E^i & -a_{EE}^i & \\
& & & \cdot & \cdot & \cdot & \cdot & \cdot \\
& & & & \cdot & \cdot & \cdot & -a_{EE}^{N-2} \\
& 0 & & & & \cdot & \cdot & -a_E^{N-1} \\
& & & & & -a_{WW}^N & -a_W^N & a_P^N
\end{bmatrix}.$$

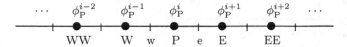

$$\cdots \quad \phi_P^{i-2} \qquad \phi_P^{i-1} \qquad \phi_P^i \qquad \phi_P^{i+1} \qquad \phi_P^{i+2} \quad \cdots$$

WW   W   w   P   e   E   EE

Fig. 4.18. CV dependencies with higher order scheme for 1-D transport problem

For the two- and three-dimensional cases fully analogous considerations can be made for the assembly of the discrete equation systems. For a two-dimensional rectangular domain with $N \times M$ CVs (see Fig. 4.19), we have, for instance, in the case of the discretization given in Sect. 4.6 equations of the form

$$a_P^{i,j} \phi_P^{i,j} - a_E^{i,j} \phi_E^{i,j} - a_W^{i,j} \phi_W^{i,j} - a_S^{i,j} \phi_S^{i,j} - a_N^{i,j} \phi_N^{i,j} = b_P^{i,j}$$

for $i = 1, \ldots, N$ and $j = 1, \ldots, M$. In the case of a lexicographical columnwise numbering of the nodal values (index j is counted up first) and a corresponding arrangement of the unknown variables $\phi_P^{i,j}$ (see Fig. 4.19), the system matrix $\mathbf{A}$ takes the following form:

$$
\mathbf{A} =
\begin{bmatrix}
a_{\mathrm{P}}^{1,1} & -a_{N}^{1,1} \cdot & 0 & \cdot -a_{\mathrm{E}}^{1,1} & & & \\
-a_{\mathrm{S}}^{1,2} & \cdot & \cdot & & \cdot & 0 & \\
0 & & \cdot & \cdot & \cdot & & \cdot \\
\cdot & & \cdot & \cdot & \cdot & & -a_{\mathrm{E}}^{N-1,M} \\
-a_{\mathrm{W}}^{2,1} & & & \cdot & \cdot & & \cdot \\
& & \cdot & & \cdot & \cdot & 0 \\
& & \cdot & & \cdot & \cdot & \cdot \\
& 0 & \cdot & & \cdot & \cdot & -a_{N}^{N,M-1} \\
& & -a_{\mathrm{W}}^{N,M} \cdot & 0 & \cdot -a_{\mathrm{S}}^{N,M} & a_{\mathrm{P}}^{N,M}
\end{bmatrix}
\cdot
$$

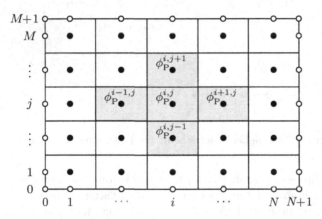

Fig. 4.19. Arrangement of CVs and nodes for 2-D transport problem

As outlined in Sect. 4.5, due to the discretization of the diffusive fluxes, in the non-Cartesian case additional coefficients can arise, whereby the number of non-zero diagonals in the system matrix increases. Using the discretization exemplarily given in Sect. 4.5, for instance, one would have additional dependencies with the points NE, NW, SE, and SW, which are required to linearly interpolate the values of ϕ in the vertices of the CV (see Fig. 4.20). Thus, in the case of a structured grid a matrix with 9 non-zero diagonals would result.

4.9 Numerical Example

As a concrete, simple (two-dimensional) example for the application of the FVM, we consider the computation of the heat transfer in a trapezoidal plate (density ρ, heat conductivity κ) with a constant heat source q all over the

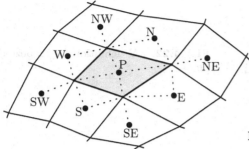

Fig. 4.20. Interpolation of vertice values for non-Cartesian CV

plate. At three sides the temperature T is prescribed and at the fourth side the heat flux is given (equal to zero). The problem data are summarized in Fig. 4.21. The problem is described by the heat conduction equation

$$-\kappa\frac{\partial^2 T}{\partial x^2} - \kappa\frac{\partial^2 T}{\partial y^2} = \rho q \qquad (4.24)$$

with the boundary conditions as indicated in Fig. 4.21 (cp. Sect. 2.3.2). For the discretization we employ a grid with only two CVs as illustrated in Fig. 4.22. The required coordinates for the distinguished points for both CVs are indicated in Table 4.2.

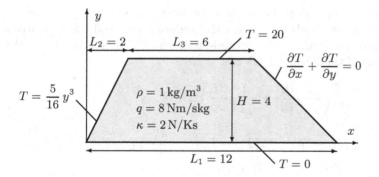

Fig. 4.21. Configuration of trapezoidal plate heat conduction example (temperature in K, length in m)

The integration of (4.24) over a control volume V and the application of the Gauß integral theorem gives:

$$\sum_c F_c = -\kappa\sum_c \int_{S_c}\left(\frac{\partial T}{\partial x}n_1 + \frac{\partial T}{\partial y}n_2\right)\mathrm{d}S_c = \int_V q\,\mathrm{d}V,$$

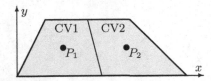

Fig. 4.22. CV definition for trapezoidal plate

Table 4.2. Coordinates of distinguished points for discretized trapezoidal plate

Point	CV1 x	y	CV2 x	y
P	13/4	2	31/4	2
e	11/2	2	10	2
w	1	2	11/2	2
n	7/2	4	13/2	4
s	3	0	9	0
nw	2	4	5	4
ne	5	4	8	4
se	6	0	12	0
sw	0	0	6	0
Volume	18		18	

where the summation has to be carried out over $c = \mathrm{s,n,w,e}$. For the approximation of the integrals we employ the midpoint rule and the derivatives at CV faces are approximated by second-order central differences. Thus, the approximations of the fluxes for CV1 is:

$$F_{\mathrm{e}} = -\kappa \int_{S_{\mathrm{e}}} \left(\frac{4}{\sqrt{17}} \frac{\partial T}{\partial x} + \frac{1}{\sqrt{17}} \frac{\partial T}{\partial y} \right) \mathrm{d}S_{\mathrm{e}} \approx$$

$$\approx D_{\mathrm{e}}(T_{\mathrm{E}} - T_{\mathrm{P}}) + N_{\mathrm{e}}(T_{\mathrm{ne}} - T_{\mathrm{se}}) = -\frac{17}{9}(T_{\mathrm{E}} - T_{\mathrm{P}}) - 10\,,$$

$$F_{\mathrm{w}} = -\kappa \int_{S_{\mathrm{w}}} \left(-\frac{2}{\sqrt{5}} \frac{\partial T}{\partial x} + \frac{1}{\sqrt{5}} \frac{\partial T}{\partial y} \right) \mathrm{d}S_{\mathrm{w}} =$$

$$= -\kappa \int_{S_{\mathrm{w}}} \left(-\frac{2}{\sqrt{5}} \frac{120}{16} x^2 + \frac{1}{\sqrt{5}} \frac{15}{16} y^2 \right) \mathrm{d}S_{\mathrm{w}} = 60\,,$$

$$F_{\mathrm{s}} = -\kappa \int_{S_{\mathrm{s}}} \left(-\frac{\partial T}{\partial y} \right) \mathrm{d}S_{\mathrm{s}} \approx -\kappa \left(\frac{\partial T}{\partial y} \right)_{\mathrm{s}} (x_{\mathrm{se}} - x_{\mathrm{sw}}) \approx$$

$$\approx -\kappa \left(\frac{T_{\mathrm{P}} - T_{\mathrm{S}}}{y_{\mathrm{P}} - y_{\mathrm{S}}} \right) (x_{\mathrm{se}} - x_{\mathrm{sw}}) = 6T_{\mathrm{P}}\,,$$

$$F_n = -\kappa \int_{S_n} \frac{\partial T}{\partial y}\, dS_n \approx -\kappa \left(\frac{\partial T}{\partial y}\right)_n (x_{ne} - x_{nw}) \approx$$

$$\approx -\kappa \left(\frac{T_N - T_P}{y_N - y_P}\right)(x_{ne} - x_{nw}) = 3T_P - 60\,.$$

The flux F_w has been computed exactly from the given boundary value function. Similarly, one obtains for CV2:

$$F_e = 0\,, \quad F_w \approx \frac{17}{9}(T_P - T_W) + 10\,, \quad F_s \approx 6T_P\,, \quad F_n \approx 3T_P - 60\,.$$

For both CVs we have $\delta V = 18$, such that the following discrete balance equations result:

$$\frac{98}{9}T_P - \frac{17}{9}T_E = 154 \quad \text{and} \quad \frac{98}{9}T_P - \frac{17}{9}T_W = 194\,.$$

We have $T_P = T_1$ and $T_E = T_2$ for CV1, and $T_P = T_2$ and $T_W = T_1$ for CV2. This gives the linear system of equations

$$98T_1 - 17T_2 = 1386 \quad \text{and} \quad 98T_2 - 17T_1 = 1746$$

for the two unknown temperatures T_1 and T_2. Its solution gives $T_1 \approx 17,77$ and $T_2 \approx 20,90$.

Exercises for Chap. 4

Exercise 4.1. Determine the leading error terms for the one-dimensional midpoint and trapezoidal rules by Taylor series expansion and compare the results.

Exercise 4.2. Let the concentration of a pollutant $\phi = \phi(x)$ in a chimney be described by the differential equation

$$-3\phi' - 2\phi'' = x\cos(\pi x) \quad \text{for} \quad 0 < x < 6$$

with the boundary conditions $\phi'(0) = 1$ and $\phi(6) = 2$. Compute the values ϕ_1 and ϕ_2 in the centers of the two control volumes CV1 = $[0, 4]$ and CV2 = $[4, 6]$ with a finite-volume discretization using the UDS method for the convective term.

Exercise 4.3. Consider the heat conduction in a square plate with the problem data given in Fig. 4.23. Compute the solution with a finite-volume method for the two grids illustarted in Fig. 4.24. Compare the results with the analytic solution $T_a(x, y) = 20 - 2y^2 + x^3y - xy^3$.

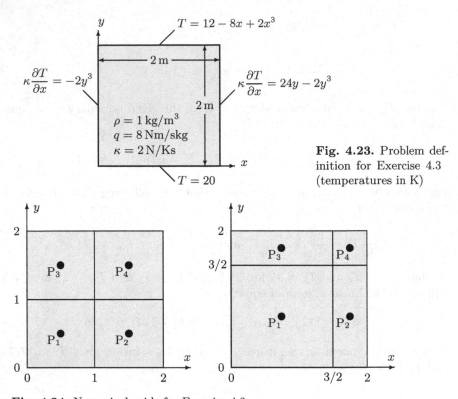

Fig. 4.23. Problem definition for Exercise 4.3 (temperatures in K)

Fig. 4.24. Numerical grids for Exercise 4.3

Exercise 4.4. Formulate a finite-volume method of 2nd order for equidistant grids for the bar equation (2.38). Use this for computing the displacement of a bar of length $L = 60\,\mathrm{m}$ with the boundary conditions (2.39) with $A(x) = 1 + x/60$, $u_0 = 0$, and $k_L = 4\,\mathrm{N}$ employing a discretization with three equidistant CVs.

Exercise 4.5. Formulate a finite-volume method of 4th order for the membrane equation (2.17) for an equidistant Cartesian grid.

Exercise 4.6. Consider the integral

$$I = \int_{S_e} \phi \, \mathrm{d}S$$

for the function $\phi = \phi(x, y)$ over the face S_e of the CV $[1,3]^2$. (i) Determine the leading error term and the order (with respect to the length Δy of S_e) for the approximation

$$I \approx \phi(3, \alpha) \Delta y$$

depending on the parameter $\alpha \in [1,3]$. (ii) Compute I for the function $\phi(x,y) = x^3 y^4$ directly (analytically) and with the approximation defined in (i) with $\alpha = 2$. Compare the two solutions.

Exercise 4.7. The velocity vector of a two-dimensional flow is given by

$$\mathbf{v} = (v_1(x,y), v_2(x,y)) = (x \cos \pi y, x^4 y).$$

Let the flux through the surface S of the control volume $V = [1,2]^2$ be defined by

$$I = \int_S v_i n_i \, \mathrm{d}S.$$

(i) Approximate the integral with the Simpson rule. (ii) Transform the integral with the Gauß integral theorem into a volume integral (over V) and approximate this with the midpoint rule.

5

Finite-Element Methods

The techniques known today as finite-element methods (FEM) date back to work conducted between 1940 and 1960 in the field of structural mechanics. The term finite element was introduced by Clough (1960). Nowadays, the FEM is widely used primarily for numerical computations in solid mechanics and can be regarded as a standard tool there. However, the FEM also has entered other application areas. We will address in this chapter the basic ideas of the method mostly by means of representative examples. The Galerkin method is employed as the general framework allowing for a universal application of the method. We remark that for the FEM there exists a comparatively elegant mathematical theory with respect to existence, convergence criteria, and error estimations. However, we will not address these aspects here (see, e.g., [3]).

5.1 Galerkin Method

The basis for a universal application of the finite-element method is provided by the *Galerkin method*, which we will introduce by means of a simple example. For this we consider the Poisson equation

$$-\frac{\partial^2 \phi}{\partial x_i \partial x_i} = f \tag{5.1}$$

for a problem domain Ω with the corresponding Dirichlet and Neumann boundary conditions

$$\phi = \phi_b \quad \text{on } \Gamma_1 \,, \tag{5.2}$$

$$\frac{\partial \phi}{\partial x_i} n_i = t_b \quad \text{on } \Gamma_2 \,, \tag{5.3}$$

where the disjoint boundary parts Γ_1 and Γ_2 together add up to the whole boundary Γ of Ω.

For the unknown function $\phi = \phi(\mathbf{x})$ we make an ansatz of the form

$$\phi(\mathbf{x}) \approx \varphi_0(\mathbf{x}) + \sum_{k=1}^{N} c_k \varphi_k(\mathbf{x}) \tag{5.4}$$

with the unknown coefficients c_k and prescribed functions φ_k. Here, the function φ_0 should satisfy the Dirichlet boundary conditions, i.e., $\varphi_0 = \phi_b$ on Γ_1, and for the other functions φ_k ($k = 1, \ldots, N$) the corresponding homogeneous Dirichlet condition $\varphi_k = 0$ on Γ_1 should be fulfilled. So, in total the ansatz satisfies the Dirichlet condition (5.2) on Γ_1. The Neumann boundary condition first remains unconsidered.

Inserting the ansatz (5.4) into the differential equation (5.1) yields:

$$-\frac{\partial^2 \varphi_0}{\partial x_i \partial x_i} - \sum_{k=1}^{N} c_k \frac{\partial^2 \varphi_k}{\partial x_i \partial x_i} = f \quad \text{in } \Omega\,.$$

The problem now is to find coefficients c_k such that this equation is fulfilled "as good as possible". As a measure for the error, the *residual R* is defined:

$$R = -\frac{\partial^2 \varphi_0}{\partial x_i \partial x_i} - \sum_{k=1}^{N} c_k \frac{\partial^2 \varphi_k}{\partial x_i \partial x_i} - f\,.$$

Conditions for the coefficients c_k are obtained by requiring that the sum of the integrals over Ω for the *weighted residuals* for N linearly independent weight functions ω_j ($j = 1, \ldots, N$) vanish:

$$\int_{\Omega} R\,\omega_j\,\mathrm{d}\Omega = 0 \quad \text{for all } j = 1, \ldots, N\,. \tag{5.5}$$

This equation system can be used for the determination of the unknown coefficients c_k (there are N equations for N unknowns).

For arbitrary weight functions this approach is called *method of weighted residuals*. Selecting as weight functions ω_j again the ansatz functions φ_j (for $j = 1, \ldots, N$) defines the *Galerkin method*. In this case, after inserting R and φ_j into (5.5), one has for the determination of the coefficients c_k the equations

$$-\int_{\Omega} \frac{\partial^2 \varphi_0}{\partial x_i \partial x_i} \varphi_j\,\mathrm{d}\Omega - \int_{\Omega} \sum_{k=1}^{N} c_k \frac{\partial^2 \varphi_k}{\partial x_i \partial x_i} \varphi_j\,\mathrm{d}\Omega = \int_{\Omega} f\,\varphi_j\,\mathrm{d}\Omega\,. \tag{5.6}$$

In the second term on the left hand side the sequence of summation and integration can be exchanged and the coefficients c_k, which do not depend on $\mathbf{x}$, can be taken out of the integral. By subsequent partial integration in the integrals on the left hand side the system (5.6) can equivalently be written as

$$\int_\Omega \frac{\partial \varphi_0}{\partial x_i} \frac{\partial \varphi_j}{\partial x_i} \, d\Omega - \int_\Gamma \frac{\partial \varphi_0}{\partial x_i} \varphi_j n_i \, d\Gamma +$$

$$\sum_{k=1}^{N} c_k \left[\int_\Omega \frac{\partial \varphi_k}{\partial x_i} \frac{\partial \varphi_j}{\partial x_i} \, d\Omega - \int_\Gamma \frac{\partial \varphi_k}{\partial x_i} \varphi_j n_i \, d\Gamma \right] = \int_\Omega f \, \varphi_j \, d\Omega \,.$$

Since the weight functions φ_j vanish on Γ_1, it is sufficient to integrate the boundary integrals over Γ_2 instead of Γ. Exchanging the sequence of integration and summation for the boundary integral terms one obtains:

$$\int_\Omega \frac{\partial \varphi_0}{\partial x_i} \frac{\partial \varphi_j}{\partial x_i} \, d\Omega + \sum_{k=1}^{N} c_k \int_\Omega \frac{\partial \varphi_k}{\partial x_i} \frac{\partial \varphi_j}{\partial x_i} \, d\Omega$$

$$= \int_\Omega f \, \varphi_j \, d\Omega + \int_{\Gamma_2} \frac{\partial}{\partial x_i} \underbrace{\left(\varphi_0 + \sum_{k=1}^{N} c_k \varphi_k \right)}_{\approx \phi} n_i \, \varphi_j \, d\Gamma \,.$$

In the boundary integral the Neumann boundary condition (5.3) on Γ_2 now can be inserted, such that finally for all $j = 1, \dots, N$ the following relation results:

$$\int_\Omega \frac{\partial \varphi_0}{\partial x_i} \frac{\partial \varphi_j}{\partial x_i} \, d\Omega + \sum_{k=1}^{N} c_k \int_\Omega \frac{\partial \varphi_k}{\partial x_i} \frac{\partial \varphi_j}{\partial x_i} \, d\Omega = \int_\Omega f \, \varphi_j \, d\Omega + \int_{\Gamma_2} t_b \, \varphi_j \, d\Gamma \,. \tag{5.7}$$

Thus, the following linear equation system with N equations and N unknowns for the unknown coefficients c_k results:

$$\sum_{k=1}^{N} S_{jk} c_k = b_j \quad \text{for } j = 1, \dots, N \quad \text{(or compactly } \mathbf{Sc} = \mathbf{b}) \tag{5.8}$$

with

$$S_{jk} = \int_\Omega \frac{\partial \varphi_k}{\partial x_i} \frac{\partial \varphi_j}{\partial x_i} \, d\Omega \,,$$

$$b_j = \int_{\Gamma_2} t_b \varphi_j \, d\Gamma + \int_\Omega f \, \varphi_j \, d\Omega - \int_\Omega \frac{\partial \varphi_0}{\partial x_i} \frac{\partial \varphi_j}{\partial x_i} \, d\Omega \,.$$

According to structural mechanics problems the matrix $\mathbf{S}$ and the vector $\mathbf{b}$ are called *stiffness matrix* and *load vector*, respectively. After solving the equation system (5.8) for the coefficients c_k one finally has an approximative solution of the problem according to the relation (5.4).

One can immediately recognize the relation to the weak formulation of the problem, which for our example is given by

$$\int\limits_{\Omega} \frac{\partial \phi}{\partial x_i} \frac{\partial \varphi}{\partial x_i} \, \mathrm{d}\Omega = \int\limits_{\Omega} f \varphi \, \mathrm{d}\Omega + \int\limits_{\Gamma_2} \frac{\partial \phi}{\partial x_i} n_i \varphi \, \mathrm{d}\Gamma$$

for all test functions φ (see Chap. 2). The Galerkin method can directly be derived from this formulation by inserting the ansatz (5.4) for ϕ and taking the ansatz functions φ_j as test functions.

The Galerkin method is determined by the choice of the ansatz functions φ_k. Chosing special piecewise polynomial functions this defines a finite-element method, i.e., the finite-element method can be interpreted as a Galerkin method with special ansatz functions. We will next deal with the question of how to select these functions.

5.2 Finite-Element Discretization

As for finite-volume methods, for the application of a finite-element method first the problem domain has to be discretized. For this one subdivides the domain into non-overlapping simple subdomains E_i for $i = 1, \ldots, N$, the so called *finite elements* (see Fig. 5.1). Depending on the spatial problem dimension usually the following element types are employed:

- 1-d: subintervals;
- 2-d: triangles, quadrilaterals, curvilinear elements;
- 3-d: tetrahedras, hexahedras, prisms, curvilinear elements.

Mixed subdivisions, e.g., into triangles and quadrilaterals in the two-dimensional case, are also possible (and sometimes also sensible). Curvilinear elements mainly are employed for a better approximation of curved boundaries. However, in this introductory text we will not address such elements.

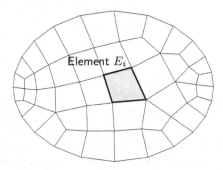

Fig. 5.1. Subdivision of problem domain into finite elements

In each subdomain (element) E_i piecewise polynomial ansatz functions for ϕ are selected. In the two-dimensional case this can be, for instance, linear polynomials of the form

$$\phi^i(x_1, x_2) = a_1^i + a_2^i x_1 + a_3^i x_2 .$$

At the element interfaces the ansatz functions have to fulfill certain (problem dependent) continuity requirements, not least so that a physically meaningful numerical solution can be obtained. These requirements can involve not only the continuity of the global solution, but also the continuity of derivatives (e.g., for beam or plate problems). Approaches fulfilling the problem specific continuity requirements are called *conforming*. In practice, in particular for beam and plate problems, also *non-conforming* elements are employed because the fulfillment of continuity is related to some computational effort. Also with such elements, under certain prerequisites, physically meaningful numerical solutions can be obtained. However, we will not address this issue further here (see [2]).

In order to fulfill the continuity requirements, it is helpful to express the ansatz functions directly in dependence on function values and/or derivatives. The values $\phi_1^i, \ldots, \phi_p^i$ employed for this are denoted as *local nodal variables* or also *degrees of freedom* of an element (thus, this can be function values and/or derivatives!). In addition, for this representation *local shape functions* $N_1^i, \ldots, N_p^i$ are introduced. With this – by a simple transformation – the ansatz functions can be represented in each element as a linear combination of these shape functions with the nodal variables as coefficients:

$$\phi^i(\mathbf{x}) = \sum_{j=1}^{p} \phi_j^i N_j^i(\mathbf{x}) . \tag{5.9}$$

If only function values are used as nodal variables, i.e.,

$$\phi_j^i = \phi^i(\mathbf{x}_j) \quad \text{for } j = 1, \ldots, p$$

at suitable locations $\mathbf{x}_1, \ldots, \mathbf{x}_p$ in the Element E_i, the local shape functions fulfill the relations

$$N_j^i(\mathbf{x}_n) = \begin{cases} 1 & \text{for } j = n, \\ 0 & \text{for } j \neq n, \end{cases} \tag{5.10}$$

since ϕ^i at the nodes $\mathbf{x}_n$ must take the nodal value ϕ_n^i.

From the elementwise representation of the unknown function by local nodal variables and local shape functions according to (5.9) a global representation of the solution (over the whole problem domain) can be assembled in a systematic way. For this, all nodal variables in the problem domain are numbered consecutively, where common local nodal variables of adjoint elements are counted only once. This way a global representation of the finite-element solution in the form

$$\phi(\mathbf{x}) \approx \varphi_0(\mathbf{x}) + \sum_{k=1}^{N} \phi_k N_k(\mathbf{x}) \qquad (5.11)$$

is obtained, where N_k is a composition of that local shape functions N_j^i, for which the local nodal variable ϕ_j^i coincides with the global nodal variable ϕ_k. In the function φ_0 the geometrical (Dirichlet) boundary conditions are subsumed.

The functions N_k are called *global shape functions*. In a finite-element method these correspond to the ansatz functions φ_k in the general Galerkin method described in the previous section. An important property of the global shape functions is that the function N_k is non-zero only in those elements which have the node P_k in common. This property ensures that in the system matrix $\mathbf{S}$ defined by (5.8) most entries are zero, i.e., $\mathbf{S}$ is a sparse matrix (as was already the case for the FVM).

In the next sections the general procedure formulated above will be exemplified and concretized by means of simple finite elements, in particular also with respect to the practical realization of the method.

5.3 One-Dimensional Linear Elements

As the simplest example for a finite element we first consider one-dimensional linear elements, which in the structural mechanics context also are denoted as *bar elements*. As example problem we take the (one-dimensional) differential equation

$$-\alpha\phi'' = f \quad \text{in} \quad \Omega = [a, b] \qquad (5.12)$$

with the boundary conditions

$$\phi(a) = \phi_a \quad \text{and} \quad \alpha\phi'(b) = h_b, \qquad (5.13)$$

where α is constant. With the corresponding interpretation of the problem quantities, for instance, the problem describes the deformation of a bar or the heat conduction in a one-dimensional medium (see Chap. 2).

5.3.1 Discretization

For the discretization of the problem domain we divide the interval $[a, b]$ into the elements (subintervals) $E_i = [x_{i-1}, x_i]$ for $i = 1, \ldots, N$, where $x_0 = a$ and $x_N = b$ (see Fig. 5.2). In the element E_i we make for ϕ the linear ansatz:

$$\phi^i(x) = a_1^i + a_2^i x. \qquad (5.14)$$

Element E_i

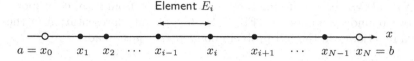

$$a = x_0 \quad x_1 \quad x_2 \quad \cdots \quad x_{i-1} \quad x_i \quad x_{i+1} \quad \cdots \quad x_{N-1} \quad x_N = b$$

Fig. 5.2. Element allocation for one-dimensional finite elements

Since a linear function is uniquely determined by the values at two points, as local nodal variables the values $\phi_1^i = \phi^i(x_{i-1})$ and $\phi_2^i = \phi^i(x_i)$ at the ends of the elements appear to be the natural choice.

From the two conditional equations

$$\phi^i(x_{i-1}) = a_1^i + a_2^i x_{i-1} = \phi_1^i \quad \text{and} \quad \phi^i(x_i) = a_1^i + a_2^i x_i = \phi_2^i$$

the ansatz coefficients result in

$$a_1^i = \frac{\phi_1^i x_i - \phi_2^i x_{i-1}}{x_i - x_{i-1}} \quad \text{and} \quad a_2^i = \frac{\phi_2^i - \phi_1^i}{x_i - x_{i-1}}.$$

Inserting these into the ansatz (5.14) one obtains for ϕ in the element E_i the representation

$$\phi^i(x) = \phi_1^i \frac{x_i - x}{x_i - x_{i-1}} + \phi_2^i \frac{x - x_{i-1}}{x_i - x_{i-1}} = \phi_1^i N_1^i + \phi_2^i N_2^i \qquad (5.15)$$

with the local shape functions

$$N_1^i(x) = \frac{x_i - x}{x_i - x_{i-1}} \quad \text{and} \quad N_2^i(x) = \frac{x - x_{i-1}}{x_i - x_{i-1}}. \qquad (5.16)$$

One can observe that these fulfill the conditions (5.10) (see Fig. 5.3).

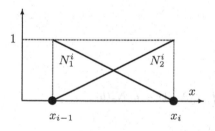

Fig. 5.3. Local shape functions for one-dimensional linear finite elements

With the local shape functions (5.16) the global shape functions N_i for $i = 0, \ldots, N$ now can be defined as follows:

$$N_i(x) = \begin{cases} N_2^i(x) & \text{for} \quad x_{i-1} < x < x_i \text{ and } i \geq 1, \\ N_1^{i+1}(x) & \text{for} \quad x_i < x < x_{i+1} \text{ and } i < N, \\ 0 & \text{otherwise,} \end{cases} \qquad (5.17)$$

where N_0 and N_N, each differing only in one element from zero, correspond to the two boundary points (see Fig. 5.4). The global representation of the unknown function ϕ in the full problem domain $\Omega = [a, b]$ reads:

$$\phi(x) \approx \phi_a N_0(x) + \sum_{k=1}^{N} \phi_k N_k(x) \tag{5.18}$$

with the global nodal variables $\phi_k = \phi(x_k)$ for $k = 1, \ldots, N$. The function $\phi_a N_0$ corresponds to the function φ_0 in the general ansatz (5.11).

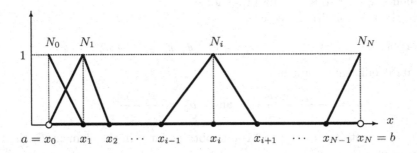

Fig. 5.4. Global shape functions for one-dimensional linear finite elements

The application of the Galerkin method for problem (5.12) with the ansatz (5.18) and the weight functions N_j leads to the global system

$$\sum_{k=1}^{N} S_{jk} \phi_k = b_j \quad \text{for all} \quad j = 1, \ldots, N \tag{5.19}$$

with

$$S_{jk} = \alpha \int_a^b N_k' N_j' \, dx \,,$$

$$b_j = \int_a^b f N_j \, dx - \phi_a \alpha \int_a^b N_0' N_j' \, dx + h_b N_j(b) \,.$$

It should be noted that the two last summands in the expression for b_j, which contain the boundary conditions, only contribute for $j = 1$ and $j = N$ to the load vector, respectively.

In principle, the components of the stiffness matrix S_{jk} and the load vector b_j could be computed from the global representation given by (5.19) (as integrals over the full interval $[a, b]$). For the systematic practical implementation of the finite-element method, however, this appears not to be convenient. A much more profitable approach is a local elementwise procedure, in which the computation of the integrals is first performed at the element level using the local shape functions and afterwards the global system is assembled from the individual element contributions.

5.3.2 Global and Local View

For the elementwise approach one exploits the composition of the global and local shape functions according to (5.17) as well as the following relations between the global and local nodal variables (see Fig. 5.5):

$$\phi_k = \phi_1^{k+1} = \phi_2^k \quad \text{for} \quad k = 1, \ldots, N-1 \quad \text{and} \quad \phi_N = \phi_2^N. \tag{5.20}$$

For the practical implementation it is convenient to retain these relations in matrix form in a so-called *coincidence matrix* (see Table 5.1).

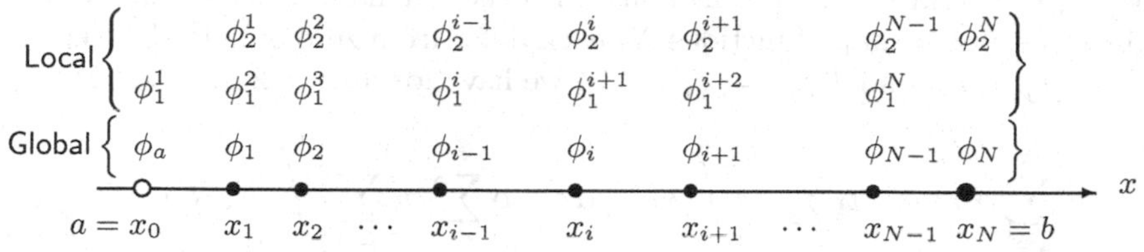

Fig. 5.5. Relation between local and global nodal variables for one-dimensional linear finite elements

Table 5.1. Allocation of global to local nodal variables for one-dimensional linear finite elements (coincidence matrix)

Local nodal variable	Element 1	2	$\cdots$	i	$\cdots$	$N-1$	N
1	ϕ_a	1	$\cdots$	$i-1$	$\cdots$	$N-2$	$N-1$
2	1	2	$\cdots$	i	$\cdots$	$N-1$	N

Analogous to the global formulation for the full problem domain $\Omega = [a, b]$ we consider a local formulation for the element $E_i = [x_{i-1}, x_i]$, where the boundary conditions are not taken into account for the time being. The local ansatz for the unknown function ϕ in the element $[x_{i-1}, x_i]$ reads (see (5.15):

$$\phi^i(x) = \sum_{k=1}^{2} \phi_k^i N_k^i(x) \tag{5.21}$$

The application of the Galerkin method for the differential equation (5.12) with the ansatz (5.21) for the element E_i and the weight functions N_j^i leads to the equation system

$$\sum_{k=1}^{2} S_{jk}^i \phi_k^i = b_j^i \quad \text{for } j = 1, 2 \quad \text{(or compactly } \mathbf{S}^i \phi^i = \mathbf{b}^i) \tag{5.22}$$

with

$$S_{jk}^i = \alpha \int_{x_{i-1}}^{x_i} (N_k^i)'(N_j^i)' \, dx \quad \text{and} \quad b_j^i = \int_{x_{i-1}}^{x_i} f N_j^i \, dx \tag{5.23}$$

The matrix $\mathbf{S}^i$ and the vector $\mathbf{b}^i$ are called *element stiffness matrix* and *element load vector*, respectively.

Next we consider the relations between the global and local equation systems (5.19) and (5.22). Let us start with the left hand sides of the systems. Since the global shape functions N_j each differ from zero only in the elements $E_j = [x_{j-1}, x_j]$ and $E_{j+1} = [x_j, x_{j+1}]$, we have for all $j = 2, \ldots, N-1$:

$$\sum_{k=1}^{N} S_{jk} \phi_k = \alpha \sum_{k=1}^{N} \phi_k \int_a^b N_k' N_j' \, dx = \alpha \sum_{k=1}^{N} \phi_k \sum_{i=1}^{N} \int_{x_{i-1}}^{x_i} N_k' N_j' \, dx =$$

$$= \alpha \sum_{k=j-1}^{j+1} \phi_k \sum_{i=j}^{j+1} \int_{x_{i-1}}^{x_i} N_k' N_j' \, dx =$$

$$= \alpha \phi_{j-1} \int_{x_{j-1}}^{x_j} N_{j-1}' N_j' \, dx + \alpha \phi_j \int_{x_{j-1}}^{x_j} N_j' N_j' \, dx +$$

$$\alpha \phi_j \int_{x_j}^{x_{j+1}} N_j' N_j' \, dx + \alpha \phi_{j+1} \int_{x_j}^{x_{j+1}} N_{j+1}' N_j' \, dx \, .$$

Using the relations (5.17) between the local and global shape functions it further holds

$$\sum_{k=1}^{N} S_{jk} \phi_k = \alpha \phi_{j-1} \int_{x_{j-1}}^{x_j} (N_1^j)'(N_2^j)' \, dx + \alpha \phi_j \int_{x_{j-1}}^{x_j} (N_2^j)'(N_2^j)' \, dx +$$

$$\alpha \phi_j \int_{x_j}^{x_{j+1}} (N_1^{j+1})'(N_1^{j+1})' \, dx + \alpha \phi_{j+1} \int_{x_j}^{x_{j+1}} (N_2^{j+1})'(N_1^{j+1})' \, dx$$

$$= \phi_{j-1} S_{21}^j + \phi_j \left(S_{22}^j + S_{11}^{j+1} \right) + \phi_{j+1} S_{12}^{j+1} \, .$$

Analogously for $j = 1$ and $j = N$ one gets the relations

$$\sum_{k=1}^{N} S_{1k}\phi_k = \phi_1\left(S_{22}^1 + S_{11}^2\right) + \phi_2 S_{12}^2,$$

$$\sum_{k=1}^{N} S_{Nk}\phi_k = \phi_{N-1}S_{21}^N + \phi_N S_{22}^N.$$

For the right hand sides of the equation systems (5.19) and (5.22) the following relations hold for $j = 1, \ldots, N$:

$$b_j = \int_a^b f N_j \, \mathrm{d}x - \phi_a\alpha \int_a^b N_0' N_j' \, \mathrm{d}x + h_b N_j(b) =$$

$$= \sum_{i=1}^{N} \int_{x_{i-1}}^{x_i} f N_j \, \mathrm{d}x - \phi_a\alpha \sum_{i=1}^{N} \int_{x_{i-1}}^{x_i} N_0' N_j' \, \mathrm{d}x + h_b N_j(b) =$$

$$= \int_{x_{j-1}}^{x_j} f N_j \, \mathrm{d}x + \int_{x_j}^{x_{j+1}} f N_j \, \mathrm{d}x - \phi_a\alpha \int_{x_{j-1}}^{x_j} N_0' N_j' \, \mathrm{d}x + h_b N_j(b) =$$

$$= \int_{x_{j-1}}^{x_j} f N_2^j \, \mathrm{d}x + \int_{x_j}^{x_{j+1}} f N_1^{j+1} \, \mathrm{d}x - \phi_a\alpha \int_{x_{j-1}}^{x_j} N_0' (N_2^j)' \, \mathrm{d}x + h_b N_j(b) =$$

$$= \begin{cases} b_2^1 + b_1^1 - \phi_a S_{21}^1 & \text{for } j = 1, \\ b_2^j + b_1^{j+1} & \text{for } j = 2, \ldots, N-1, \\ b_2^j + h_b & \text{for } j = N. \end{cases}$$

Thus, the global equation system (5.19) is composed of contributions from the element stiffness matrices and load vectors as follows:

$$\underbrace{\begin{bmatrix} S_{22}^1+S_{11}^2 & S_{12}^2 & & & \\ S_{21}^2 & S_{22}^2+S_{11}^3 & S_{12}^3 & & 0 \\ & \ddots & \ddots & \ddots & \\ 0 & & S_{21}^{N-1} & S_{22}^{N-1}+S_{11}^N & S_{12}^N \\ & & & S_{21}^N & S_{22}^N \end{bmatrix}}_{S} \underbrace{\begin{bmatrix} \phi_1 \\ \phi_2 \\ \vdots \\ \\ \phi_N \end{bmatrix}}_{\phi} = \underbrace{\begin{bmatrix} b_2^1+b_1^2 - \phi_a S_{21}^1 \\ b_2^2+b_1^3 \\ \vdots \\ b_2^{N-1}+b_1^N \\ b_2^N + h_b \end{bmatrix}}_{b}.$$

One can observe the typical band structure of the stiffness matrix, which results due to the locality of the element ansatz functions.

Note that with a corresponding polynomial ansatz, one-dimensional elements of arbitrary order can be defined where an ansatz of order p requires $p + 1$ nodal variables. The principal procedure runs completely analogous to that outlined for the linear case.

5.4 Practical Realization

The correlations identified in the previous section, which can be similarly formulated for other problems and other finite-elements, now suggest the following general procedure for the practical realization of the finite-element method:

(i)   Subdivision of problem domain into elements and selection of element ansatz type.
(ii)   Computation of element stiffness matrices and load vectors for all elements.
(iii) Assembling of global stiffness matrix and load vectors from the element stiffness matrices and load vectors.
(iv)   Consideration of boundary conditions.
(v)   Solution of global equation system.

In the next few sections we will go into more detail for some of these steps.

5.4.1 Assembling of Equation Systems

The assembling of the global equation system in step (iii) can be performed very systematically elementwise according to the following procedure:

- *Stiffness matrix:* if the k-th and j-th nodal variable of the i-th element is equal to n and m then S_{kj}^i is added to S_{nm}.
- *Load vector:* if the j-th nodal variable of the i-th element is equal to m than b_j^i is added to b_m.

This can be realized in a computer code very easily by employing the information stored in the coincidence matrix.

When assembling the global system it is advantageous first not to take into account the boundary conditions – let the resulting system be denoted by $\tilde{S}\tilde{\phi} = \tilde{b}$ – and to do the necessary modifications for their consideration subsequently (step (iv)). In the case of Neumann boundary conditions for this just the respective contributions have to be added to the corresponding components of the load vector. For Dirichlet boundary conditions it has to be ensured that in the global system the nodal variables for which certain values are prescribed really get these values. This can be realized as follows: if for the k-th nodal variable the value ϕ_r is prescribed, first the ϕ_r-fold of the k-th column of $\tilde{S}$ is subtracted from the load vector $\tilde{b}$. Afterwards the k-th row and column in $\tilde{\mathbf{S}}$ is set to zero, the k-th main diagonal element is set to 1, and the k-th component of the load vector is set to ϕ_r.

Normally, the equation system is then solved in this form, although due to the prescribed boundary values it contains trivial equations. Of course, the already known nodal variables could be eliminated from the system by simply deleting the corresponding rows and columns from the system. Since in practice the number of prescribed values compared to the total number of unknowns is relatively small, this is usually not done. This has the advantage that then all nodal variables are contained in the computed solution vector and can be used directly for further processing (e.g., for a visualization).

For the example in Sect. 5.3 the above procedure means that the global equation system is set up first in the form

$$
\underbrace{\begin{bmatrix}
S_{22}^1 & S_{12}^1 \\
S_{21}^1 & S_{22}^1+S_{11}^2 & S_{12}^2 \\
 & S_{21}^2 & S_{22}^2+S_{11}^3 & S_{12}^3 & & 0 \\
 & & \ddots & \ddots & \ddots \\
 & 0 & & S_{21}^{N-1} & S_{22}^{N-1}+S_{11}^N & S_{12}^N \\
 & & & & S_{21}^N & S_{22}^N
\end{bmatrix}}_{\tilde{\mathbf{S}}}
\underbrace{\begin{bmatrix}
\phi_0 \\ \phi_1 \\ \phi_2 \\ \vdots \\ \\ \phi_N
\end{bmatrix}}_{\tilde{\phi}}
=
\underbrace{\begin{bmatrix}
b_1^1 \\ b_2^1+b_1^2 \\ b_2^2+b_1^3 \\ \vdots \\ b_2^{N-1}+b_1^N \\ b_2^N
\end{bmatrix}}_{\tilde{\mathbf{b}}}.
$$

Note that the matrix $\tilde{\mathbf{S}}$ is singular, such that the corresponding equation system in this form possesses no unique solution. Only after the consideration of the boundary conditions is this uniquely determined. The Neumann boundary condition $\alpha\phi'(b) = h_b$ requires the modification $b_N = \tilde{b}_N + h_b$ and the integration of the Dirichlet boundary condition $\phi_0 = \phi_a$ leads to the system

$$
\begin{bmatrix}
1 & 0 \\
0 & S_{22}^1+S_{11}^2 & S_{12}^2 \\
 & S_{21}^2 & S_{22}^2+S_{11}^3 & S_{12}^3 & & 0 \\
 & & \ddots & \ddots & \ddots \\
 & 0 & & S_{21}^{N-1} & S_{22}^{N-1}+S_{11}^N & S_{12}^N \\
 & & & & S_{21}^N & S_{22}^N
\end{bmatrix}
\begin{bmatrix}
\phi_0 \\ \phi_1 \\ \phi_2 \\ \vdots \\ \\ \phi_N
\end{bmatrix}
=
\begin{bmatrix}
\phi_a \\ b_2^1+b_1^2 - \phi_a S_{21}^1 \\ b_2^2+b_1^3 \\ \vdots \\ b_2^{N-1}+b_1^N \\ b_2^N + h_b
\end{bmatrix}.
$$

Deleting the first row and first column from this system, the system $\mathbf{S}\phi = \mathbf{b}$ is recovered again.

5.4.2 Computation of Element Contributions

For the necessary evaluation of the element integrals in step (ii) for the computation of the element stiffness matrices and load vectors – with respect to a universal treatment of all elements – it is useful first to transform the integrals by means of a corresponding variable substitution to a unit domain.

For the example from Sect. 5.3 the integrals in (5.23) over the elements $E_i = [x_{i-1}, x_i]$ can be transformed via the following variable substitution to the unit interval $[0, 1]$:

$$x = x_{i-1} + \Delta x_i \xi \quad \text{with} \quad \Delta x_i = x_i - x_{i-1}. \tag{5.24}$$

In the unit element $[0, 1]$ one obtains for ϕ the representation

$$\phi^e(\xi) = \phi_1^e(1 - \xi) + \phi_2^e \xi = \phi_1^e N_1^e + \phi_2^e N_2^e \tag{5.25}$$

with the local shape functions (see Fig. 5.6)

$$N_1^e = 1 - \xi \quad \text{and} \quad N_2^e = \xi,$$

and the local nodal variables $\phi_1^e = \phi^e(0)$ and $\phi_2^e = \phi^e(1)$.

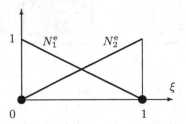

Fig. 5.6. Local shape functions for linear one-dimensional unit element

For the integrals in (5.23), because of

$$\frac{d}{d\xi} = \frac{d}{dx}\frac{dx}{d\xi} = \Delta x_i \frac{d}{dx},$$

one has the transformation rules

$$S_{jk}^i = \alpha \int_{x_{i-1}}^{x_i} \frac{dN_k^i}{dx}\frac{dN_j^i}{dx}\, dx = \frac{\alpha}{\Delta x_i} \int_0^1 \frac{dN_k^e}{d\xi}\frac{dN_j^e}{d\xi}\, d\xi, \tag{5.26}$$

$$b_j^i = \int_{x_{i-1}}^{x_i} f N_j^i \, dx = \Delta x_i \int_0^1 f N_j^e \, d\xi. \tag{5.27}$$

The integrals over the unit interval normally are evaluated numerically (usually with Gauß quadrature, see Sect. 5.7). Note that in the present example the corresponding integral in (5.26) no longer depends on the element i, such that it has to be evaluated only once (this is generally valid if α is elementwise constant). Therefore, in this case for the computation of the element matrices first the corresponding integrals for the unit interval can be computed (once):

$$
\mathbf{S}^e = \alpha
\begin{bmatrix}
\int\limits_0^1 \dfrac{dN_1^e}{d\xi}\dfrac{dN_1^e}{d\xi}\,d\xi & \int\limits_0^1 \dfrac{dN_1^e}{d\xi}\dfrac{dN_2^e}{d\xi}\,d\xi \\[2mm]
\int\limits_0^1 \dfrac{dN_2^e}{d\xi}\dfrac{dN_1^e}{d\xi}\,d\xi & \int\limits_0^1 \dfrac{dN_2^e}{d\xi}\dfrac{dN_2^e}{d\xi}\,d\xi
\end{bmatrix}
= \alpha
\begin{bmatrix} 1 & -1 \\ -1 & 1 \end{bmatrix},
\tag{5.28}
$$

from which, by using the transformation rule (5.26), the element matrices for all elements E_i can be evaluated in a simple way (without further integral evaluations):

$$
\mathbf{S}^i = \frac{1}{\Delta x_i}\mathbf{S}^e = \frac{\alpha}{\Delta x_i}\begin{bmatrix} 1 & -1 \\ -1 & 1 \end{bmatrix}.
\tag{5.29}
$$

The same applies for the computation of the element load vector according to (5.27), if f is elementwise constant.

5.4.3 Numerical Example

To exemplify the procedure described in the preceding sections we consider a concrete numerical example for the problem (5.12) and (5.13) with the data $[a, b] = [0, 8]$, $\alpha = 1/2$, $\phi_a = 0$, $h_b = 255$, and $f(x) = -3x^2/2$. We subdivide the problem domain $[0, 8]$ equidistantly into 4 elements, such that $\Delta x_i = \Delta x = 2$ for all elements $i = 1, 2, 3, 4$ (see Fig. 5.7).

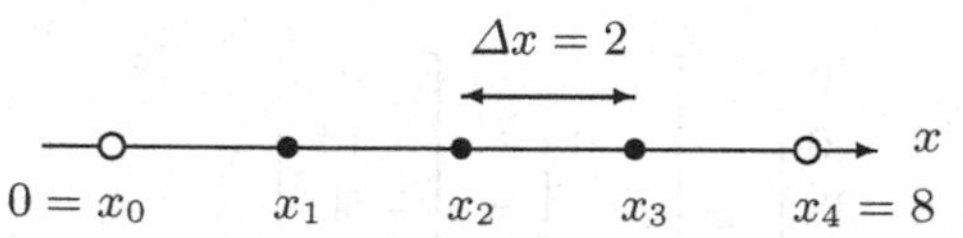

Fig. 5.7. Element subdivision for one-dimensional numerical example

According to (5.29) the element stiffness matrices result in

$$
\mathbf{S}^i = \frac{1}{4}\begin{bmatrix} 1 & -1 \\ -1 & 1 \end{bmatrix}
$$

for all elements $i = 1, 2, 3, 4$. With this for the global stiffness matrix (for the time being without boundary conditions) one obtains

$$
\tilde{\mathbf{S}} = \frac{1}{4}
\begin{bmatrix}
1 & -1 & 0 & 0 & 0 \\
-1 & 2 & -1 & 0 & 0 \\
0 & -1 & 2 & -1 & 0 \\
0 & 0 & -1 & 2 & -1 \\
0 & 0 & 0 & -1 & 1
\end{bmatrix}.
$$

For clarification we note that the assembling of $\tilde{\mathbf{S}}$ corresponds to the addition of the four 5×5 matrices

$$
\tilde{\mathbf{S}}^1 = \frac{1}{4}
\begin{bmatrix}
1 & -1 & 0 & 0 & 0 \\
-1 & 1 & 0 & 0 & 0 \\
0 & 0 & 0 & 0 & 0 \\
0 & 0 & 0 & 0 & 0 \\
0 & 0 & 0 & 0 & 0
\end{bmatrix},
\quad
\tilde{\mathbf{S}}^2 = \frac{1}{4}
\begin{bmatrix}
0 & 0 & 0 & 0 & 0 \\
0 & 1 & -1 & 0 & 0 \\
0 & -1 & 1 & 0 & 0 \\
0 & 0 & 0 & 0 & 0 \\
0 & 0 & 0 & 0 & 0
\end{bmatrix},
$$

$$
\tilde{\mathbf{S}}^3 = \frac{1}{4}
\begin{bmatrix}
0 & 0 & 0 & 0 & 0 \\
0 & 0 & 0 & 0 & 0 \\
0 & 0 & 1 & -1 & 0 \\
0 & 0 & -1 & 1 & 0 \\
0 & 0 & 0 & 0 & 0
\end{bmatrix},
\quad
\tilde{\mathbf{S}}^4 = \frac{1}{4}
\begin{bmatrix}
0 & 0 & 0 & 0 & 0 \\
0 & 0 & 0 & 0 & 0 \\
0 & 0 & 0 & 0 & 0 \\
0 & 0 & 0 & 1 & -1 \\
0 & 0 & 0 & -1 & 1
\end{bmatrix},
$$

which contain the respective element stiffness matrices at locations corresponding to the relations between the local and global nodal variables.

For the element load vectors one obtains from (5.27) the relation

$$
\mathbf{b}^i = -\frac{3 \Delta x}{2}
\begin{bmatrix}
\int_0^1 (x_{i-1} + \Delta x\, \xi)^2 (1 - \xi)\, d\xi \\
\int_0^1 (x_{i-1} + \Delta x\, \xi)^2 \xi\, d\xi
\end{bmatrix},
$$

which results in the following values for the four elements:

$$
\mathbf{b}^1 = \begin{bmatrix} -1 \\ -3 \end{bmatrix}, \quad
\mathbf{b}^2 = \begin{bmatrix} -11 \\ -17 \end{bmatrix}, \quad
\mathbf{b}^3 = \begin{bmatrix} -33 \\ -43 \end{bmatrix}, \quad
\mathbf{b}^4 = \begin{bmatrix} -67 \\ -81 \end{bmatrix}.
$$

This yields the global load vector (without boundary conditions)

$$
\tilde{\mathbf{b}} =
\begin{bmatrix} -1 \\ -3 \\ 0 \\ 0 \\ 0 \end{bmatrix} +
\begin{bmatrix} 0 \\ -11 \\ -17 \\ 0 \\ 0 \end{bmatrix} +
\begin{bmatrix} 0 \\ 0 \\ -33 \\ -43 \\ 0 \end{bmatrix} +
\begin{bmatrix} 0 \\ 0 \\ 0 \\ -67 \\ -81 \end{bmatrix} =
\begin{bmatrix} -1 \\ -14 \\ -50 \\ -110 \\ -81 \end{bmatrix}.
$$

Next, the two boundary conditions have to be considered. For the Neumann condition $\phi'(8)/2 = 255$ the value $h_b = 255$ has to be added to the last component of the load vector $\tilde{\mathbf{b}}$. The incorporation of the Dirichlet boundary condition $\phi(0) = 0$ finally leads to the global equation system

$$
\frac{1}{4}
\begin{bmatrix}
1 & 0 & 0 & 0 & 0 \\
0 & 2 & -1 & 0 & 0 \\
0 & -1 & 2 & -1 & 0 \\
0 & 0 & -1 & 2 & -1 \\
0 & 0 & 0 & -1 & 1
\end{bmatrix}
\begin{bmatrix} \phi_0 \\ \phi_1 \\ \phi_2 \\ \phi_3 \\ \phi_4 \end{bmatrix} =
\begin{bmatrix} 0 \\ -14 \\ -50 \\ -110 \\ 174 \end{bmatrix}.
$$

The solution of the system yields:

$$\phi_0 = 0 \, , \quad \phi_1 = 0 \, , \quad \phi_2 = 56 \, , \quad \phi_3 = 312 \, , \quad \phi_4 = 1008 \, .$$

The values of the nodal variables in this case coincide with the corresponding values of the exact solution $\phi(x) = x^4/4 - 2x$, which, of course, can easily be determined for the considered example.

5.5 One-Dimensional Cubic Elements

As a further example of a one-dimensional element, which also represents a conforming ansatz for the beam bending, we consider an element with a cubic ansatz. Simultaneously, this element, which also is called *beam element*, should serve as an example for the procedure if derivatives are employed as nodal variables. As an example problem we consider the (one-dimensional) beam equation

$$B\phi^{(4)} = -f_q \, , \tag{5.30}$$

for the problem domain $\Omega = [a, b]$, where at the interval ends corresponding geometric boundary conditions for ϕ are prescribed (see Sect. 2.4.2).

5.5.1 Discretization

Let the subdivision of the problem domain $\Omega = [a, b]$ into elements again be given according to Fig. 5.2. We can directly start our considerations for the unit element and afterwards transfer it by the variable transformation (5.24) to the individual elements. This results in a simplified presentation.

A cubic ansatz for ϕ on $[0, 1]$ is:

$$\phi^e(\xi) = a_1^e + a_2^e \xi + a_3^e \xi^2 + a_4^e \xi^3 \, . \tag{5.31}$$

This cubic function is uniquely determined by four values. As nodal variables we select the function values and the first derivatives of ϕ at the ends of the interval:

$$\phi_1^e = \phi(0), \quad \phi_2^e = \frac{d\phi}{d\xi}(0), \quad \phi_3^e = \phi(1), \quad \phi_4^e = \frac{d\phi}{d\xi}(1) \, . \tag{5.32}$$

From (5.31) and (5.32) one obtains the following relations between the nodal variables ϕ_i^e and the coefficients a_i^e:

$$\begin{bmatrix} a_1^e \\ a_2^e \\ a_3^e \\ a_4^e \end{bmatrix} = \begin{bmatrix} 1 & 0 & 0 & 0 \\ 0 & 1 & 0 & 0 \\ -3 & -2 & 3 & -1 \\ 2 & 1 & -2 & 1 \end{bmatrix} \begin{bmatrix} \phi_1^e \\ \phi_2^e \\ \phi_3^e \\ \phi_4^e \end{bmatrix} \, .$$

Inserting this into the ansatz (5.31) one obtains the representation

$$\phi^e(\xi) = \phi_1^e N_1^e + \phi_2^e N_2^e + \phi_3^e N_3^e + \phi_4^e N_4^e \tag{5.33}$$

with the local shape functions

$$N_1^e(\xi) = (1-\xi)^2(1+2\xi), \quad N_2^e(\xi) = \xi(1-\xi)^2,$$
$$N_3^e(\xi) = \xi^2(3-2\xi), \quad\quad\quad N_4^e(\xi) = \xi^2(\xi-1),$$

which again are characterized by the property of taking the value 1 at the assigned nodal variable and zero at the other ones (see Fig. 5.8).

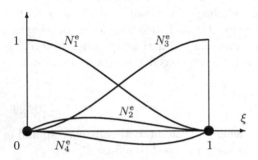

Fig. 5.8. Local shape functions for one-dimensional cubic finite element

The weak form of the beam equation (5.30) for the unit element $[0, 1]$ with the test functions φ reads:

$$B \int_0^1 \phi'' \varphi'' \, dx = -\int_0^1 f_q \varphi \, dx . \tag{5.34}$$

With the ansatz (5.33) this leads via the Galerkin method to the following equation system for the unit element:

$$\sum_{k=1}^{4} S_{jk}^e \phi_k^e = b_j^e \quad \text{for} \quad j = 1, \ldots, 4 \tag{5.35}$$

with the element stiffness matrix and element load vector

$$S_{jk}^e = B \int_0^1 (N_k^e)''(N_j^e)'' \, dx \quad \text{and} \quad b_j^e = -\int_0^1 f_q N_j^e \, dx .$$

For the element stiffness matrix and the element load vector for the element $E_i = [x_{i-1}, x_i]$ we obtain with the variable transformation (5.24) the following relations:

$$S^i_{jk} = B \int_{x_{i-1}}^{x_i} \frac{d^2 N^i_k}{dx^2} \frac{d^2 N^i_j}{dx^2}\, dx \;=\; \frac{B}{\Delta x_i^3} \int_0^1 \frac{d^2 N^e_k}{d\xi^2} \frac{d^2 N^e_j}{d\xi^2}\, d\xi \;=\; \frac{1}{\Delta x_i^3} S^e_{jk}\,,$$

$$b^i_j = -\int_{x_{i-1}}^{x_i} f_q N^i_j\, dx \;=\; -\Delta x_i \int_0^1 f_q N^e_j\, d\xi \;=\; \Delta x_i\, b^e_j\,.$$

When using an ansatz with derivatives as nodal variables, the derivatives in the integrals transformed to the unit interval refer to the variable ξ. For the later coupling with neighboring elements, however, it is more practical that only derivatives with respect to x appear as nodal variables, since otherwise derivatives of neighboring elements have a different meaning. This must be taken into account by the transformation to the original integrals via the relation

$$\frac{d\phi}{d\xi} = \frac{d\phi}{dx} \frac{dx}{d\xi} = \Delta x_i \frac{d\phi}{dx}$$

between the derivatives. In this way, for the beam element the following expressions for the element stiffness matrix $\mathbf{S}^i$ and the element load vector $\mathbf{b}^i$ result:

$$\mathbf{S}^i = \frac{B}{\Delta x_i^3} \begin{bmatrix} 12 & 6\Delta x_i & -12 & 6\Delta x_i \\ 6\Delta x_i & 4\Delta x_i^2 & -6\Delta x_i & 2\Delta x_i^2 \\ -12 & -6\Delta x_i & 12 & -6\Delta x_i \\ 6\Delta x_i & 2\Delta x_i^2 & -6\Delta x_i & 4\Delta x_i^2 \end{bmatrix}, \tag{5.36}$$

$$\mathbf{b}^i = \frac{f^i_q \Delta x_i}{12} \begin{bmatrix} -6 \\ -\Delta x_i \\ -6 \\ \Delta x_i \end{bmatrix}, \tag{5.37}$$

where we have assumed that f_q takes the constant value f^i_q in the element E_i (if this is not the case, the value in the center of the element can be taken as mean value or the integrals must be evaluated numerically). The corresponding local nodal variables are

$$\phi^i_1 = \phi(x_{i-1})\,, \quad \phi^i_2 = \frac{d\phi}{dx}(x_{i-1})\,, \quad \phi^i_3 = \phi(x_i)\,, \quad \phi^i_4 = \frac{d\phi}{dx}(x_i)\,.$$

The global equation system can be assembled in an analogous way as for the linear element by using the relations between local and global nodal variables. For N elements we have $2N + 2$ global nodal variables, including the ones which possibly are already prescribed due to geometric boundary conditions. Numbering the global nodal variables successively from 1 to $2N+2$, the corresponding coincidence matrix takes the form given in Table 5.2. We will exemplify the assembling of the global system in the next section by means of a concrete example.

Table 5.2. Relation of global and local nodal variables for one-dimensional cubic element (coincidence matrix)

Local nodal variable	Element 1	2	$\cdots$	i	$\cdots$	$N-1$	N
1	1	3	$\cdots$	$2i-1$	$\cdots$	$2N-3$	$2N-1$
2	2	4	$\cdots$	$2i$	$\cdots$	$2N-2$	$2N$
3	3	5	$\cdots$	$2i+1$	$\cdots$	$2N-1$	$2N+1$
4	4	6	$\cdots$	$2i+2$	$\cdots$	$2N$	$2N+2$

5.5.2 Numerical Example

As an example of a concrete application of the beam element we consider the deflection of a continuous beam of length $L = 7$ m with constant cross-section, which is clamped at one end and simply supported at two other locations $x = 3L/7$ and $x = L$. The beam is loaded by a continuously distributed load $f_q = -12$ N/m and a concentrated force $F = -6$ N at $x = 2L/7$. The flexural stiffness is $B = 4$ Nm². The unknown is the deflection $w = w(x)$ of the beam (in the vertical direction) for $0 \leq x \leq 7$. Figure 5.9 shows a sketch of the problem situation. For details about the problem modeling we refer to Sect. 2.4.2.

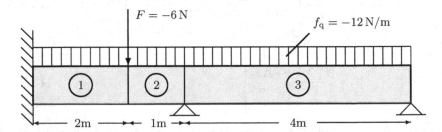

Fig. 5.9. Twofold simply supported, laterally clamped continuous beam under vertical load with subdivision into finite elements

We divide the beam into three elements, as indicated in Fig. 5.9. The lengths of the elements are $h_1 = 2$ m, $h_2 = 1$ m, and $h_3 = 4$ m. The global nodal variables, i.e., the respective function values and first derivatives of w at the ends of the elements, are:

$$
\begin{bmatrix} \phi_1 \\ \phi_2 \\ \phi_3 \\ \phi_4 \\ \phi_5 \\ \phi_6 \\ \phi_7 \\ \phi_8 \end{bmatrix} = \begin{bmatrix} w(0) \\ w'(0) \\ w(2) \\ w'(2) \\ w(3) \\ w'(3) \\ w(7) \\ w'(7) \end{bmatrix} .
$$

Inserting the concrete numerical values into formula (5.36) gives the element stiffness matrices

$$
\mathbf{S}^1 = \frac{B}{4} \begin{bmatrix} 6 & 6 & -6 & 6 \\ 6 & 8 & -6 & 4 \\ -6 & -6 & 6 & -6 \\ 6 & 4 & -6 & 8 \end{bmatrix}, \quad \mathbf{S}^2 = 2B \begin{bmatrix} 6 & 3 & -6 & 3 \\ 3 & 2 & -3 & 1 \\ -6 & -3 & 6 & -3 \\ 3 & 1 & -3 & 2 \end{bmatrix},
$$

$$
\mathbf{S}^3 = \frac{B}{32} \begin{bmatrix} 6 & 12 & -6 & 12 \\ 12 & 32 & -12 & 16 \\ -6 & -12 & 6 & -12 \\ 12 & 16 & -12 & 32 \end{bmatrix} .
$$

With the coincidence matrix in Table 5.2 (with N=3) the global stiffness matrix (without boundary conditions) becomes

$$
\tilde{\mathbf{S}} = \frac{B}{16} \begin{bmatrix} 24 & 24 & -24 & 24 & 0 & 0 & 0 & 0 \\ 24 & 32 & -24 & 16 & 0 & 0 & 0 & 0 \\ -24 & -24 & 216 & 72 & -192 & 96 & 0 & 0 \\ 24 & 16 & 72 & 96 & -96 & 32 & 0 & 0 \\ 0 & 0 & -192 & -96 & 195 & -90 & -3 & 6 \\ 0 & 0 & 96 & 32 & -90 & 80 & -6 & 8 \\ 0 & 0 & 0 & 0 & -3 & -6 & 3 & -6 \\ 0 & 0 & 0 & 0 & 6 & 8 & -6 & 16 \end{bmatrix} .
$$

For the element load vectors of the three elements we obtain with formula (5.37):

$$
\mathbf{b}^1 = \frac{f_q}{6} \begin{bmatrix} 6 \\ 2 \\ 6 \\ -2 \end{bmatrix}, \quad \mathbf{b}^2 = \frac{f_q}{12} \begin{bmatrix} 6 \\ 1 \\ 6 \\ -1 \end{bmatrix}, \quad \mathbf{b}^3 = \frac{f_q}{3} \begin{bmatrix} 6 \\ 4 \\ 6 \\ -4 \end{bmatrix} .
$$

With this the global load vector results in

$$\tilde{b} = \frac{f_q}{6}\begin{bmatrix} 6 \\ 2 \\ 6 \\ -2 \\ 0 \\ 0 \\ 0 \end{bmatrix} + \frac{f_q}{12}\begin{bmatrix} 0 \\ 0 \\ 6 \\ 1 \\ 6 \\ -1 \\ 0 \end{bmatrix} + \frac{f_q}{3}\begin{bmatrix} 0 \\ 0 \\ 0 \\ 0 \\ 6 \\ 4 \\ 6 \\ -4 \end{bmatrix} + \begin{bmatrix} 0 \\ 0 \\ F \\ 0 \\ 0 \\ 0 \\ 0 \end{bmatrix} = \begin{bmatrix} -12 \\ -4 \\ -24 \\ 3 \\ -30 \\ -15 \\ -24 \\ 16 \end{bmatrix},$$

where the concentrated force at $x = 2$, affecting the nodal variable $\phi_3 = w(2)$, is already considered.

According to the prescribed geometric boundary conditions

$$w(0) = w'(0) = w(3) = w(7) = 0$$

we have:

$$\phi_1 = \phi_2 = \phi_5 = \phi_7 = 0.$$

By just deleting the corresponding rows and columns (possible in this case because all prescribed values are 0), the following equation system remains to be solved:

$$\frac{1}{4}\begin{bmatrix} 216 & 72 & 96 & 0 \\ 72 & 96 & 32 & 0 \\ 96 & 32 & 80 & 8 \\ 0 & 0 & 8 & 16 \end{bmatrix}\begin{bmatrix} \phi_3 \\ \phi_4 \\ \phi_6 \\ \phi_8 \end{bmatrix} = \begin{bmatrix} -24 \\ 3 \\ -15 \\ 16 \end{bmatrix}$$

The solution of the system gives:

$$\phi_3 = 0.009625 , \quad \phi_4 = 0.611 , \quad \phi_6 = -1.48 , \quad \phi_8 = 4.74 .$$

Thus, for instance, the deflection of the beam at the point of application of the concentrated force is $w(2) = 0.009625\,\text{m}$.

5.6 Two-Dimensional Elements

The techniques introduced for the one-dimensional case can be employed in a very similar way for the definition of finite elements in two spatial dimensions (and also for three-dimensional elements, which, however, will not be addressed in this introductory text). As a representative example we consider the Poisson equation for the unknown function $\phi = \phi(x, y)$ with homogeneous Dirichlet boundary conditions $\phi = 0$ on the boundary Γ (to save indices here and in the following we write $x = x_1$ and $y = x_2$). As already outlined in Sect. 5.1, in this case the application of the Galerkin method with the global shape functions N_k $(k = 1, \ldots, N)$ as ansatz functions leads to the following linear equation system for the unknown coefficients (nodal variables) ϕ_k:

$$\sum_{k=1}^{N} \phi_k \int_{\Omega} \left(\frac{\partial N_k}{\partial x} \frac{\partial N_j}{\partial x} + \frac{\partial N_k}{\partial y} \frac{\partial N_j}{\partial y} \right) \mathrm{d}\Omega = \int_{\Omega} f N_j \, \mathrm{d}\Omega \quad \text{for } j = 1, \ldots, N.$$

Assuming again that f is constant within an element (with value f^i), for the considered problem integrals of the following types over the elements E_i $(i = 1, \ldots, N)$ have to be computed:

$$\int_{E_i} \left(\frac{\partial N_k^i}{\partial x} \frac{\partial N_j^i}{\partial x} + \frac{\partial N_k^i}{\partial y} \frac{\partial N_j^i}{\partial y} \right) \mathrm{d}\Omega \quad \text{and} \quad \int_{E_i} N_j^i \, \mathrm{d}\Omega, \tag{5.38}$$

where N_k^i again denotes the local shape function in the element.

For two-dimensional problems in practice usually element subdivisions into triangles or quadrilaterals are employed. As in the one-dimensional case the elementwise computation of the contributions for the stiffness matrix and the load vector can be put down to the computation of integrals over a unit domain. We will exemplify this for simple triangular and quadrilateral elements.

5.6.1 Variable Transformation for Triangular Elements

A triangle D_i in general location with the vertices (x_1, y_1), (x_2, y_2), and (x_3, y_3) can uniquely be mapped by the variable transformation

$$\begin{aligned} x &= x_1 + (x_2 - x_1)\xi + (x_3 - x_1)\eta, \\ y &= y_1 + (y_2 - y_1)\xi + (y_3 - y_1)\eta \end{aligned} \tag{5.39}$$

to the isosceles unit triangle D_0 with edge length 1 (see Fig. 5.10). The two integrals in (5.38) transform according to the relations

$$\int_{D_i} \left(\frac{\partial N_k^i}{\partial x} \frac{\partial N_j^i}{\partial x} + \frac{\partial N_k^i}{\partial y} \frac{\partial N_j^i}{\partial y} \right) \mathrm{d}x\mathrm{d}y = k_1^i \int_{D_0} \frac{\partial N_k^e}{\partial \xi} \frac{\partial N_j^e}{\partial \xi} \, \mathrm{d}\xi\mathrm{d}\eta +$$

$$k_2^i \int_{D_0} \frac{\partial N_k^e}{\partial \eta} \frac{\partial N_j^e}{\partial \eta} \, \mathrm{d}\xi\mathrm{d}\eta + k_3^i \int_{D_0} \left(\frac{\partial N_k^e}{\partial \xi} \frac{\partial N_j^e}{\partial \eta} + \frac{\partial N_k^e}{\partial \eta} \frac{\partial N_j^e}{\partial \xi} \right) \mathrm{d}\xi\mathrm{d}\eta \tag{5.40}$$

and

$$\int_{D_i} N_j^i \, \mathrm{d}x\mathrm{d}y = J \int_{D_0} N_j^e \, \mathrm{d}\xi\mathrm{d}\eta, \tag{5.41}$$

where

$$k_1^i = \left[(x_3 - x_1)^2 + (y_3 - y_1)^2\right]/J,$$
$$k_2^i = \left[(x_2 - x_1)^2 + (y_2 - y_1)^2\right]/J, \qquad (5.42)$$
$$k_3^i = \left[(x_3 - x_1)(x_1 - x_2) + (y_3 - y_1)(y_1 - y_2)\right]/J.$$

Here,

$$J = (x_2 - x_1)(y_3 - y_1) - (x_3 - x_1)(y_2 - y_1)$$

denotes the *Jacobi determinant* of the coordinate transformation (5.39). The value of J corresponds to the double area of the triangle D_i.

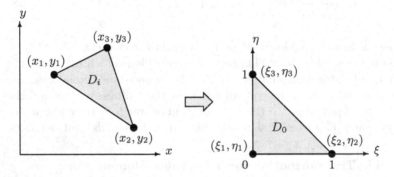

Fig. 5.10. Transformation of triangle in arbitrary location to unit triangle

Note that with the same transformation (5.39) also a parallelogram in arbitrary location can be transformed to the unit square $Q_0 = [0,1]^2$ (see Fig. 5.11). This becomes obvious if the triangle in Fig. 5.10 is complemented to a parallelogramm.

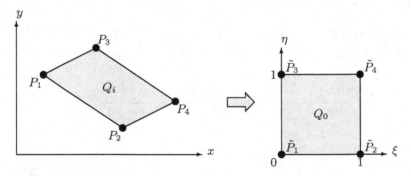

Fig. 5.11. Transformation of parallelogram in arbitrary location to unit square

5.6.2 Linear Triangular Elements

The most simple triangular element is obtained with the following linear ansatz for ϕ (in the unit triangle):

$$\phi^e(\xi, \eta) = a_1^e + a_2^e \xi + a_3^e \eta. \tag{5.43}$$

This function is uniquely determined by the prescription of the three function values ϕ_1^e, ϕ_2^e, and ϕ_3^e at the vertices of the triangle, which are selected as nodal variables. Between the nodal variables and the ansatz coefficients we have the relation

$$\begin{bmatrix} a_1^e \\ a_2^e \\ a_3^e \end{bmatrix} = \begin{bmatrix} 1 & 0 & 0 \\ -1 & 1 & 0 \\ -1 & 0 & 1 \end{bmatrix} \begin{bmatrix} \phi_1^e \\ \phi_2^e \\ \phi_3^e \end{bmatrix}.$$

Inserting the coefficients a_i^e in the ansatz (5.43) one obtains for ϕ on the unit triangle the representation

$$\phi^e(\xi, \eta) = \phi_1^e N_1^e + \phi_2^e N_2^e + \phi_3^e N_3^e \tag{5.44}$$

with the local shape functions

$$N_1^e(\xi, \eta) = 1 - \xi - \eta, \quad N_2^e(\xi, \eta) = \xi, \quad N_3^e(\xi, \eta) = \eta.$$

The shape function N_1^e is illustrated in Fig. 5.12. The other two run analogously (with value 1 in the points P_2 and P_3, respectively).

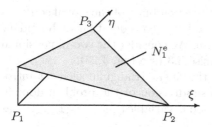

Fig. 5.12. Shape function N_1^e for linear triangular finite elements

For the integrals over the unit triangle appearing in (5.40) and (5.41) one gets:

$$S_{jk}^{1e} = \int_{D_0} \frac{\partial N_k^e}{\partial \xi} \frac{\partial N_j^e}{\partial \xi} \, d\xi d\eta = \frac{1}{2} \begin{bmatrix} 1 & -1 & 0 \\ -1 & 1 & 0 \\ 0 & 0 & 0 \end{bmatrix},$$

$$S_{jk}^{2e} = \int_{D_0} \frac{\partial N_k^e}{\partial \eta} \frac{\partial N_j^e}{\partial \eta} \, d\xi d\eta = \frac{1}{2} \begin{bmatrix} 1 & 0 & -1 \\ 0 & 0 & 0 \\ -1 & 0 & 1 \end{bmatrix},$$

$$S_{jk}^{3e} = \int\limits_{D_0} \left(\frac{\partial N_k^e}{\partial \xi} \frac{\partial N_j^e}{\partial \eta} + \frac{\partial N_k^e}{\partial \eta} \frac{\partial N_j^e}{\partial \xi} \right) \mathrm{d}\xi \mathrm{d}\eta = \frac{1}{2} \begin{bmatrix} 2 & -1 & -1 \\ -1 & 0 & 1 \\ -1 & 1 & 0 \end{bmatrix},$$

$$b_j^e = \int\limits_{D_0} N_j^e \, \mathrm{d}\xi \mathrm{d}\eta = \frac{1}{6} \begin{bmatrix} 1 \\ 1 \\ 1 \end{bmatrix}.$$

Thus, the element stiffness matrix results in

$$\mathbf{S}^i = k_1^i \mathbf{S}^{1e} + k_2^i \mathbf{S}^{2e} + k_3^i \mathbf{S}^{3e} = \frac{1}{2} \begin{bmatrix} k_1^i + k_2^i + 2k_3^i & -k_1^i - k_3^i & -k_2^i - k_3^i \\ -k_1^i - k_3^i & k_1^i & k_3^i \\ -k_2^i - k_3^i & k_3^i & k_2^i \end{bmatrix}$$

with the (element dependent) constants k_1^i, k_2^i, and k_3^i defined in (5.42). The element load vector is given by

$$\mathbf{b}^i = f^i J \mathbf{b}^e = \frac{f^i J}{6} \begin{bmatrix} 1 \\ 1 \\ 1 \end{bmatrix}.$$

If terms other than those in the considered Poisson problem appear in the underlying differential equations, the corresponding element stiffness matrices and load vectors can be determined in a completely analogous way.

5.6.3 Numerical Example

We will now exemplify the procedure for the assembling of the global equation system from the element contributions and the consideration of boundary conditions also for the two-dimensional case. We consider the two-dimensional heat conduction problem defined in Exercise 4.3 (see Fig. 4.23).

For the problem solution we employ the linear triangular element introduced in the preceding section with the subdivision of the problem domain into triangles D_i $(i = 1, \ldots, 8)$ and the nodal variables ϕ_k $(k = 1, \ldots, 9)$ as indicated in Fig. 5.13.

The application of the Galerkin method leads to the following global problem formulation:

$$\sum_{k=1}^{9} \phi_k \int\limits_0^2 \int\limits_0^2 \frac{\partial N_k}{\partial x} \frac{\partial N_j}{\partial x} + \frac{\partial N_k}{\partial y} \frac{\partial N_j}{\partial y} \, \mathrm{d}x \mathrm{d}y =$$

$$\int\limits_0^2 y^3 N_j \, \mathrm{d}y + \int\limits_0^2 (12y - y^3) N_j \, \mathrm{d}y + 4 \int\limits_0^2 \int\limits_0^2 N_j \, \mathrm{d}x \mathrm{d}y \qquad (5.45)$$

for $j = 1, \ldots, 9$ with the global shape functions N_j.

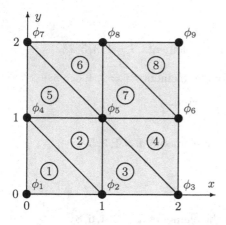

Fig. 5.13. Finite-element discretization with triangular elements for 2-d example problem

The assembling of the global equation system follows the procedure described in Sect. 5.4. For the corresponding computations the coordinates of the element vertices and the allocation of the global to the local nodal variables of the individual elements (coincidence matrix) are summarized in Tables 5.3 and 5.4, respectively. Note that although in the coincidence matrix the starting point of the local numbering is arbitrary, the sequence of the numbering (here counter-clockwise) should be the same in each element (see Fig. 5.12).

Table 5.3. Node coordinates of triangular elements for 2-d example problem

Coordinates	Node								
	1	2	3	4	5	6	7	8	9
x	0	1	2	0	1	2	0	1	2
y	0	0	0	1	1	1	2	2	2

Table 5.4. Allocation of global and local nodal variables for 2-d example problem (coincidence matrix)

Local nodal variable	Element							
	1	2	3	4	5	6	7	8
1	1	2	2	3	4	5	5	6
2	2	5	3	6	5	8	6	9
3	4	4	5	5	7	7	8	8

For the constants k_i^i, k_2^i, and k_3^i in (5.42) one obtains:

$$k_1^i = 1, \quad k_2^i = 1, \quad k_3^i = 0 \quad \text{for the elements } i = 1, 3, 5, 7,$$

$$k_1^i = 2, \quad k_2^i = 1, \quad k_3^i = -1 \quad \text{for the elements } i = 2, 4, 6, 8.$$

With this the element stiffness matrices become

$$\mathbf{S}^i = \frac{1}{2} \begin{bmatrix} 2 & -1 & -1 \\ -1 & 1 & 0 \\ -1 & 0 & 1 \end{bmatrix} \quad \text{for the elements } i = 1, 3, 5, 7,$$

$$\mathbf{S}^i = \frac{1}{2} \begin{bmatrix} 1 & -1 & 0 \\ -1 & 2 & -1 \\ 0 & -1 & 1 \end{bmatrix} \quad \text{for the elements } i = 2, 4, 6, 8.$$

For the element load vector one gets for all elements $i = 1, \ldots, 8$:

$$\mathbf{b}^i = \frac{4}{6} \begin{bmatrix} 1 \\ 1 \\ 1 \end{bmatrix}.$$

Next, the assembling of the global system can be done according to the methodology introduced in Sect. 5.4. Due to the symmetry of the global stiffness matrix it is sufficient if only the coefficients on and above the main diagonal are considered. Employing the information of the coincidence matrix in Table 5.4 we obtain the entries for the global stiffness matrix and the global load vector according to the schemes given in Tables 5.5 and 5.6, respectively. All components which are not indicated are zero.

The global stiffness matrix and load vector are obtained by summation of the individual contributions for the coefficients. For our example this yields:

$$\tilde{\mathbf{S}} = \frac{1}{2} \begin{bmatrix} 2 & -1 & 0 & -1 & 0 & 0 & 0 & 0 & 0 \\ -1 & 4 & -1 & 0 & -2 & 0 & 0 & 0 & 0 \\ 0 & -1 & 2 & 0 & 0 & -1 & 0 & 0 & 0 \\ -1 & 0 & 0 & 4 & -2 & 0 & -1 & 0 & 0 \\ 0 & -2 & 0 & -2 & 8 & -2 & 0 & -2 & 0 \\ 0 & 0 & -1 & 0 & -2 & 4 & 0 & 0 & -1 \\ 0 & 0 & 0 & -1 & 0 & 0 & 2 & -1 & 0 \\ 0 & 0 & 0 & 0 & -2 & 0 & -1 & 4 & -1 \\ 0 & 0 & 0 & 0 & 0 & -1 & 0 & -1 & 2 \end{bmatrix} \quad \text{and} \quad \tilde{\mathbf{b}} = \frac{4}{6} \begin{bmatrix} 1 \\ 3 \\ 2 \\ 3 \\ 6 \\ 3 \\ 2 \\ 3 \\ 1 \end{bmatrix}.$$

Again, one can recognize the typical band structure of the matrix.

Next, the boundary conditions have to be taken into account. Let us start with the Neumann boundary conditions, which require the evaluation of the boundary integrals in (5.45) resulting in additional contributions to the load vector. For the elementwise computation of these integrals one can employ

Table 5.5. Composition of global stiffness matrix for 2-d example problem

Global stiffness matrix	Element 1	2	3	4	5	6	7	8
S_{11}	S^1_{11}							
S_{12}	S^1_{12}							
S_{14}	S^1_{13}							
S_{22}	S^1_{22}	S^2_{11}	S^3_{11}					
S_{23}			S^3_{12}					
S_{24}	S^1_{23}	S^2_{13}						
S_{25}		S^2_{12}	S^3_{13}					
S_{33}			S^3_{22}	S^4_{11}				
S_{35}			S^3_{23}	S^4_{13}				
S_{36}				S^4_{12}				
S_{44}	S^1_{33}	S^2_{33}			S^5_{11}			
S_{45}		S^2_{23}			S^5_{12}			
S_{47}					S^5_{13}			
S_{55}		S^2_{22}	S^3_{33}	S^4_{33}	S^5_{22}	S^6_{11}	S^7_{11}	
S_{56}				S^4_{23}			S^7_{12}	
S_{57}					S^5_{23}	S^6_{13}		
S_{58}						S^6_{12}	S^7_{13}	
S_{66}				S^4_{22}			S^7_{22}	S^8_{11}
S_{68}							S^7_{23}	S^8_{13}
S_{69}								S^8_{12}
S_{77}					S^5_{33}	S^6_{33}		
S_{78}						S^6_{23}		
S_{88}						S^6_{22}	S^7_{33}	S^8_{33}
S_{89}								S^8_{23}
S_{99}								S^8_{22}

Table 5.6. Composition of global load vector for 2-d example problem

Global load vector	Element 1	2	3	4	5	6	7	8
b_1	b^1_1							
b_2	b^1_2	b^2_1	b^3_1					
b_3			b^3_2	b^4_1				
b_4	b^1_3	b^2_3			b^5_1			
b_5		b^2_2	b^3_3	b^4_3	b^5_2	b^6_1	b^7_1	
b_6				b^4_2			b^7_2	b^8_1
b_7					b^5_3	b^6_3		
b_8						b^6_2	b^7_3	b^8_3
b_9								b^8_2

(one-dimensional) boundary shape functions for the element edges located at boundaries where a Neumann condition is prescribed. For the employed linear ansatz the boundary shape functions for the unit interval in y direction read:

$$N_1^r(\eta) = 1 - \eta \quad \text{and} \quad N_2^r(\eta) = \eta \,.$$

The assignment of the boundary nodal variables to the elements involving a Neumann condition is given in Table 5.7. The boundary shape function (now dependent on y) for the corresponding elements are

$$N_1^1(y) = N_1^4(y) = 1 - y \,, \quad N_2^1(y) = N_2^4(y) = y \,,$$

$$N_1^5(y) = N_1^8(y) = 2 - y \,, \quad N_2^5(y) = N_2^8(y) = y - 1 \,.$$

The superscripts and subscripts relate to the element and the boundary nodal variable, respectively. Thus, $N_i^k = N_i^k(y)$ is the boundary shape function in the k-th element, which has the value 1 at the i-th boundary nodal variable.

Table 5.7. Allocation of boundary nodal variables and elements for example problem

Local nodal variable	Element 1	4	5	8
1	1	3	4	6
2	4	6	7	9

The computation of the corresponding boundary integral contributions yields:

$$r_1^1 = \int_0^1 y^3 N_1^1(y)\, dy = \int_0^1 y^3 (1 - y)\, dy = \frac{1}{20} \,,$$

$$r_2^1 = \int_0^1 y^3 N_2^1(y)\, dy = \int_0^1 y^3 y\, dy = \frac{1}{5} \,,$$

$$r_1^4 = \int_0^1 (12y - y^3) N_1^4(y)\, dy = \int_0^1 (12y - y^3)(1 - y)\, dy = \frac{39}{20} \,,$$

$$r_2^4 = \int_0^1 (12y - y^3) N_2^4(y)\, dy = \int_0^1 (12y - y^3) y\, dy = \frac{19}{5} \,,$$

$$r_1^5 = \int_0^1 y^3 N_1^5(y)\, dy = \int_1^2 y^3 (2 - y)\, dy = \frac{13}{10} \,,$$

$$r_2^5 = \int\limits_0^1 y^3 N_2^5(y)\, dy = \int\limits_1^2 y^3(y-1)\, dy = \frac{49}{20}\,,$$

$$r_1^8 = \int\limits_0^1 (12y - y^3)N_1^8(y)\, dy = \int\limits_1^2 (12y - y^3)(2-y)\, dy = \frac{67}{10}\,,$$

$$r_2^8 = \int\limits_0^1 (12y - y^3)N_2^8(y)\, dy = \int\limits_1^2 (12y - y^3)(y-1)\, dy = \frac{151}{20}\,.$$

According to the allocation given in Table 5.7, the coefficients of the load vector have to be modified as follows (not strictly mathematically, we use the same notation for the modified coefficients):

$$\tilde{b}_1 \leftarrow \tilde{b}_1 + r_1^1\,,$$
$$\tilde{b}_3 \leftarrow \tilde{b}_3 + r_1^4\,,$$
$$\tilde{b}_4 \leftarrow \tilde{b}_4 + r_2^1 + r_1^5\,,$$
$$\tilde{b}_6 \leftarrow \tilde{b}_6 + r_2^4 + r_1^8\,,$$
$$\tilde{b}_7 \leftarrow \tilde{b}_7 + r_2^5\,,$$
$$\tilde{b}_9 \leftarrow \tilde{b}_9 + r_2^8\,.$$

The other coefficients remain unchanged. Inserting the concrete values, we obtain for the load vector:

$$\tilde{\mathbf{b}} = \frac{4}{6}\begin{bmatrix} 1 \\ 3 \\ 2 \\ 3 \\ 6 \\ 3 \\ 2 \\ 3 \\ 1 \end{bmatrix} + \frac{1}{20}\begin{bmatrix} 1 \\ 0 \\ 39 \\ 30 \\ 0 \\ 210 \\ 49 \\ 0 \\ 151 \end{bmatrix} = \frac{1}{60}\begin{bmatrix} 43 \\ 120 \\ 197 \\ 210 \\ 240 \\ 750 \\ 227 \\ 120 \\ 493 \end{bmatrix}.$$

According to the Dirichlet boundary conditions we have:

$$\phi_1 = \phi_2 = \phi_3 = 20\,,\quad \phi_7 = 12\,,\quad \phi_8 = 6\,,\quad \phi_9 = 12\,.$$

Proceeding in the way as outlined in Sect. 5.4 for considering these conditions we finally obtain for the stiffness matrix and the load vector:

$$S = \begin{bmatrix} 1 & 0 & 0 & 0 & 0 & 0 & 0 & 0 & 0 \\ 0 & 1 & 0 & 0 & 0 & 0 & 0 & 0 & 0 \\ 0 & 0 & 1 & 0 & 0 & 0 & 0 & 0 & 0 \\ 0 & 0 & 0 & 4 & -2 & 0 & 0 & 0 & 0 \\ 0 & 0 & 0 & -2 & 8 & -2 & 0 & 0 & 0 \\ 0 & 0 & 0 & 0 & -2 & 4 & 0 & 0 & 0 \\ 0 & 0 & 0 & 0 & 0 & 0 & 1 & 0 & 0 \\ 0 & 0 & 0 & 0 & 0 & 0 & 0 & 1 & 0 \\ 0 & 0 & 0 & 0 & 0 & 0 & 0 & 0 & 1 \end{bmatrix} \quad \text{and} \quad b = \begin{bmatrix} 20 \\ 20 \\ 20 \\ 39 \\ 60 \\ 57 \\ 12 \\ 6 \\ 12 \end{bmatrix}.$$

Eliminating the already known nodal variables, the equation system that remains to be solved is

$$\begin{bmatrix} 4 & -2 & 0 \\ -2 & 8 & -2 \\ 0 & -2 & 4 \end{bmatrix} \begin{bmatrix} \phi_4 \\ \phi_5 \\ \phi_6 \end{bmatrix} = \frac{1}{2} \begin{bmatrix} 39 \\ 60 \\ 75 \end{bmatrix},$$

which has the solution

$$\phi_4 = 18.75, \quad \phi_5 = 18.0, \quad \phi_6 = 23.25.$$

For the elements employed the computed coefficients directly correspond to the temperatures at the corresponding locations. Comparing these with the respective analytical solution (see Table 5.8), already a relatively good agreement of the results is achieved – despite the rather coarse element subdivision and the low order of the polynomial ansatz.

Table 5.8. Analytical and numerical solution with linear triangular elements for 2-d example problem

	$T(0,1)$	$T(1,1)$	$T(2,1)$
Analytical	18.00	18.00	24.00
Numerical	18.75	18.00	23.25

5.6.4 Bilinear Parallelogram Elements

As an example of a quadrilateral element we consider the simplest element out of this class, i.e., the *bilinear parallelogram element*. For this, we use the following bilinear ansatz in the unit square $Q_0 = [0,1]^2$ for ϕ:

$$\phi^e(\xi, \eta) = a_1^e + a_2^e \xi + a_3^e \eta + a_4^e \xi \eta. \tag{5.46}$$

As nodal variables we chose the four function values ϕ_1^e, ϕ_2^e, ϕ_3^e, and ϕ_4^e at the vertices of the square (in successive counter-clockwise numbering), by which the function is uniquely determined in the element. At the edges of the square one has a linear course of the ansatz function. Thus, the function values at the vertices uniquely determine the linear course along an edge, such that the continuity to neighboring elements is ensured. Consequently, the bilinear parallelogram element (by preserving the continuity) also can be combined easily with the linear triangular element.

Between the nodal variables ϕ_i^e and the ansatz coefficients a_i^e we have the following correlation:

$$
\begin{bmatrix} a_1^e \\ a_2^e \\ a_3^e \\ a_4^e \end{bmatrix} = \begin{bmatrix} 1 & 0 & 0 & 0 \\ -1 & 1 & 0 & 0 \\ -1 & 0 & 0 & 1 \\ 1 & -1 & 1 & -1 \end{bmatrix} \begin{bmatrix} \phi_1^e \\ \phi_2^e \\ \phi_3^e \\ \phi_4^e \end{bmatrix} .
$$

Inserting this into the ansatz (5.46) we obtain the shape function representation

$$
\phi^e(\xi, \eta) = \phi_1^e N_1^e + \phi_2^e N_2^e + \phi_3^e N_3^e + \phi_4^e N_4^e \tag{5.47}
$$

with the local shape functions

$$
\begin{aligned}
N_1^e(\xi, \eta) &= (1 - \xi)(1 - \eta) , \\
N_2^e(\xi, \eta) &= \xi(1 - \eta) , \\
N_3^e(\xi, \eta) &= \xi\eta , \\
N_4^e(\xi, \eta) &= (1 - \xi)\eta .
\end{aligned}
$$

The function N_1^e is illustrated in Fig. 5.14. The other ones are obtained by rotations by 90°, 180°, and 270°.

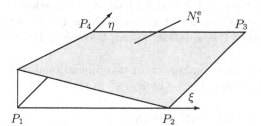

Fig. 5.14. Shape function N_1^e for bilinear parallelogram element

For the integrals in (5.40) and (5.41) over the unit square one obtains

$$
S_{jk}^{1e} = \int_{Q_0} \frac{\partial N_k^e}{\partial \xi} \frac{\partial N_j^e}{\partial \xi} \, d\xi d\eta = \frac{1}{6} \begin{bmatrix} 2 & -2 & -1 & 1 \\ -2 & 2 & 1 & -1 \\ -1 & 1 & 2 & -2 \\ 1 & -1 & -2 & 2 \end{bmatrix} ,
$$

$$S_{jk}^{2e} = \int_{Q_0} \frac{\partial N_k^e}{\partial \eta} \frac{\partial N_j^e}{\partial \eta} \, d\xi d\eta = \frac{1}{6} \begin{bmatrix} 2 & 1 & -1 & -2 \\ 1 & 2 & -2 & -1 \\ -1 & -2 & 2 & 1 \\ -2 & -1 & 1 & 2 \end{bmatrix},$$

$$S_{jk}^{3e} = \int_{Q_0} \left(\frac{\partial N_k^e}{\partial \xi} \frac{\partial N_j^e}{\partial \eta} + \frac{\partial N_k^e}{\partial \eta} \frac{\partial N_j^e}{\partial \xi} \right) d\xi d\eta = \frac{1}{6} \begin{bmatrix} 3 & 0 & -3 & 0 \\ 0 & -2 & 0 & 3 \\ -3 & 0 & 2 & 0 \\ 0 & 3 & 0 & -2 \end{bmatrix},$$

$$b_j^e = \int_{Q_0} N_j^e \, d\xi d\eta = \frac{1}{4} \begin{bmatrix} 1 \\ 1 \\ 1 \\ 1 \end{bmatrix}.$$

With this the element stiffness matrix and the element load vector for the bilinear parallelogram element can be computed as

$$\mathbf{S}^e = k_1^i \mathbf{S}^{1e} + k_2^i \mathbf{S}^{2e} + k_3^i \mathbf{S}^{3e} \quad \text{and} \quad \mathbf{b}^i = f^i J \mathbf{b}^e,$$

where k_1, k_2, and k_3 again are the (element dependent) constants defined in (5.42).

5.6.5 Other Two-Dimensional Elements

With a complete polynomial ansatz (all terms up to a certain degree appear), principally elements of arbitrary high order can be constructed, also in the two-dimensional case. However, polynomials with a degree higher than three are rarely used, since with the polynomial degree the number of nodal variables increases quite rapidly, i.e., $(p+1)(p+2)/2$ nodal variables are needed for degree p. The corresponding system matrices then possess many non-zero entries and the numerical solution of the equation system becomes relatively costly (see Chap. 7).

The complete ansatz polynomial of second order for two-dimensional problems, for instance, is (for simplicity we omit the index "e" in the following)

$$\phi(\xi, \eta) = a_1 + a_2 \xi + a_3 \eta + a_4 \xi^2 + a_5 \xi \eta + a_6 \eta^2.$$

As nodal variables for corresponding triangular elements, for instance, the function values of ϕ in the vertices and in the midpoints of the edges of the triangles can be taken. In the cubic case already 10 nodal variables per element are required. Also incomplete (reduced) polynomial ansatz functions, in which some terms in the complete polynomial are omitted, are possible.

In Table 5.9 an overview of several common triangular elements are given together with the corresponding polynomial ansatz functions. Here, the notation indicated in Table 5.10 is used for the definition of the nodal variables.

Thus, the symbol "⊚", for instance, means that at the corresponding location the following 6 nodal variables are defined:

$$\phi,\ \frac{\partial \phi}{\partial \xi},\ \frac{\partial \phi}{\partial \eta},\ \frac{\partial^2 \phi}{\partial \xi^2},\ \frac{\partial^2 \phi}{\partial \eta^2},\ \frac{\partial^2 \phi}{\partial \xi \partial \eta}.$$

The point in the interior of the triangle for the two cubic elements corresponds to the barycenter of the triangle. Elements with derivatives as nodal variables, in particular, are well suited for plate problems. For instance, the last three elements in Table 5.9 also are conforming for these kind of problems because the continuity of the normal derivatives along the triangle edges at the interfaces of neighboring elements is ensured.

By selecting different polynomial ansatz functions and nodal variables also quadrilateral elements of different order – either conforming or non-conforming – can be defined. In Table 5.11 the most important quadrilateral elements are summarized together with the corresponding polynomial ansatz functions, where for the definition of the nodal variables again the notation introduced in Table 5.10 is employed.

The ansatz polynomial of the biquadratic or the bicubic elements can be seen as a product of quadratic or cubic polynomials in ξ and η, respectively. The corresponding elements are denoted as *Lagrange elements* because they are in direct relation to Lagrange interpolation. The bilinear element described in the preceding section also belongs to this element class.

The *Serendipity elements* are characterized by ansatz functions, which on each edge of the quadrilateral represent a complete polynomial that is uniquely determined by the respective prescribed nodal values along the edges. This way the continuity of neighboring elements is ensured. However, the internal nodes that result from the corresponding product ansatz are omitted, such that the ansatz polynomial is incomplete.

The question of which finite element is best suited for a concrete problem (i.e., achievement of a desired accuracy with lowest possible computational effort), cannot be answered universally because this strongly depends on the specific problem. The higher the degree of the ansatz, the more accurate the elementwise approximation of the solution. Simultaneously, however, the effort for the assembling of the equation system and its numerical solution increases quickly. As already mentioned there is an elaborated mathematical theory on the method also addressing aspects of approximation errors (see, e.g., [3]). We will not go into further detail here, but note that under certain regularity assumptions for the finite-element solution ϕ_h error estimates of the form

$$\|\phi(\mathbf{x}) - \phi_h(\mathbf{x})\| \leq C h^{p+1} \tag{5.48}$$

can be derived. Here, h is a measure of the element size, C is a constant not depending on h (but on ϕ), p is the degree of the ansatz polynomial, and $\| \cdot \|$ denotes a suitable norm.

Table 5.9. Overview of common two-dimensional triangular elements

Nodal variables	Description
	Linear triangular element 3 degrees of freedom, continuous $\phi(\xi, \eta) = a_1 + a_2\xi + a_3\eta$
	Quadratic triangular element 6 degrees of freedom, continuous $\phi(\xi, \eta) = a_1 + a_2\xi + a_3\eta + a_4\xi^2 + a_5\xi\eta + a_6\eta^2$
	Cubic triangular element 10 degrees of freedom, continuous $\phi(\xi, \eta) = a_1 + a_2\xi + a_3\eta + a_4\xi^2 + a_5\xi\eta + a_6\eta^2 +$ $\qquad a_7\xi^3 + a_8\xi^2\eta + a_9\xi\eta^2 + a_{10}\eta^3$
	Cubic triangular element 10 degrees of freedom, continuous, continuous 1st derivatives $\phi(\xi, \eta) = a_1 + a_2\xi + a_3\eta + a_4\xi^2 + a_5\xi\eta + a_6\eta^2 +$ $\qquad a_7\xi^3 + a_8\xi^2\eta + a_9\xi\eta^2 + a_{10}\eta^3$
	Bell triangle 18 degrees of freedom, continuous, continuous 1st derivatives $\phi(\xi, \eta) =$ reduced 5th order polynomial
	Argyris triangle 21 degrees of freedom, continuous, continuous 1st derivatives $\phi(\xi, \eta) =$ complete 5th order polynomial

Table 5.10. Notation for defining nodal variables

Symbol	Prescribed nodal variables
●	Function value
◉	Function value and 1st derivatives
◎	Function value, 1st and 2nd derivatives
+	Normal derivative

Table 5.11. Overview of common two-dimensional quadrilateral elements

Nodal variables	Description
	Bilinear quadrilateral element 4 degrees of freedom, continuous $\phi(\xi, \eta) = a_1 + a_2\xi + a_3\eta + a_4\xi\eta$
	Quadratic quadrilateral element (Serendipity) 8 degrees of freedom, continuous $\phi(\xi, \eta) = a_1 + a_2\xi + a_3\eta + a_4\xi^2 + a_5\xi\eta + a_6\eta^2 +$ $\quad a_7\xi^2\eta + a_8\xi\eta^2$
	Biquadratic quadrilateral element (Lagrange) 9 degrees of freedom, continuous $\phi(\xi, \eta) = a_1 + a_2\xi + a_3\eta + a_4\xi^2 + a_5\xi\eta + a_6\eta^2 + a_7\xi^2\eta +$ $\quad a_8\xi\eta^2 + a_9\xi^2\eta^2$
	Cubic quadrilateral element (Serendipity) 12 degrees of freedom, continuous $\phi(\xi, \eta) = a_1 + a_2\xi + a_3\eta + a_4\xi^2 + a_5\xi\eta + a_6\eta^2 + a_7\xi^3 +$ $\quad a_8\xi^2\eta + a_9\xi\eta^2 + a_{10}\eta^3 + a_{11}\xi^3\eta + a_{12}\xi\eta^3$
	Cubic quadrilateral element 12 degrees of freedom, continuous, continuous 1st derivatives $\phi(\xi, \eta) = a_1 + a_2\xi + a_3\eta + a_4\xi^2 + a_5\xi\eta + a_6\eta^2 + a_7\xi^3 +$ $\quad a_8\xi^2\eta + a_9\xi\eta^2 + a_{10}\eta^3 + a_{11}\xi^3\eta + a_{12}\xi\eta^3$
	Bicubic quadrilateral element (Lagrange) 16 degrees of freedom, continuous $\phi(\xi, \eta) = a_1 + a_2\xi + a_3\eta + a_4\xi^2 + a_5\xi\eta + a_6\eta^2 + a_7\xi^3 +$ $\quad a_8\xi^2\eta + a_9\xi\eta^2 + a_{10}\eta^3 + a_{11}\xi^3\eta + a_{12}\xi^2\eta^2 +$ $\quad a_{13}\xi\eta^3 + a_{14}\xi^2\eta^3 + a_{15}\xi^3\eta^2 + a_{16}\xi^3\eta^3$

5.7 Numerical Integration

Already, for simpler ansatz functions relatively complex integrals have to be evaluated. For the considered simple examples it has been possible to evaluate the integrals for the computation of the various element contributions simply "by hand". For more complex elements (e.g., isoparametric elements, which we will discuss in Sect. 9.2) an exact integration usually is no longer possible and numerical integration techniques have to be employed for this purpose. An exact computation of the integrals also is not necessary because by the discretization already an error is introduced and a numerical integration scheme of sufficiently high order does not significantly deteriorate the accuracy of the numerical solution.

We already dealt with numerical integration methods in connection with finite-volume methods (see Sect. 4.2). In that case it was necessary that the integration points coincide with the locations at which the corresponding variables are defined. This is not necessary for the evaluation of the integrals within the finite-element method because of the use of shape functions within the elements. Thus, more efficient methods can be employed here.

In general, a formula for the numerical integration of an arbitrary (scalar) function $\phi = \phi(\mathbf{x})$ over a domain Ω can be written in the form

$$\int_\Omega \phi \, d\Omega \approx \sum_{i=1}^{p} \phi(\mathbf{x}_i) w_i$$

with suitable integration weights w_i and nodal points $\mathbf{x}_i$ in Ω. The idea for the construction of more efficient numerical integration formulas is to select the locations of nodal points – for a given number of points – in such a way that the error becomes minimal. Optimal in this sense are the *Gauß quadature formulas* with which a polynomial of degree $2p-1$ still can be integrated exactly with p nodal points. Thus, these formulas for a given number of nodal points are much more accurate than the ones we have discussed in the context of finite-volume methods. For instance, with the Simpson rule (with 3 nodal points) only polynomials up to degree 2 can be integrated exactly, while the corresponding Gaussian quadrature formula still yields exact values for polynomials up to degree 5.

We will not go into the details of the derivation of the Gauß quadrature formulas, which can be found in the corresponding literature (e.g., [2] and the references given there). Suffice it to say in the one-dimensional case the nodal points are defined by roots of *Legendre polynomials* and that the corresponding integration weights are obtained by integration of *Lagrange polynomials*. In Table 5.12 the nodal points and the weights of the Gauß quadrature formulas for the unit interval up to the order $p = 3$ are given. The method with $p = 1$ corresponds to the midpoint rule.

For the integration error of the one-dimensional Gauß quadrature of order p one has the following general expression (under certain regularity assumptions on ϕ):

$$\int_0^1 \phi \, d\xi - \sum_{i=1}^{p} \phi(\xi_i) w_i = \frac{2^{2p+1}(p!)^4}{[(2p)!]^3(2p+1)} \, \phi^{(2p)}(\xi_\mathrm{m}) \,,$$

where ξ_m is located in the interval $(0, 1)$ (where exactly is usually not known).

The one-dimensional Gauß quadrature formulas can also be used for the unit square and unit cube by just applying the corresponding formula successively for each spatial direction (productwise application). This corresponds

Table 5.12. Nodal points and weights for one-dimensional Gauß quadrature up to order 3

Order	i	ξ_i	w_i
1	1	$1/2$	1
2	1	$(3-\sqrt{3})/6$	$1/2$
	2	$(3+\sqrt{3})/6$	$1/2$
3	1	$(5-\sqrt{15})/6$	$5/18$
	2	$1/2$	$4/9$
	3	$(5+\sqrt{15})/6$	$5/18$

to a distribution of the nodal points and weights according to the one-dimensional case in each coordinate direction (see Fig. 5.15). However, the resulting formulas are no longer optimal in the above mentioned sense. It is also possible to derive directly corresponding optimal formulas for a specific domain. For the unit square, for instance, the nodal points are then located on circles around the point $(\xi, \eta) = (1/2, 1/2)$.

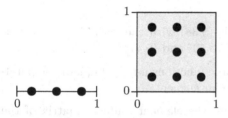

Fig. 5.15. Distribution of nodal points for Gauß quadrature with $p=3$ for unit interval and unit square (for productwise application)

Corresponding integration formulas are available also for the unit triangle. As an example, in Table 5.13 the coordinates of the nodal points and the corresponding weights for the formula with 7 nodal points are given. Figure 5.16 shows the corresponding distribution of the nodal points within the unit triangle. This formula yields exact integrals for polynomials up to degree 5.

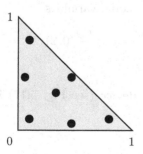

Fig. 5.16. Distribution of nodal points for exact integration of polynomials up to degree 5 for unit triangle

Table 5.13. Nodal points and integration weights for Gauß quadrature with 7 points for unit triangle

i	ξ_i	η_i	w_i
1	$1/3$	$1/3$	$9/80$
2	$(6 + \sqrt{15})/21$	$(6 + \sqrt{15})/21$	$(155 + \sqrt{15})/2400$
3	$(9 - 2\sqrt{15})/21$	$(6 + \sqrt{15})/21$	$(155 + \sqrt{15})/2400$
4	$(6 + \sqrt{15})/21$	$(9 - 2\sqrt{15})/21$	$(155 + \sqrt{15})/2400$
5	$(6 - \sqrt{15})/21$	$(6 - \sqrt{15})/21$	$(155 - \sqrt{15})/2400$
6	$(9 + 2\sqrt{15})/21$	$(6 - \sqrt{15})/21$	$(155 - \sqrt{15})/2400$
7	$(6 - \sqrt{15})/21$	$(9 + 2\sqrt{15})/21$	$(155 - \sqrt{15})/2400$

The nodal points and weights for the unit domains in the different spatial dimensions for the Gauß quadrature formulas of different orders can be found in tabular form in the literature (e.g., [20] and the references given therein).

Exercises for Chap. 5

Exercise 5.1. Transform the integrals (5.38) with the variable transformation (5.39) to the unit triangle D_0 (cf. (5.40) and (5.41)).

Exercise 5.2. Determine the local shape functions for the biquadratic parallelogram element and the cubic triangular element.

Exercise 5.3. Derive an expression for the element stiffness matrix of the quadratic triangular element when applied to the two-dimensional scalar transport equation (4.1).

Exercise 5.4. In the problem domain $0 \leq x \leq 1$ the one-dimensional differential equation $\phi'(x) - \phi(x) = 0$ with the initial condition $\phi(0) = 1$ is given. Compute a finite-element solution with 4 equidistant linear elements.

Exercise 5.5. In the triangle with the vertices $(0,0)$, $(1,0)$, and $(0,2)$ a finite element is defined by a bilinear ansatz and the local nodal variables

$$\phi_1 = \phi(0,0) , \quad \phi_2 = \phi(1,0) , \quad \phi_3 = \phi(0,2) , \quad \phi_4 = \frac{\partial \phi}{\partial x}(0,2) .$$

Determine the corresponding local shape functions.

Exercise 5.6. The temperature $\phi = \phi(x,y)$ in a device (see Fig. 5.17) is described by the differential equation

$$-2\frac{\partial^2 \phi}{\partial x^2} - 2\frac{\partial^2 \phi}{\partial y^2} = xy .$$

On the whole boundary the value $\phi = 1$ is prescribed. The device is subdivided into 4 bilinear parallelogram elements $E_1, \ldots, E_4$ with the global nodal variables $\phi_1, \ldots, \phi_{10}$ according to Fig. 5.17. The local nodal variables are numbered clockwise starting at the right bottom corner of the elements. (i) Give the corresponding coincidence matrix. (ii) Compute the coefficient b_2^2 of the element load vector for the element E_2. (iii) Express the components $\tilde{S}_{87}$ of the global stiffness matrix and b_5 of the global load vector in dependence on the element stiffness matrices and load vectors S_{jk}^i and b_j^i (for $i, j, k = 1, \ldots, 4$).

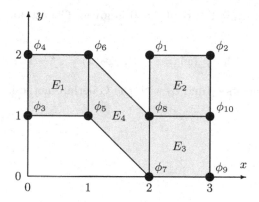

Fig. 5.17. Finite-element discretization with bilinear parallelogram elements for Exercise 5.6

Exercise 5.7. Compute the heat transfer example from Sect. 5.6.3 with the discretization into 4 bilinear parallelogram elements shown in Fig. 5.18. Compare the results with that for the linear triangular elements (see Table 5.8).

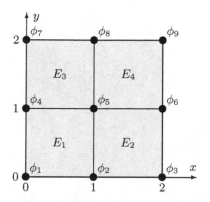

Fig. 5.18. Finite-element discretization with bilinear parallelogram elements for Exercise 5.7

Exercise 5.8. A finite-element discretization for a device consists of four bilinear quadrangular elements with counterclockwise numbering of the local nodal variables. The following components of the global stiffness matrix are given:

$$S_{85} = S_{41}^1 + S_{21}^3, \quad S_{46} = S_{32}^4, \quad S_{12} = S_{34}^3,$$

$$S_{53} = S_{43}^2 + S_{41}^4 , \quad S_{49} = S_{23}^1 , \quad S_{72} = S_{21}^2$$

(i) Give the corresponding coincidence matrix. (ii) Sketch a suitable finite-element discretization with the corresponding numberings of elements and global nodal variables.

Exercise 5.9. For the function $\phi = \phi(x)$ the differential equation

$$5\phi' - \phi'' = 4x \quad \text{for} \quad 0 < x < 1$$

with the boundary conditions $\phi'(0) = 1$ and $\phi(1) = 0$ is given. Choose for ϕ the ansatz

$$\phi(x) \approx c_1 \varphi_1(x) + c_2 \varphi_2(x) \quad \text{with} \quad \varphi_1(x) = x - 1 , \quad \varphi_2(x) = x^2 - 1$$

and determine the ansatz coefficients c_1 und c_2 with the Galerkin method.

6

Time Discretization

In many practical applications the processes under consideration are unsteady and thus require for their numerical simulation the solution of time-dependent model equations. The time has a certain exceptional role in the differential equations because, unlike for spatial coordinates, there is a distinguished direction owing to the principle of causality. This fact has to be taken into account for the discretization techniques employed for time. In this chapter the most important aspects with respect to this issue are discussed.

6.1 Basics

For unsteady processes the physical quantities – in addition to the spatial dependence – also depend on the time t. In the applications considered here mainly two types of time-dependent problems appear: transport and vibration processes. Examples of corresponding processes are, for instance, the Kármán vortex street formed when fluids flow around bodies (see Fig. 6.1), or vibrations of a structure (see Fig. 6.2), respectively.

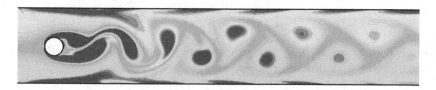

Fig. 6.1. Kármán vortex street (instantaneous vorticity)

While the equations for unsteady transport processes only involve first derivatives with respect to time, for vibration processes second time derivatives also appear. In the first case the problem is called *parabolic*, in the second case *hyperbolic*. Since we do not need the underlying concepts in the following,

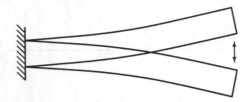

Fig. 6.2. Vibrations of a clamped beam

we will disperse with a more precise definition of the terms that can be used for a general classification of second-order partial differential equations (see, e.g., [9] or [12]).

An example of a parabolic problem is the general unsteady scalar transport equation (cf. Sect. 2.3.2)

$$\frac{\partial(\rho\phi)}{\partial t} + \frac{\partial}{\partial x_i}\left(\rho v_i \phi - \alpha \frac{\partial \phi}{\partial x_i}\right) = f. \tag{6.1}$$

An example of the hyperbolic type are the equations of linear elastodynamics (cf. Sect. 2.4.1). For a vibrating beam, as illustrated in Fig. 6.2, one has, for instance:

$$\rho A \frac{\partial^2 w}{\partial t^2} + \frac{\partial^2}{\partial x^2}\left(B \frac{\partial^2 w}{\partial x^2}\right) + f_q = 0. \tag{6.2}$$

Compared to the corresponding steady problems the time is an additional coordinate, i.e., $\phi = \phi(\mathbf{x}, t)$ or $w = w(\mathbf{x}, t)$. Also, all involved prescribed quantities may depend on time. Note that vibration processes frequently can be formulated by means of a separation ansatz, i.e., $\phi(\mathbf{x}, t) := \phi_1(\mathbf{x})\,\phi_2(t)$, in the form of eigenvalue problems. However, we will not go into further detail with this here (see, e.g., [2]).

In order to fully define time-dependent problems, initial conditions are required in addition to the boundary conditions (which may also depend on time). For transport problems an initial distribution of the unknown function has to be prescribed, e.g.,

$$\phi(\mathbf{x}, t_0) = \phi^0(\mathbf{x})$$

for problem (6.1), while vibration problems require an additional initial distribution for the first time derivative, e.g.,

$$w(x, t_0) = w^0(x) \quad \text{und} \quad \frac{\partial w}{\partial t}(x, t_0) = w^1(x)$$

for problem (6.2).

For the numerical solution of time-dependent problems usually first a spatial discretization with one of the techniques described in the preceding sections is performed. This results in a system of ordinary differential equations (with respect to time). In a finite-difference method setting this approach is referred to as *method of lines*. For instance, the spatial discretization of (6.1) with a finite-volume method yields for each control volume the equation

$$\frac{\partial \phi_P}{\partial t} = \frac{1}{\rho \, \delta V} \left[-a_P(t)\phi_P + \sum_c a_c(t)\phi_c + b_P(t) \right], \tag{6.3}$$

where, for the sake of simplicity we assume the density ρ and the volume δV is temporally constant (and we will also do so in the following). Globally, i.e., for all control volumes, (6.3) corresponds to a (coupled) system of ordinary differential equations for the unknown functions $\phi_P^i = \phi_P^i(t)$ for $i = 1, \ldots, N$, where N is the number of control volumes.

When employing a finite-element method for the spatial discretization of a time-dependent problem, in the Galerkin method a corresponding ansatz with time-dependent coefficients is made:

$$\phi(\mathbf{x}, t) = \varphi_0(\mathbf{x}, t) + \sum_{k=1}^N c_k(t)\varphi_k(\mathbf{x}) \, .$$

For temporally varying boundary conditions also the function φ_0 has to be time-dependent, because it has to fulfil the inhomogeneous boundary conditions within the whole time interval. Applying the Galerkin method with this ansatz in an analogous way as in the steady case leads to a system of ordinary differential equations for the unknown functions $c_k = c_k(t)$ (see Exercise 6.2).

For ease of notation in the following the right hand side of the equation resulting from the spatial discretization (either obtained by finite-volume or finite-element methods) is expressed by the operator $\mathcal{L}$:

$$\frac{\partial \phi}{\partial t} = \mathcal{L}(\phi) \, ,$$

where $\phi = \phi(t)$ denotes the vector of the unknown functions. For instance, in the case of a finite-volume space discretization of (6.1) according to (6.3), the components of $\mathcal{L}(\phi)$ are defined by the right hand side of (6.3).

For the time discretization, i.e., for the discretization of the systems of ordinary differential equations, techniques similar to those for the spatial coordinates can be employed (i.e., finite-difference, finite-volume, or finite-element methods). Since the application of the different methods does not result in principal differences in the resulting discrete systems, we restrict ourselves to the (most simple) case of finite-difference approximations.

First, the time interval $[t_0, T]$ under consideration is divided into individual, generally non-equidistant, subintervals Δt_n:

$$t_{n+1} = t_n + \Delta t_n \, , \quad n = 0, 1, 2, \ldots$$

For a further simplification of notation the variable value at time t_n is indicated with an index n, e.g.:

$$\mathcal{L}(\phi(t_n)) = \mathcal{L}(\phi^n).$$

According to the principle of causality the solution at time t_{n+1} only can depend on *previous* points in time $t_n, t_{n-1}, \ldots$ Since the time in this sense is a "one-way" coordinate, the solution for t_{n+1} has to be determined as a function of the boundary conditions and the solutions at earlier times. Thus, the time discretization always consists in an extrapolation. Starting from the prescribed initial conditions at t_0, the unknown variable ϕ is successively computed at the points of time $t_1, t_2, \ldots$ (see Fig. 6.3).

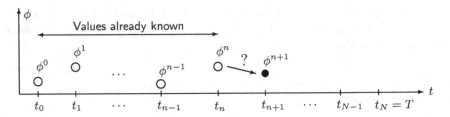

Fig. 6.3. Time-stepping process

Finally, the temporal developing of ϕ can be represented as a sequence of different spatial values at discrete points in time (see Fig. 6.4 for a spatially two-dimensional problem). It should be noted that for transport problems steady solutions are often also computed with a time discretization method as a limit for $t \to \infty$ from the time-dependent equations. However, this method, known as *pseudo time stepping*, does not usually result in an efficient method, but may be useful when, owing to stability problems, the direct solution of the steady problem is hard to obtain (the time-stepping acts as a relaxation, see Sect. 10.3.3).

There must be at least *one* already known time level to discretize the time derivative. If only one time level is used, i.e., the values at t_n, one speaks of *one-step methods*, while if more known time levels are employed, i.e., values for

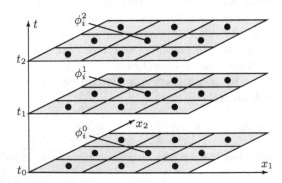

Fig. 6.4. Relation between spatial and temporal discretization

the time levels $t_n, t_{n-1}, \ldots$, one talks about *multi-step methods*. Furthermore
– and more importantly – the methods for the time discretization generally
are divided into two classes according to the choice of points of time, at which
the right hand side is evaluated:

- *Explicit methods*: discretization of the right hand side only at previous
  (already known) time levels:

$$\phi^{n+1} = \mathcal{F}(\phi^n, \phi^{n-1}, \ldots).$$

- *Implicit methods*: discretization of the right hand side also at new (unknown) time level:

$$\phi^{n+1} = \mathcal{F}(\phi^{n+1}, \phi^n, \phi^{n-1}, \ldots).$$

Here, $\mathcal{F}$ denotes some discretization rule for the choice of which we will give
examples later. The distinction into explicit and implicit methods is a very
important attribute because far reaching differences with respect to the properties of the numerical schemes arise (we will go into more detail in Sect. 8.1.2).

In the next two sections some important and representative variants for the
above classes of methods will be introduced. We restrict our considerations to
problems of the parabolic type (only first time derivative). However, it should
be mentioned that the described methods in principle also can be applied to
problems with second time derivative, i.e., problems of the type

$$\frac{\partial^2 \phi}{\partial t^2} = \mathcal{L}(\phi), \qquad (6.4)$$

by introducing the first time derivatives

$$\psi = \frac{\partial \phi}{\partial t}$$

as additional unknowns (order reduction). According to (6.4) one has

$$\frac{\partial \psi}{\partial t} = \mathcal{L}(\phi)$$

and with the definitions

$$\tilde{\phi} = \begin{bmatrix} \psi \\ \phi \end{bmatrix} \quad \text{and} \quad \tilde{\mathcal{L}}(\tilde{\phi}) = \begin{bmatrix} \mathcal{L}(\phi) \\ \psi \end{bmatrix}$$

a system of the form

$$\frac{\partial \tilde{\phi}}{\partial t} = \tilde{\mathcal{L}}(\tilde{\phi})$$

results, which is equivalent to (6.4) and involves only first time derivatives.
However, this way the number of unknowns doubles so that methods that
solve the system (6.4) directly (for instance the so-called *Newmark methods*,
see, e.g., [2]) usually are more efficient.

6.2 Explicit Methods

We start with the most simple example of a time discretization method, the *explicit Euler method*, which is obtained by approximating the time derivative at time level t_n by means of a forward differencing scheme:

$$\frac{\partial \phi}{\partial t}(t_n) \approx \frac{\phi^{n+1} - \phi^n}{\Delta t_n} = \mathcal{L}(\phi^n). \tag{6.5}$$

This corresponds to an approximation of the time derivative of the components ϕ_i of ϕ at the time t_n by means of the slope of the straight line through the points ϕ_i^n and ϕ_i^{n+1} (see Fig. 6.5). The method is first-order accurate (with respect to time) and is also known as the *Euler polygon method*.

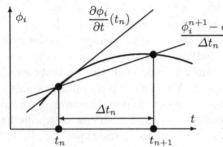

Fig. 6.5. Approximation of time derivative with explicit Euler method

The relation (6.5) can be resolved explicitly for ϕ^{n+1}:

$$\phi^{n+1} = \phi^n + \Delta t_n \mathcal{L}(\phi^n).$$

On the right hand side there are only values from the already known time level, such that the equations for the values at the point of time t_{n+1} at the different spatial grid points are fully decoupled and can be computed independently from each other. This is characteristic for explicit methods.

Let us consider as an example the unsteady one-dimensional diffusion equation (with constant material parameters):

$$\frac{\partial \phi}{\partial t} = \frac{\alpha}{\rho}\frac{\partial^2 \phi}{\partial x^2}. \tag{6.6}$$

A finite-volume space discretization with the central differencing scheme for the diffusion term for an equidistant grid with grid spacing Δx yields for each control volume the ordinary differential equation:

$$\frac{\partial \phi_P}{\partial t}(t)\,\Delta x = \frac{\alpha}{\rho}\frac{\phi_E(t) - \phi_P(t)}{\Delta x} - \frac{\alpha}{\rho}\frac{\phi_P(t) - \phi_W(t)}{\Delta x} \tag{6.7}$$

The explicit Euler method with fixed time step size Δt gives the following approximation:

$$\frac{\phi_P^{n+1} - \phi_P^n}{\Delta t} \Delta x = \frac{\alpha}{\rho} \frac{\phi_E^n - \phi_P^n}{\Delta x} - \frac{\alpha}{\rho} \frac{\phi_P^n - \phi_W^n}{\Delta x}$$

Resolving for ϕ_P^{n+1} yields

$$\phi_P^{n+1} = \frac{\alpha \Delta t}{\rho \Delta x^2}(\phi_E^n + \phi_W^n) + (1 - \frac{2\alpha \Delta t}{\rho \Delta x^2})\phi_P^n \,.$$

The procedure is illustrated graphically in Fig. 6.6. One observes that a number of time steps is necessary until changes at the boundary affect the interior of the spatial problem domain. The finer the grid, the slower the spreading of the information (for the same time step size). As we will see in Sect. 8.1.2, this leads to a limitation of the time step size (stability condition), which depends quadratically on the spatial resolution and, with a finer spatial grid, becomes more and more restrictive. This limitation is purely due to numerical reasons and independent from the actual temporal developing of the problem solution.

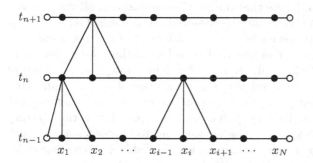

Fig. 6.6. Procedure and flow of information for explicit Euler method

There are numerous other explicit one-step methods that differ from the explicit Euler method in the approximation of the right hand sides. The *modified explicit Euler method* formulated by Collatz (1960) is

$$\frac{\phi^{n+1} - \phi^n}{\Delta t_n} = \mathcal{L}(\phi^n + \frac{\Delta t_n}{2}\mathcal{L}(\phi^n)) \,,$$

which, compared to the explicit Euler method, requires just one additional evaluation of the right hand side, but is second order accurate on equidistant grids.

Another class of explicit one-step methods are the *Runge-Kutta methods*, which are frequently used in practice particularly for aerodynamical flow simulations. These methods can be defined for arbitrary order. As an example, the classical Runge-Kutta method of fourth order is defined by:

$$\frac{\phi^{n+1} - \phi^n}{\Delta t_n} = \frac{1}{6}(f_1 + 2f_2 + 2f_3 + f_4) \,,$$

where

$$f_1 = \mathcal{L}(\phi^n), \qquad\qquad f_2 = \mathcal{L}\left(\phi^n + \frac{\Delta t_n}{2} f_1\right),$$
$$f_3 = \mathcal{L}\left(\phi^n + \frac{\Delta t_n}{2} f_2\right), \quad f_4 = \mathcal{L}(\phi^n + \Delta t_n f_3).$$

Combinations of Runge-Kutta methods of the orders p and $p+1$ frequently are employed to obtain procedures with an automatic time step size control. The resulting methods are known as *Runge-Kutta-Fehlberg methods* (see, e.g., [24]).

In multi-step methods more than two time levels are employed to approximate the time derivative. A corresponding discretization scheme can, for instance, be defined by assuming a piecewise polynomial course of the unknown function with respect to time (e.g., quadratically for three time levels) or by a suitable Taylor series expansion (see, e.g., [12]).

The computation of the solution with a multi-step method must always be started with a single-step method because initially only the solution at t_0 is available. Having computed the solutions for $t_1, \ldots, t_{p-2}$, one can continue with a p time level method. Note that during the computation all variable values from the involved time levels have to be stored. In the case of large systems and many time levels this results in a relatively large memory requirement.

Depending on the number of involved time levels, the approximation of the time derivative, and the evaluation of the right hand side, a variety of multi-step methods can be defined. An important class of explicit multi-step methods that are frequently employed in practice are the *Adams-Bashforth methods*. These can be derived by polynomial interpolation with arbitrary orders. However, in practice, only the methods up to the order 4 are used. For equdistant time steps these are summarized in Table 6.1 (the first order Adams-Bashforth method is again the explicit Euler method).

Table 6.1. Adams-Bashforth methods up to the order 4

Formula	Order
$\dfrac{\phi^{n+1} - \phi^n}{\Delta t} = \mathcal{L}(\phi^n)$	1
$\dfrac{\phi^{n+1} - \phi^n}{\Delta t} = \dfrac{1}{2}\left[3\mathcal{L}(\phi^n) - \mathcal{L}(\phi^{n-1})\right]$	2
$\dfrac{\phi^{n+1} - \phi^n}{\Delta t} = \dfrac{1}{12}\left[23\mathcal{L}(\phi^n) - 16\mathcal{L}(\phi^{n-1}) + 5\mathcal{L}(\phi^{n-2})\right]$	3
$\dfrac{\phi^{n+1} - \phi^n}{\Delta t} = \dfrac{1}{24}\left[55\mathcal{L}(\phi^n) - 59\mathcal{L}(\phi^{n-1}) + 37\mathcal{L}(\phi^{n-2}) - 9\mathcal{L}(\phi^{n-3})\right]$	4

6.3 Implicit Methods

Approximating the time derivative at time t_{n+1} by a first order backward difference formula (see Fig. 6.7) results in the *implicit Euler method*:

$$\frac{\partial \phi}{\partial t}(t_{n+1}) \approx \frac{\phi^{n+1} - \phi^n}{\Delta t_n} = \mathcal{L}(\phi^{n+1})$$

This differs from the explicit variant only in the evaluation of the right hand side, which now is computed at the new (unknown) time level. Consequently, explicitly solving for ϕ_P^{n+1} is no longer possible because all variables of the new time level are coupled to each other. Thus, for the computation of *each* new time level – as in the steady case – the solution of an equation system is necessary. This is characteristic for implicit methods.

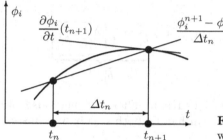

Fig. 6.7. Approximation of time derivative with implicit Euler method

For instance, discretizing the one-dimensional diffusion equation (6.6) with the spatial discretization (6.7) using the implicit Euler method gives:

$$(1 + \frac{2\alpha \Delta t}{\rho \Delta x^2})\phi_P^{n+1} = \frac{\alpha \Delta t}{\rho \Delta x^2}(\phi_E^{n+1} + \phi_W^{n+1}) + \phi_P^n . \qquad (6.8)$$

Regarded over all control volumes, this represents a tridiagonal linear equation system that has to be solved for each time step.

In the implicit case changes at the boundary in the actual time step spread in the whole spatial problem domain (see Fig. 6.8) so that the stability problems indicated for the explicit Euler method do not occur in this form. The implicit Euler method turns out to be stable independently of Δx and Δt (see Sect. 8.1.2).

As the explicit method, the implicit Euler method is first order accurate in time. It is more costly than the explicit variant because more computational effort (for the solution the equation system) and more memory (for the coefficients and the source terms) is required per time step. However, there is no limitation for the time step size due to stability reasons. The higher effort of the method usually is more than compensated for by the possibility of selecting larger time steps. Thus, in most cases it is in total much more efficient than the explicit variant.

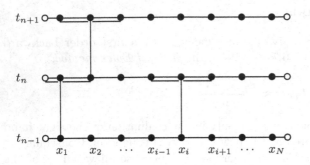

Fig. 6.8. Procedure and flow of information for implicit Euler method

The algebraic equations resulting with the implicit Euler method for an unsteady transport problem differ from the corresponding steady case (when using the same spatial discretizations) only by two additional terms in the coefficients a_P and b_P:

$$\underbrace{\left(a_P^{n+1} + \boxed{\frac{\delta V \rho}{\Delta t_n}}\right)}_{\tilde{a}_P^{n+1}} \phi_P^{n+1} = \sum_c a_c^{n+1} \phi_c^{n+1} + \underbrace{b_P^{n+1} + \boxed{\frac{\delta V \rho}{\Delta t_n}} \phi_P^n}_{\tilde{b}_P^{n+1}} .$$

In the limit $\Delta t_n \to \infty$ the steady equations result. Thus, the methods can be easily combined for a single code that can handle both steady and unsteady cases.

An important implicit one-step method frequently used in practice is the *Crank-Nicolson method*, which is obtained when for each component ϕ_i of ϕ the time derivative at time $t_{n+1/2} = (t_n + t_{n+1})/2$ is approximated by the straight line connecting ϕ_i^{n+1} and ϕ_i^n (see Fig. 6.9):

$$\frac{\partial \phi}{\partial t}(t_{n+1/2}) \approx \frac{\phi^{n+1} - \phi^n}{\Delta t_n} = \frac{1}{2}\left[\mathcal{L}(\phi^{n+1}) + \mathcal{L}(\phi^n)\right] .$$

This corresponds to a central difference approximation of the time derivative at time $t_{n+1/2}$. The scheme has a second order temporal accuracy. The method is also called trapezoidal rule because the application of the latter for numerical integration of the equivalent integral equation yields the same formula.

The computational effort for the Crank-Nicolson method is only slightly higher than for the implicit Euler method because only $\mathcal{L}(\phi^n)$ has to be computed additionally and the solution of the resulting equation systems usually is a bit more "difficult". However, due to the higher order the accuracy is much better. One can show that the Crank-Nicolson method is the most accurate second-order method.

For problem (6.6) with the spatial discretization (6.7), the Crank-Nicolson method results in the following approximation:

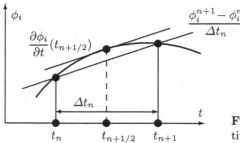

Fig. 6.9. Approximation of time derivative with Crank-Nicolson method

$$2(1 + \frac{\alpha \Delta t}{\rho \Delta x^2})\phi_{\mathrm{P}}^{n+1} = \frac{\alpha \Delta t}{\rho \Delta x^2}(\phi_{\mathrm{E}}^{n+1} + \phi_{\mathrm{W}}^{n+1}) + \frac{\alpha \Delta t}{\rho \Delta x^2}(\phi_{\mathrm{E}}^{n} + \phi_{\mathrm{W}}^{n}) +$$

$$2(1 - \frac{\alpha \Delta t}{\rho \Delta x^2})\phi_{\mathrm{P}}^{n} .$$

Although it is implicit, the Crank-Nicolson method may suffer from stability problems for cases where the problem solution is spatially not "smooth" (no *strong A-stability*, see, e.g., [12]). By interspersing some steps of the implicit Euler method at regular intervals, a damping of the corresponding (non-physical) oscillations can be achieved while preserving the second-order accuracy of the scheme.

Note that the explicit and implicit Euler methods as well as the Crank-Nicolson method can be integrated into a single code in a simple way by introducing a control parameter θ as follows:

$$\frac{\phi^{n+1} - \phi^n}{\Delta t_n} = \theta \mathcal{L}(\phi^{n+1}) + (1 - \theta)\mathcal{L}(\phi^n) .$$

This approach in the literature is often called θ-*method*. For $\theta = 0$ and $\theta = 1$ the explicit and implicit Euler methods, respectively, result. $\theta = 1/2$ gives the Crank-Nicolson method. Valid time discretizations are also obtained for all other values of θ in the interval $[0, 1]$. However, for $\theta \neq 1/2$ the method is only of first order.

As in the explicit case, implicit multi-step methods of different order can be defined depending on the number of involved time levels, the approximation of the time derivative, and the evaluation of the right hand side An important class of methods are the *BDF-methods* (backward-differencing formula). These can be derived with arbitrary order by approximating the time derivative at t_{n+1} with backward-differencing formulas involving a corresponding number of previous time levels. The corresponding methods for equidistant time steps up to the order 4 are indicated in Table 6.2.

The first order BDF-method corresponds to the implicit Euler method. In particular, the second order BDF-method is frequently used in practice. With this the unknown function is approximated by the parabola defined by the function values at the time levels t_{n-1}, t_n, and t_{n+1} (see Fig. 6.10). Having comparably good stability properties this method only involves slightly

Table 6.2. BDF-methods up to the order 4

Formula	Order
$\dfrac{\phi^{n+1} - \phi^n}{\Delta t} = \mathcal{L}(\phi^{n+1})$	1
$\dfrac{3\phi^{n+1} - 4\phi^n + \phi^{n-1}}{2\Delta t} = \mathcal{L}(\phi^{n+1})$	2
$\dfrac{11\phi^{n+1} - 18\phi^n + 9\phi^{n-1} + 2\phi^{n-2}}{6\Delta t} = \mathcal{L}(\phi^{n+1})$	3
$\dfrac{25\phi^{n+1} - 48\phi^n + 36\phi^{n-1} - 16\phi^{n-2} + 3\phi^{n-3}}{2\Delta t} = \mathcal{L}(\phi^{n+1})$	4

more computational effort per time step and is much more accurate than the implicit Euler method. Only the values ϕ^{n-1} have to be stored additionally. From order three on the stability properties deteriorate with increasing order such that the application of a BDF-method with order higher than 4 is not recommended.

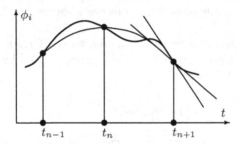

Fig. 6.10. Approximation of time derivative with second-order BDF-method

Another class of implicit multi-step methods are the *Adams-Moulton methods* – the implicit counterparts to the (explicit) Adams-Bashforth methods. The corresponding formulas up to the order 4 for equidistant grids are summarized in Table 6.3. The Adams-Moulton methods of first and second orders correspond to the implicit Euler and Crank-Nicolson methods, respectively.

Adams-Moulton methods can be used together with Adams-Bashforth methods of the same order as *predictor-corrector methods*. Here, the idea is to determine with the explicit predictor method in a "cheap" way a good starting value for the implicit corrector method. For instance, the corresponding predictor-corrector method of fourth order is given by:

$$\phi^{*} = \phi^n + \frac{\Delta t}{24} \left[55\mathcal{L}(\phi^n) - 59\mathcal{L}(\phi^{n-1}) + 37\mathcal{L}(\phi^{n-2}) - 9\mathcal{L}(\phi^{n-3})\right],$$

$$\phi^{n+1} = \phi^n + \frac{\Delta t}{24} \left[9\mathcal{L}(\phi^{*}) + 19\mathcal{L}(\phi^n) - 5\mathcal{L}(\phi^{n-1}) + \mathcal{L}(\phi^{n-2})\right].$$

Table 6.3. Adams-Moulton methods up to the order 4

Formula	Order
$\dfrac{\phi^{n+1} - \phi^n}{\Delta t} = \mathcal{L}(\phi^{n+1})$	1
$\dfrac{\phi^{n+1} - \phi^n}{\Delta t} = \dfrac{1}{2}\left[\mathcal{L}(\phi^{n+1}) + \mathcal{L}(\phi^n)\right]$	2
$\dfrac{\phi^{n+1} - \phi^n}{\Delta t} = \dfrac{1}{12}\left[5\mathcal{L}(\phi^{n+1}) + 8\mathcal{L}(\phi^n) - \mathcal{L}(\phi^{n-1})\right]$	3
$\dfrac{\phi^{n+1} - \phi^n}{\Delta t} = \dfrac{1}{24}\left[9\mathcal{L}(\phi^{n+1}) + 19\mathcal{L}(\phi^n) - 5\mathcal{L}(\phi^{n-1}) + \mathcal{L}(\phi^{n-2})\right]$	4

The error for such a combined method is equal to that of the implicit method, which is always the smaller one.

6.4 Numerical Example

As a more complex application example for the numerical simulation of time-dependent processes and for comparison of different time discretization methods we consider the unsteady flow around a circular cylinder in a channel with time dependent inflow condition. The problem configuration is shown in Fig. 6.11. The problem can be described by the two-dimensional incompressible Navier-Stokes equations as given in Sect. 2.5.1. The kinematic viscosity is defined as $\nu = 10^{-3}\,\mathrm{m}^2/\mathrm{s}$, and the fluid density is $\rho = 1.0\,\mathrm{kg/m}^3$. The inflow condition for the velocity component v_1 in x_1-direction is

$$v_1(0, x_2, t) = 4v_{\max}x_2(H - x_2)\sin(\pi t/8)/H^2, \quad \text{for} \quad 0 \le t \le 8\,\mathrm{s}$$

with $v_{\max} = 1.5\,\mathrm{m/s}$. This corresponds to the velocity profile of a fully developed channel flow, where the Reynolds number $Re = \bar{v}D/\nu$ based on the cylinder diameter $D = 0.1\,\mathrm{m}$ and the mean velocity $\bar{v}(t) = 2v_{\max}(0, H/2, t)/3$ varies in the range $0 \le Re \le 100$.

As reference quantities the drag and lift coefficients

$$c_{\mathrm{D}} = \frac{2F_w}{\rho\bar{v}^2 D} \quad \text{and} \quad c_{\mathrm{L}} = \frac{2F_a}{\rho\bar{v}^2 D}$$

for the cylinder are considered. Hereby, the drag and lift forces are defined by

$$F_{\mathrm{D}} = \int_S \left(\rho\nu\frac{\partial v_t}{\partial x_i}n_i n_2 - pn_1\right)\mathrm{d}S \quad \text{and} \quad F_{\mathrm{L}} = -\int_S \left(\rho\nu\frac{\partial v_t}{\partial x_i}n_i n_1 + pn_2\right)\mathrm{d}S,$$

where S denotes the cylinder surface (circle), $\mathbf{n} = (n_1, n_2)$ is the normal vector on S, and v_t is the tangential velocity on S.

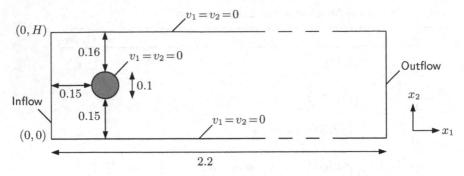

Fig. 6.11. Configuration for two-dimensional flow around cylinder (lengths in m)

The temporal development of the flow is illustrated in Fig. 6.12 and shows the vorticity at different points of time. First, there are two counterrotating vortices behind the cylinder, that become unstable after a certain time (with increasing oncoming flow). A Kármán vortex street forms and finally decays with decreasing oncoming flow. The problem effectively involves two kinds of time dependencies: an "outer" one due to the time-dependent boundary condition and an "inner" one due to the vortex separation by physical instability (bifurcation).

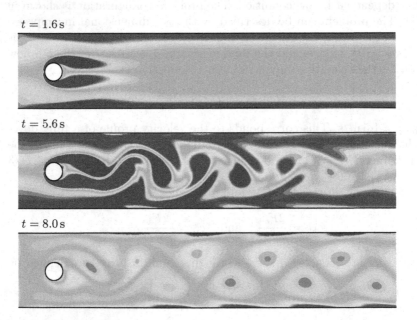

Fig. 6.12. Temporal development of vorticity for unsteady flow around cylinder

The spatial discretization uses a finite-volume method with central differencing scheme on a grid with 24 576 CVs, a typical grid size for this kind of problem. The grid can be seen in Fig. 6.13, where for visibility only every fourth grid line is shown, i.e., the real number of CVs is 16 times larger. For the time discretization the implicit Euler method, the Crank-Nicolson method, and the second-order BDF-method are compared. We only consider implicit methods because explicit methods for this kind of problems are orders of magnitude slower and, therefore, are out of discussion here.

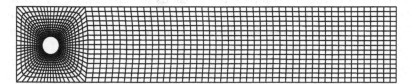

Fig. 6.13. Numerical grid for flow around cylinder (every fourth grid line is shown)

Figure 6.14 shows the temporal development of the lift coefficient c_L obtained with the different time discretization methods, where for each case the same time step size $\Delta t = 0.02$ s is employed. Thus 400 time steps for the given time interval of 8 s are necessary. The "exact" solution, which has been obtained by a computation with a very fine grid and a very small time step size, is also indicated.

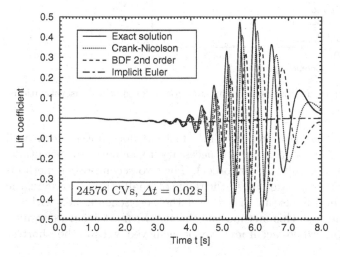

Fig. 6.14. Temporal development of lift coefficient for different time discretization schemes for flow around cylinder

One can see quite significant differences in the results obtained with the different methods. With the implicit Euler method the oscillations with the given time step size are not captured at all. The discretization error in this case is so large, that the oscillations are damped completely. The BDF-method is able to resolve the oscillations to some extent, but the amplitude is clearly too small. The Crank-Nicolson method, as one would expect from a corresponding analysis of the discretization error, gives the best result.

One of the most important practical aspects for a numerical method is how much computing time the method needs to compute the solution with a certain accuracy. In order to compare the methods in this respect, Fig. 6.15 shows the relative error for the maximum of the lift coefficient against the computing time, which is needed for different time step sizes, for the different time discretization methods.

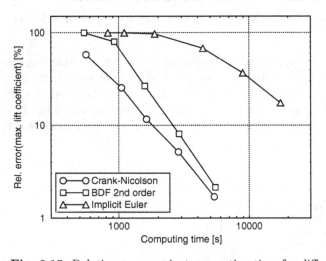

Fig. 6.15. Relative error against computing time for different time discretizations

One can observe that the implicit Euler method does not perform well because very small time step sizes are necessary to achieve an acceptable accuracy (the method is only of first order). The two second-order methods don't differ that much, in particular, in the range of small errors. However, the Crank-Nicolson method is also in this respect the best scheme, i.e., within a given computing time with this method one obtains the most accurate results. In other words, a precribed accuracy can be achieved within the shortest computing time.

In order to point out that not all quantities of a problem react with the same sensitivity to the discretization employed, in Fig. 6.16 the temporal development of the drag coefficient c_D that results with the different time discretization schemes is given. In contrast to the corresponding lift coefficients one can observe only minor differences in the results for the different methods.

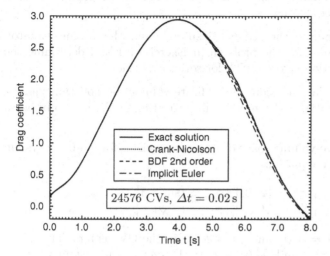

Fig. 6.16. Temporal development of drag coefficient with different time discretizations

In summary, one can conclude that the time discretization method together with the time step size has to be chosen according to the accuracy requirements of the underlying problem, where the stability and approximation properties of the method have to be taken into account. This is not always an easy undertaking. In the case of strong temporal variations in the solution, the method should in any event be at least of second order.

Exercises for Chap. 6

Exercise 6.1. The temperature distribtion $T = T(t,x)$ in a bar of length L with constant material properties is decribed by the differential equation

$$\frac{\partial T}{\partial t} - \alpha \frac{\partial^2 T}{\partial x^2} = 0 \quad \text{with} \quad \alpha = \frac{\kappa}{\rho c_p}$$

for $0 < x < L$ and $t > 0$ (cf. Sect. 2.3.2). As initial and boundary conditions $T(0,x) = \sin(\pi x) + x$ and $T(t,0) = 0$, and $T(t,L) = 1$ are given (in K). The problem parameters are $L = 1\,\text{m}$ and $\alpha = 1\,\text{m}^2/\text{s}$. (i) Use the FVM with two equidistant CVs and second-order central differences for the spatial discretization and formulate the resulting ordinary differential equations for the two CVs. (ii) Compute the temperature until the time $t = 0,4\,\text{s}$ with the implicit and explicit Euler methods, each with $\Delta t = 0,1\,\text{s}$ and $0,2\,\text{s}$. (iii) Discuss the results in comparison to the analytic solution $T_a(t, x) = e^{-\alpha t \pi^2} \sin(\pi x) + x$.

Exercise 6.2. Discretize the unsteady transport equation (2.24) with the finite-element method and formulate the θ-method for the resulting system of ordinary differential equations.

Exercise 6.3. Formulate the Adams-Bashforth and the Adams-Moulton methods of fourth order for the problem in Exercise 6.1 and determine the corresponding truncation errors by Taylor series expansion.

Exercise 6.4. Formulate a second-order finite-volume method (for equidistant grids) for the spatial and temporal discretization of the unsteady beam equation (6.2).

Exercise 6.5. A finite-volume space discretization yields for $t > 0$ the system of ordinary differential equations

$$\begin{bmatrix} \phi_1' \\ \phi_2' \end{bmatrix} = \begin{bmatrix} 2 & 0 \\ t & \sqrt{t} \end{bmatrix} \begin{bmatrix} \phi_1 \\ \phi_2 \end{bmatrix} + \begin{bmatrix} \sin(\pi t) \\ \sqrt{t} \end{bmatrix}$$

for the two values $\phi_1 = \phi_1(t)$ and $\phi_2 = \phi_2(t)$ in the CV centers. The initial conditions are $\phi_1(0) = 2$ and $\phi_2(0) = 1$. (i) Discretize the system with the θ-method. (ii) Compute $\phi_1^1 = \phi_2(\Delta t)$ with the time step size $\Delta t = 2$.

7

Solution of Algebraic Systems of Equations

The discretization of steady or unsteady problems with implicit time integration, either by finite-volume or finite-element methods, results in large sparse systems of algebraic equations. The solution procedure for these equation systems is an important part of a numerical method. Frequently, much more than 50% of the total computation time is required for the numerical solution of these systems. Therefore, in this chapter we will deal in some detail with solution methods for such systems, considering first the linear and afterwards the nonlinear case.

We will give examples of some typical solution procedures in order to highlight the characteristic properties of such methods, particularly with respect to their computational efficiency. A large variety of methods exist, some of which are specially designed for certain classes of systems. For these we refer to the corresponding literature (e.g., [11]).

7.1 Linear Systems

The linear equation systems to be solved are of the following general form:

$$a_P^i \phi_P^i - \sum_c a_c^i \phi_c^i = b_P^i \quad \text{for} \quad i = 1, \ldots, K, \tag{7.1}$$

where the summation index c runs over the nodes in the neighborhood of node P, which are involved in the corresponding discretization. In matrix notation we denote the system (7.1) as

$$\mathbf{A}\phi = \mathbf{b}. \tag{7.2}$$

The structure of the system matrix $\mathbf{A}$ is determined mainly from the numerical grid and from the discretization scheme employed (see, for instance, the examples in Sect. 4.8). The dimension K of the matrix is defined by the number of unknown nodal values.

The system matrices resulting from the discretization methods that were discussed are, after consideration of boundary conditions, usually non-singular (under certain requirements for the discretization), such that the equation systems possess a unique solution. In principle there are two possibilities for the numerical solution of the systems:

- direct methods,
- iterative methods.

The characteristic of direct methods is that the *exact* solution of the system (neglecting rounding errors) is obtained by a *nonrecurring* application of an algorithm, whereas with iterative methods an *approximative* solution is successively improved by a *repeated* application of a certain iteration rule. In the following we will exemplify typical variants for both classes of methods.

7.1.1 Direct Solution Methods

The only method for the direct solution of matrix equations that is of interest for the problems in the present context is the *Gauß elimination* or variants of it. Also in practice, especially for structural mechanics FEM computations, these are still frequently employed. Applied to general matrices the Gauß elimination has an important interpretation as a multiplicative decomposition of the given matrix $\mathbf{A}$ into lower and upper triangular matrices $\mathbf{L}$ and $\mathbf{U}$ (LU-decomposition):

$$
\mathbf{A} = \underbrace{\begin{bmatrix} 1 & 0 & \cdot & \cdot & \cdot & 0 \\ l_{2,1} & \cdot & & & & \cdot \\ \cdot & \cdot & \cdot & & & \\ \cdot & \cdot & \cdot & \cdot & & 0 \\ \cdot & & & \cdot & \cdot & \\ l_{K,1} & \cdot & \cdot & \cdot & l_{K,K-1} & 1 \end{bmatrix}}_{\mathbf{L}} \underbrace{\begin{bmatrix} u_{1,1} & u_{1,2} & \cdot & \cdot & \cdot & u_{1,K} \\ 0 & \cdot & \cdot & & & \cdot \\ \cdot & & \cdot & \cdot & & \\ \cdot & & & \cdot & \cdot & u_{K-1,K} \\ 0 & \cdot & \cdot & \cdot & 0 & u_{K,K} \end{bmatrix}}_{\mathbf{U}},
$$

where the coefficients $l_{i,j}$ and $u_{i,j}$ have to be computed anyway in the course of the usual elimination procedure (see, e.g., [20] for details). Having computed the decomposition, the solution of the system – if necessary, also for different right hand side vectors – can be determined by a simple forward and backward substitution process. Here, first the system

$$\mathbf{L}\mathbf{h} = \mathbf{b}$$

is solved by forward substitution for $\mathbf{h}$ and afterwards the final solution ϕ is determined from

$$\mathbf{U}\phi = \mathbf{h}$$

by backward substitution.

If $\mathbf{A}$ is a band matrix, as is usually the case for the discretizations of the problems of interest here (see, e.g., Sect. 4.8), the matrices $\mathbf{L}$ and $\mathbf{U}$ are also band matrices (with the same band width as $\mathbf{A}$). The property that $\mathbf{A}$ is only sparsely filled within the band (sparse matrices), however, does not transfer to $\mathbf{L}$ and $\mathbf{U}$, which usually are full within the band. This is a crucial disadvantage of direct methods when they are applied to two- and, in particular, three-dimensional problems. We will return to the consequences of this "fill-in" effect with respect to the efficiency of the Gauß elimination in Sect. 7.1.7. For one-dimensional problems, for which the discretization result in for instance tridiagonal or pentadiagonal matrices, this effect plays no role (already the band of $\mathbf{A}$ is full, but narrow), such that in this case the direct solution with the Gauß elimination is quite efficient. For tridiagonal matrices the Gauß elimination in the literature is known as *TDMA (tridiagonal matrix algorithm)* or *Thomas algorithm*.

It should be noted that for symmetric positive definite systems there exists a symmetric variant of the Gauß elimination called *Cholesky method*. Here, all operations can be performed on, for instance, the lower triangular matrix, which reduces the memory requirements by a factor of two. Additionally, the Cholesky method is less sensitive to rounding errors than the usual Gauß elimination (see, e.g., [24]).

7.1.2 Basic Iterative Methods

Iterative methods start from an estimated solution ϕ^0 of the equation system, which then is successively improved by the multiple application of an iteration process $\mathcal{P}$:

$$\phi^{k+1} \leftarrow \mathcal{P}(\phi^k) \,, \quad k = 0, 1, \ldots$$

The *Jacobi method* and the *Gauß-Seidel method* belong to the simplest iterative solution methods. In the Jacobi method an improved solution is achieved by inserting the ("old") values ϕ^k in the sum term of the algebraic equation (7.1), i.e., the "new" value at node P is determined by an averaging of the "old" values in the neighboring nodes involved in the discretization. The corresponding iteration rule reads:

$$\phi_{\mathrm{P}}^{i,k+1} = \frac{1}{a_{\mathrm{P}}^i} \left(\sum_c a_c^i \phi_c^{i,k} + b_{\mathrm{P}}^i \right).$$

Figure 7.1 illustrates the procedure for a problem resulting from a two-dimensional finite-volume discretization. Contrary to many other methods (see below), the properties of the Jacobi method are completely independent from the numbering of the unknowns.

The convergence of the Jacobi method can be accelerated if one exploits the fact that when computating the new value at node P for some of the

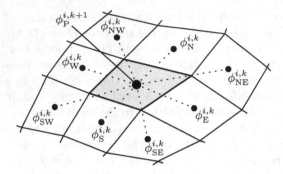

Fig. 7.1. Iteration rule for Jacobi method

neighboring nodes involved in the discretization (depending on the numbering of the nodes) already some new (improved) values are available. This leads to the *Gauß-Seidel method*:

$$\phi_P^{i,k+1} = \frac{1}{a_P^i}\left(\sum_{c_1} a_{c_1}^i \phi_{c_1}^{i,k} + \sum_{c_2} a_{c_2}^i \phi_{c_2}^{i,k+1} + b_P^i\right),$$

where the index c_1 runs over the nodes that are not yet computed and the index c_2 over the nodes for which already "new" values are available. Figure 7.2 illustrates the procedure for the case of a lexigraphical numbering of the nodes (first from west to east, then from south to north).

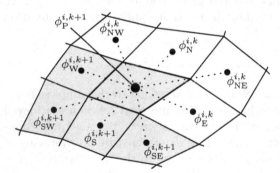

Fig. 7.2. Iteration rule for Gauß-Seidel method with lexicographical numbering of nodes

A sufficient (but not necessary) criterion for the convergence of the Jacobi and Gauß-Seidel methods is the *essential diagonal dominance* of the coefficient matrix, i.e., for all $i = 1, \dots, K$ the inequality

$$|a_P^i| \geq \sum_c |a_c^i|$$

is fulfilled, where the strict inequality ">" has to be fulfilled at least for one index i. This property is also known as *weak row sum criterion*. For instance, for a conservative discretization the convergence is ensured if all coefficients a_c^i have the same sign. In this case "=" is fulfilled for all inner nodes and ">" for the nodes in the vicinity of boundaries with Dirichlet conditions.

The rate of convergence of the Gauß-Seidel method can be further improved by an underrelaxation, resulting in the *SOR method (successive overrelaxation)*. Here, first auxiliary values ϕ_*^{k+1} are computed according to the Gauß-Seidel method. They are then linearly combined with the "old" values ϕ^k to give the new iterate:

$$\phi^{k+1} = \phi^k + \omega(\phi_*^{k+1} - \phi^k),$$

where the relaxation parameter ω should be in the interval $[1, 2)$ (in this case, the convergence is ensured if the matrix $\mathbf{A}$ fulfils certain requirements, see, e.g., [11]). For $\omega = 1$ again the Gauß-Seidel method is recovered. In summary the SOR method can be written as follows:

$$\phi_P^{i,k+1} = (1 - \omega)\phi_P^{i,k} + \frac{\omega}{a_P^i} \left(\sum_{c_1} a_{c_1}^i \phi_{c_1}^{i,k} + \sum_{c_2} a_{c_2}^i \phi_{c_2}^{i,k+1} + b_P^i \right).$$

For general equation systems it is not possible to specify an optimal value ω_{opt} for the relaxation parameter ω explicitly. Such values can be theoretically derived only for simple linear model problems For these it can be shown that the asymptotic computational effort for the SOR method is significantly lower than for the Gauß-Seidel method, i.e., with the relatively simple modification the computational effort may be reduced considerably (we come back to this issue in Sect. 7.1.7). Frequently, the optimal value for ω for the model problems also provides reasonable convergence rates for more complex problems.

There are numerous variants of the SOR method, mainly differing in the order in which the individual nodes are treated: e.g., red-black SOR, block SOR (line SOR, plane SOR, ...), symmetric SOR (SSOR). For different applications (on different computer architectures) different variants can be advantageous. We will not discuss this issue further here, but refer to the corresponding literature (e.g., [11])

7.1.3 ILU Methods

Another class of iterative methods, which became popular due to good convergence and robustness properties, is based on what is called *incomplete LU decompositions* of the system matrix. These methods, which meanwhile have been proposed in numerous variants (see, e.g., [11]), are known in literature as *ILU methods* (Incomplete LU). An incomplete LU decomposition – like a complete LU decomposition (see Sect. 7.1.1) – consists in a multiplicative splitting of the coefficient matrix into lower and upper triangular matrices, which, however, compared to a complete decomposition, are only sparsely filled within the band (as the matrix $\mathbf{A}$). The product of the two triangular matrices – again denoted by $\mathbf{L}$ and $\mathbf{U}$ – should approximate the matrix $\mathbf{A}$ as best as possible:

$$\mathbf{A} \approx \mathbf{LU},$$

i.e., $\mathbf{LU}$ should closely resemble a complete decomposition.

Since the product matrix $\mathbf{LU}$ only constitutes an approximation of the matrix $\mathbf{A}$, an iteration process has to be introduced for solving the equation system (7.2). For this the system can be equivalently written as

$$\mathbf{LU}\phi - \mathbf{LU}\phi = \mathbf{b} - \mathbf{A}\phi$$

and the iteration process can be defined by

$$\mathbf{LU}\phi^{k+1} - \mathbf{LU}\phi^{k} = \mathbf{b} - \mathbf{A}\phi^{k}.$$

The right hand side is the residual vector

$$\mathbf{r}^{k} = \mathbf{b} - \mathbf{A}\phi^{k}$$

for the k-th iteration, which (in a suitable norm) is a measure for the accuracy of the iterative solution. The actual computation of ϕ^{k+1} can be done via the correction $\Delta\phi^{k} = \phi^{k+1} - \phi^{k}$, which can be determined as the solution of the system

$$\mathbf{LU}\Delta\phi^{k} = \mathbf{r}^{k}.$$

This system can easily be solved directly by forward and backward substitution as described in Sect. 7.1.1.

The question that remains is how the triangular matrices $\mathbf{L}$ and $\mathbf{U}$ should be determined in order for the iteration process to be as efficient as possible. We will exemplify the principal idea by means of a 5-point finite-volume discretization in the two-dimensional case for a structured quadrilateral grid with lexicographical node numbering (see Sect. 4.8). In this case one can demand, for instance, that $\mathbf{L}$ and $\mathbf{U}$ possess the same structure as $\mathbf{A}$:

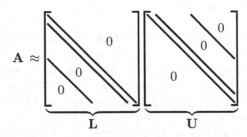

The simplest possibility to determine $\mathbf{L}$ and $\mathbf{U}$ would be to set the corresponding coefficients to those of $\mathbf{A}$. This, however, usually does not lead to a good approximation. When formally carrying out the corresponding matrix multiplication one can see that in comparison to **A** the product matrix **LU** contains two additional diagonals:

$$\mathbf{LU} = \begin{bmatrix} & & & 0 \\ & & 0 & \\ & 0 & & \\ 0 & & & \end{bmatrix} = \begin{bmatrix} & & & 0 \\ & & 0 & \\ & 0 & & \\ 0 & & & \end{bmatrix} + \begin{bmatrix} & & & 0 \\ & & 0 & \\ & 0 & & \\ 0 & & & \end{bmatrix}. \qquad (7.3)$$

$$\underbrace{}_{\mathbf{A}} \qquad \underbrace{}_{\mathbf{N}}$$

The additional coefficients correspond (for our example) to the nodes NW and SE in the numerical grid (see Fig. 7.3).

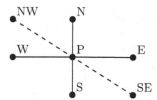

Fig. 7.3. Correspondance of grid points and coefficients in product matrix of ILU decomposition

The objective should be to determine **L** and **U** such that the contribution of the matrix **N**, which can be interpreted as the deviation of **LU** from the complete decomposition, becomes (in a suitable sense) as small as possible. A (recursive) computational procedure for the coefficients of **L** and **U** can be defined by prescribing coefficients of **N**, for which there are a variety of possibilities. The simplest one is to require that the two additional diagonals in **N** vanish, i.e., the diagonals of **A** correspond to that of **LU**, which results in the standard ILU method (see [11]). The corresponding procedure is similar to that for a complete LU decomposition, the only difference being that all diagonals within the band, which are zero in **A**, simply are set to zero also in **L** and **U**.

Stone (1968) proposed approximating the contributions of the additional diagonals by neighboring nodes:

$$\phi_{\mathrm{SE}}^{i} = \alpha(\phi_{\mathrm{S}}^{i} + \phi_{\mathrm{E}}^{i} - \phi_{\mathrm{P}}^{i}) \quad \text{and} \quad \phi_{\mathrm{NW}}^{i} = \alpha(\phi_{\mathrm{W}}^{i} + \phi_{\mathrm{N}}^{i} - \phi_{\mathrm{P}}^{i}). \qquad (7.4)$$

Again the coefficients of the matrices **L** and **U** can be recursively computed from (7.3) by taking into account the approximation (7.4) (see, e.g., [8]). The resulting ILU variant, which is frequently employed within flow simulation programs, is known as *SIP (strongly implicit procedure)*. The approximations defined by the expressions (7.4) can be influenced by the choice of the parameter α, which has to be in the interval $[0, 1)$. For $\alpha = 0$ again the standard ILU method results.

There are many more variants to define the incomplete decomposition. These variants are also for matrices with more than 5 diagonals as well as for three-dimensional problems, and even for system matrices from discretizations on unstructured grids. These differ in, for instance, the approximation of the nodes corresponding to the additional diagonals in the product matrix or by the consideration of additional diagonals in the decomposition (see, e.g., [11]).

7.1.4 Convergence of Iterative Methods

All the iterative methods described so far can be put into a general framework for iteration schemes. We will briefly outline this concept, since it also provides the basis for an insight into the convergence properties of the schemes.

With some non-singular matrix $\mathbf{B}$ the linear system (7.2) can be equivalently written as:

$$\mathbf{B}\phi + (\mathbf{A} - \mathbf{B})\phi = \mathbf{b}.$$

An iteration rule can be defined by

$$\mathbf{B}\phi^{k+1} + (\mathbf{A} - \mathbf{B})\phi^k = \mathbf{b},$$

which, solving for ϕ^{k+1}, gives the relation

$$\phi^{k+1} = \phi^k - \mathbf{B}^{-1}(\mathbf{A}\phi^k - \mathbf{b}) = (\mathbf{I} - \mathbf{B}^{-1}\mathbf{A})\phi^k + \mathbf{B}^{-1}\mathbf{b}. \qquad (7.5)$$

Depending on the choice of $\mathbf{B}$ different iterative methods can be defined. For instance, the methods introduced above can be obtained by

- Jacobi method: $\mathbf{B}^{\mathrm{JAC}} = \mathbf{A_D}$,

- Gauß-Seidel method: $\mathbf{B}^{\mathrm{GS}} = \mathbf{A_D} + \mathbf{A_L}$,

- SOR method: $\mathbf{B}^{\mathrm{SOR}} = (\mathbf{A_D} + \omega\mathbf{A_L})/\omega$,

- ILU method: $\mathbf{B}^{\mathrm{ILU}} = \mathbf{LU}$,

where the matrices $\mathbf{A_D}$, $\mathbf{A_L}$, and $\mathbf{A_U}$ are defined according to the following additive decomposition of $\mathbf{A}$ into a diagonal and lower and upper triangular matrices:

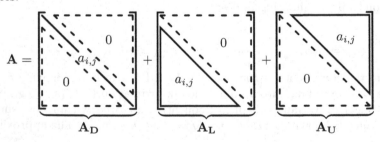

The convergence rate of iterative methods defined by (7.5) is determined by the absolute largest eigenvalue $\lambda_{\max}$ (also denoted as *spectral radius*) of the iteration matrix

$$\mathbf{C} = \mathbf{I} - \mathbf{B}^{-1}\mathbf{A}.$$

One can show that the number of iterations N_{it}, which are required to reduce the initial error by a factor ϵ, i.e., to achieve

$$\|\phi^k - \phi\| \leq \epsilon \|\phi^0 - \phi\|,$$

is given by

$$N_{\text{it}} = \frac{C(\epsilon)}{1 - \lambda_{\max}} \tag{7.6}$$

with a constant $C(\epsilon)$, which does not depend on the number of unknowns. With $\| \cdot \|$ some norm in $\mathbb{R}^K$ is denoted, e.g., the usual Euclidian distance, which for a vector $\mathbf{a}$ with the components a_i is defined by

$$\|\mathbf{a}\| = \left(\sum_{i=1}^{K} a_i^2 \right)^{1/2}.$$

If (and only if) $\lambda_{\max} < 1$ the method converges, the smaller $\lambda_{\max}$ is, the faster it converges. Due to the above properties, the matrix $\mathbf{B}$ should fulfil the following requirements:

- $\mathbf{B}$ should approximate $\mathbf{A}$ as "good" as possible in order to have a small spectral radius of $\mathbf{C}$, i.e., a low number of iterations,
- the system (7.5) should be "easily" solvable for ϕ^{k+1} in order to have a low effort for the individual iterations.

Since both criteria cannot be fulfilled optimally simultaneously, a compromise has to be found. This is the case for all the methods described above. For instance, for the Jacobi method, since $\mathbf{B}^{\text{JAC}}$ is a diagonal matrix, the solution of (7.5) just needs a division by the diagonal elements. However, the approximation of $\mathbf{A}$ just by its diagonal might be rather poor.

In Sect. 7.1.7 we will use the above considerations for an investigation of the convergence properties and the computational effort of the iterative methods introduced above for a model problem.

Let us mention here an approach which can be used in the context of iterative equation solvers to obtain a simpler structure of the system matrix or to increase its diagonal dominance. The idea is to treat matrix entries which should not be considered in the solution algorithm (e.g., contributions due to non-orthogonality of the grid, see Sect. 4.5, or flux-blending parts of higher order, see Sect. 4.3.3) as source terms (explicit treatment) by allocating the corresponding variable values with values from the preceding iteration.

Formally, this approach can be formulated by an additive decomposition of the system matrix of the form

$$\mathbf{A} = \mathbf{A}_\mathrm{I} + \mathbf{A}_\mathrm{E}$$

into an "implicit" part $\mathbf{A}_\mathrm{I}$ and an "explicit" part $\mathbf{A}_\mathrm{E}$. The corresponding iteration rule reads:

$$\mathbf{B}_\mathrm{I}\phi^{k+1} + (\mathbf{A}_\mathrm{I} - \mathbf{B}_\mathrm{I})\phi^k = \mathbf{b} - \mathbf{A}_\mathrm{E}\phi^k \,,$$

where $\mathbf{B}_\mathrm{I}$ is a suitable approximation of $\mathbf{A}_\mathrm{I}$. For instance, when using the ILU method from Sect. 7.1.3, not $\mathbf{A}$, but $\mathbf{A}_\mathrm{I}$ is (incompletely) decomposed, such that the matrices $\mathbf{L}$ and $\mathbf{U}$ get the (simpler) structure of $\mathbf{A}_\mathrm{I}$. A consequence of such an approach is that it usually results in an increase in the number of iterations, but the individual iterations are less expensive.

7.1.5 Conjugate Gradient Methods

A further important class of iterative linear system solvers is *gradient methods*. The basic idea of these methods is to solve a minimization problem that is equivalent to the original equation system. There are numerous variants of gradient methods which mainly differ in how the minimization problem is formulated and in which way the minimum is sought. We will address here the most important basics of the *conjugate gradient methods (CG methods)*, which can be considered for our applications as the most important gradient method (for details we refer to the corresponding literature, e.g., [11]).

Conjugate gradient methods were first developed for symmetric positive definite matrices (Hestenes and Stiefel, 1952). In this case the following equivalence holds:

$$\text{Solve } \mathbf{A}\phi = \mathbf{b} \quad \Leftrightarrow \quad \text{Minimize } F(\phi) = \frac{1}{2}\phi \cdot \mathbf{A}\phi - \mathbf{b} \cdot \phi.$$

The equivalence of both formulations is easily seen since the gradient of F is $\mathbf{A}\phi - \mathbf{b}$, and the vanishing of the gradient is a necessary condition for a minimum of F.

For the iterative soluton of the minimization problem, starting from a prescribed starting value ϕ^0, the functional F is minimized successively in each iteration ($k = 0, 1, \ldots$) in a certain direction $\mathbf{y}^k$:

$$\text{Minimize } F(\phi^k + \alpha\mathbf{y}^k) \text{ for all real } \alpha.$$

The value α^k, for which the functional attains its minimum, results according to

$$\frac{\mathrm{d}}{\mathrm{d}\alpha}F(\phi^k + \alpha\mathbf{y}^k) = 0 \quad \Rightarrow \quad \alpha^k = \frac{\mathbf{y}^k \cdot \mathbf{r}^k}{\mathbf{y}^k \cdot \mathbf{A}\mathbf{y}^k} \quad \text{with} \quad \mathbf{r}^k = \mathbf{b} - \mathbf{A}\phi^k$$

and defines the new iterate $\phi^{k+1} = \phi^k + \alpha^k \mathbf{y}^k$. The direction $\mathbf{y}^k$, in which the minimum is sought, is characterized by the fact that it is conjugated with respect to $\mathbf{A}$ to all previous directions, i.e.,

$$\mathbf{y}^k \cdot \mathbf{A}\mathbf{y}^i = 0 \text{ for all } i = 0, \ldots, k-1.$$

For the efficient implementation of the CG method it is mandatory that $\mathbf{y}^{k+1}$ can be determined by the following simple recursion formula (see, e.g., [11]):

$$\mathbf{y}^{k+1} = \mathbf{r}^{k+1} + \frac{\mathbf{r}^{k+1} \cdot \mathbf{r}^{k+1}}{\mathbf{r}^k \cdot \mathbf{r}^k} \mathbf{y}^k.$$

With this, in summary, the CG algorithm can be recursively formulated as follows:

- Initialization:

$$\mathbf{r}^0 := \mathbf{b} - \mathbf{A}\phi^0,$$
$$\mathbf{y}^0 := \mathbf{r}^0,$$
$$\beta^0 := \mathbf{r}^0 \cdot \mathbf{r}^0.$$

- For $k = 0, 1, \ldots$ until convergence:

$$\alpha^k = \beta^k / (\mathbf{y}^k \cdot \mathbf{A}\mathbf{y}^k),$$
$$\phi^{k+1} = \phi^k + \alpha^k \mathbf{y}^k,$$
$$\mathbf{r}^{k+1} = \mathbf{r}^k - \alpha^k \mathbf{A}\mathbf{y}^k,$$
$$\beta^{k+1} = \mathbf{r}^{k+1} \cdot \mathbf{r}^{k+1},$$
$$\mathbf{y}^{k+1} = \mathbf{r}^{k+1} + \beta^{k+1} \mathbf{y}^k / \beta^k.$$

The CG method is parameter-free and theoretically (i.e., without taking into account rounding errors) yields the exact solution after K iterations, where K is the number of unknowns of the system. Thus, for the systems of interest here, the method would, of course, be useless because K usually is very large. In practice, however, a sufficiently accurate solution is obtained with fewer iterations. One can show that the number of iterations required to reduce the absolute value of the residual to ϵ is given by

$$N_{\text{it}} \leq 1 + 0.5\sqrt{\kappa(\mathbf{A})} \ln(2/\epsilon), \tag{7.7}$$

where $\kappa(\mathbf{A})$ is the *condition number* of $\mathbf{A}$, i.e., for symmetric positive definite matrices the ratio of the largest and smallest eigenvalue of $\mathbf{A}$.

The CG method in the above form is restricted to the application for symmetric positive definite matrices, a condition that is not fulfilled for a number of problems (e.g., transport problems with convection). However, there are different generalizations of the method developed for non-symmetric matrices. Examples are:

- generalized minimal-residual method (GMRES),
- conjugate gradient squared (CGS) method,
- bi-conjugate gradient method (BICG),
- stabilized BICG method (BICGSTAB), ...

We will omit details of these methods here and refer to the corresponding literature (e.g., [11]). It should be noted that for these generalized methods the type of algebraic operations in each iteration is the same as for the CG method except that the number is larger (approximately twice as high). A complete theory concerning the convergence properties, available for the classical CG method, does not exist so far. However, experience has shown that the convergence behavior is usually close to that of the CG method.

7.1.6 Preconditioning

The convergence of CG methods can be further improved by a *preconditioning* of the equation system. The idea here is to transform the original system (7.1) by means of an (invertible) matrix $\mathbf{P}$ into an equivalent system

$$\mathbf{P}^{-1}\mathbf{A}\phi = \mathbf{P}^{-1}\mathbf{b} \tag{7.8}$$

and to apply the CG algorithm (or correspondingly generalized variants in the non-symmetric case) to the transformed system (7.8). $\mathbf{P}$ is called the *preconditioning matrix* and the resulting method is known as *preconditioned CG (PCG) method*.

The preconditioning can be integrated into the original CG algorithm as follows:

- Initialization:

$$\mathbf{r}^0 := \mathbf{b} - \mathbf{A}\phi^0 \,,$$
$$\mathbf{z}^0 := \mathbf{P}^{-1}\mathbf{r}^0 \,,$$
$$\mathbf{y}^0 := \mathbf{z}^0 \,,$$
$$\beta^0 := \mathbf{z}^0 \cdot \mathbf{r}^0 \,.$$

- For $k = 0, 1, \ldots$ until convergence:

$$\alpha^k = \beta^k / (\mathbf{y}^k \cdot \mathbf{A}\mathbf{y}^k) \,,$$
$$\phi^{k+1} = \phi^k + \alpha^k \mathbf{y}^k \,,$$
$$\mathbf{r}^{k+1} = \mathbf{r}^k - \alpha^k \mathbf{A}\mathbf{y}^k \,,$$
$$\mathbf{z}^{k+1} = \mathbf{P}^{-1}\mathbf{r}^{k+1} \,,$$
$$\beta^{k+1} = \mathbf{z}^{k+1} \cdot \mathbf{r}^{k+1} \,,$$
$$\mathbf{y}^{k+1} = \mathbf{z}^{k+1} + \beta^{k+1}\mathbf{y}^k / \beta^k \,.$$

One can observe that the additional effort within one iteration consists in the solution of an equation system with the coefficient matrix $\mathbf{P}$. The number of iterations for PCG methods according to the estimate (7.7) is given by:

$$N_{\mathrm{it}} \leq 1 + 0.5\sqrt{\kappa(\mathbf{P}^{-1}\mathbf{A})}\ln(2/\epsilon).$$

According to the above considerations, for the choice of $\mathbf{P}$ the following criteria can be formulated:

- the condition number of $\mathbf{P}^{-1}\mathbf{A}$ should be as small as possible in order to have a low number of iterations,
- the computation of $\mathbf{P}^{-1}\phi$ should be as efficient as possible in order to have a low effort for the individual iterations.

As is the case in the definition of iterative methods by the matrix $\mathbf{B}$ (see Sect. 7.1.4), a compromise between these two conflicting requirements has to be found.

Examples of frequently employed preconditioning techniques are:

- classical iterative methods with $\mathbf{P} = \mathbf{B}$ (Jacobi, Gauß-Seidel, ILU, ...),
- domain decomposition methods,
- polynomial approximations of $\mathbf{A}^{-1}$,
- multigrid methods,
- hierarchical basis methods.

For details we refer to the special literature (in particular [1] and [11]).

7.1.7 Comparison of Solution Methods

For an estimation and comparison of the computational efforts that the different solution methods require, we employ the model problem

$$-\frac{\partial^2\phi}{\partial x_i\partial x_i} = f \quad \text{in } \Omega,$$
$$\phi = \phi_S \quad \text{on } \Gamma, \tag{7.9}$$

which we consider for the one-, two-, and three-dimensional cases, $i = 1, \ldots, d$ with d equal to 1, 2, or 3. The problem domain Ω is the unit domain corresponding to the spatial dimension (i.e., unit interval, square, or cube). The discretization is performed with a central differencing scheme of second order for an equidistant grid with grid spacing h and N inner grid points in each spatial direction. The result of using the usual lexicographical numbering of the unknowns is the known matrix with dimension N^d having the value $2d$ in the main diagonal and the value -1 in all occupied subdiagonals. The model problem is well suited for the present purpose because all important properties of the solution methods can also be determined analytically (e.g., [11]). The eigenvalues of the system matrix $\mathbf{A}$ are given by

$$\lambda_j = 4d \sin^2 \left(\frac{j\pi h}{2} \right) \quad \text{for} \quad j = 1, \ldots, N .\tag{7.10}$$

For the three considered classical iterative methods one obtains from (7.10) for the spectral radii of the iteration matrices:

$$\lambda_{\max}^{\text{JAC}} = 1 - 2\sin^2 \left(\frac{\pi h}{2} \right) = 1 - \frac{\pi^2}{2} h^2 + O(h^4) ,$$

$$\lambda_{\max}^{\text{GS}} = 1 - \sin^2(\pi h) = 1 - \pi^2 h^2 + O(h^4) ,$$

$$\lambda_{\max}^{\text{SOR}} = \frac{1 - \sin(\pi h)}{1 + \sin(\pi h)} = 1 - 2\pi h + O(h^2) .$$

The last equation in each case is obtained by Taylor expansion of the sinus functions. In the case of the SOR method the optimal relaxation parameter is employed. For the model problem this results in

$$\omega_{\text{opt}} = \frac{2}{1 + \sin(\pi h)} = 2 - 2\pi h + O(h^2) .$$

Since $N \sim 1/h$, the outcome of the spectral radii together with (7.6) is that with the Jacobi and Gauß-Seidel methods the number of iterations required for the solution of the model problem are proportional to N^2 and with the SOR method are proportional to N. One can also observe that the Jacobi method needs twice as many iterations as the Gauß-Seidel method. For ILU methods a convergence behavior similar to that of the SOR method is achieved (see, e.g., [11]).

The condition number of $\mathbf{A}$ for the model problem (7.9) results from (7.10) in

$$\kappa(\mathbf{A}) \approx \frac{4d}{4d \sin^2(\pi h/2)} \approx \frac{4}{\pi^2 h^2} .$$

Together with the estimate (7.7) it follows that the number of iterations required to solve the problem with the CG method is proportional to N.

The asymptotic computational effort for the solution of the model problem with the iterative methods simply is given by the product of the number of iterations with the number of unknowns. In Table 7.1 the corresponding values are indicated together with memory requirements depending on the number of grid points for the different spatial dimensions. For comparison, the corresponding values for the direct LU decomposition are also given. One can observe the enormous increase in effort with increasing spatial dimension for the LU decomposition, which is mainly caused by the mentioned "fill-in" of the band. For instance, if we compare the effort of the SOR method with that for the direct solution with the LU decomposition, one can clearly observe the advantages that iterative methods may provide for multi-dimensional problems (in particular in the three-dimensional case). Note that also for the various variants of the Gauß elimination (e.g., the Cholesky method) the asymptotic effort remains the same.

Table 7.1. Asymptotic memory requirement and computational effort with different linear system solvers for model problem for different spatial dimensions

Dim.	Unknowns	Memory requirements		Computational effort		
		Iterative	Direct	JAC/GS	SOR/ILU/CG	Direct
1-d	N	$O(N)$	$O(N)$	$O(N^3)$	$O(N^2)$	$O(N)$
2-d	N^2	$O(N^2)$	$O(N^3)$	$O(N^4)$	$O(N^3)$	$O(N^4)$
3-d	N^3	$O(N^3)$	$O(N^5)$	$O(N^5)$	$O(N^4)$	$O(N^7)$

While the asymptotic memory requirements are the same for all iterative methods, a significant improvement can be achieved with regard to the computational effort when using the SOR, CG, or ILU methods instead of the Jacobi or Gauß-Seidel methods. An advantage of ILU methods is that the "good" convergence properties apply to a wide class of problems (robustness). However, owing to the deteriorating convergence rate, the computational effort also increases disproportionately with grid refinement for the SOR, CG, or ILU methods.

Let us look at a concrete numerical example to point out the different computing times that the different solvers may need. For this we consider the model problem (7.9) in the two-dimensional case with $N^2 = 256 \times 256$ CVs. In Table 7.2 a comparison of the computing times for different solvers is given. Here, SSOR-PCG denotes the CG method preconditioned with the symmetric SOR method (see, e.g., [11]). The multigrid method, for which also a result is given, will be explained in some more detail in Sect. 12.2. This two-dimensional problem already has significant differences in computing times of the methods. For three-dimensional problems these are even more pronounced. Multigrid methods belong to the most efficient methods for the solution of linear equation systems. However, iterative methods, such as the ones discussed in this chapter, constitute an important constituent also for multigrid methods (see Sect. 12.2).

From the above considerations some significant disadvantages of using direct methods for the solution of the linear systems, which arise from the discretization of continuum mechanics problems, become obvious and can be summarized as follows:

- In order to achieve an adequate discretization accuracy the systems usually are rather large. In particular, this is the case for three-dimensional problems and generally for all multi-dimensional problems of fluid mechanics. With direct methods the effort increases strongly disproportionately with

Table 7.2. Asymptotic computational effort and computing time for different linear system solvers applied to two-dimensional model problem

Method	Operations	$\approx$ Computing time
Gauß elimination	$O(N^4)$	1 d
Jacobi	$O(N^4)$	5 h
SOR,ILU,CG	$O(N^3)$	30 m
SSOR-PCG	$O(N^{5/2})$	5 m
Multigrid	$O(N^2)$	8 s

the problem size (see Table 7.1), such that extremely high computing times and huge memory requirements result (see Table 7.2).

- Due to the relatively high number of arithmetic operations, rounding errors may cause severe problems in direct methods (also with 64-bit word length, i.e., ca. 14-digits accuracy). The influence of rounding errors depends on the condition number $\kappa(\mathbf{A})$ of the coefficient matrix, which for the matrices under consideration is usually quite unfavorable (the finer the grid, the worse condition number).

- In the case of non-linearity in the problem, an iteration process for the solution anyway is necessary. Therefore, no exact solution of the equation system (as obtained with direct methods) is demanded, but a solution which is accurate to some tolerance is sufficient. The same applies if the matrix needs to be corrected due to a coupling (see Sect. 7.2).

So, for larger multi-dimensional problems it is usually much more efficient, to solve the linear systems by iterative methods.

7.2 Non-Linear and Coupled Systems

For the numerical solution of non-linear algebraic equation systems, as they arise, for instance, for fluid mechanics or geometrically and/or physically non-linear structural mechanics problems an iteration process is basically required. The most frequent approaches are *successive iteration* (or *Picard iteration*), *Newton methods*, or *quasi-Newton methods*, which we will briefly outline. As an example we consider a non-linear system of the form

$$\mathbf{A}(\phi)\phi = \mathbf{b}, \tag{7.11}$$

to which an arbitrary non-linear system can be transformed by a suitable definition of $\mathbf{A}$ and $\mathbf{b}$.

The iteration rule for the Newton method for (7.11) is defined by

$$\phi^{k+1} = \phi^k - \left[\frac{\partial \mathbf{r}(\phi^k)}{\partial \phi}\right]^{-1} \mathbf{r}(\phi^k) \quad \text{with} \quad \mathbf{r}(\phi^k) = \mathbf{A}(\phi^k)\phi^k - \mathbf{b}.$$

Thus, in each iteration the Jacobian matrix $\partial r/\partial \phi$ must be computed and inverted. For a quasi-Newton method this is not carried out in each iteration. Instead, the Jacobi matrix is kept constant for a certain number of iterations. On the one hand, this reduces the effort per iteration, but on the other hand, the convergence rate may deteriorate. The Newton method possesses a quadratic convergence behavior, i.e., in each iteration the error $\|\phi^k - \phi\|$ is reduced by a factor of four. A possible problem when using the Newton method is that the convergence is ensured only if the starting value ϕ^0 already is sufficiently close to the exact solution (which, of course, is not known). In order to circumvent this problem (at least partially), frequently an incremental approach is applied. For instance, this can be done by solving problem (7.11) first for a "smaller" right hand side, which is then increased step by step to the original value $\mathbf{b}$.

The Picard iteration for a system of the type (7.11) is defined by an iteration procedure of the form:

$$\phi^{k+1} = \phi^k - \tilde{\mathbf{A}}(\phi^k)\phi^{k+1} + \tilde{\mathbf{b}}$$

with suitable iteration matrix $\tilde{\mathbf{A}}(\phi)$ and right hand side $\tilde{\mathbf{b}}$ (these can be equal to $\mathbf{A}(\phi)$ and $\mathbf{b}$, for instance). In this method no Jacobian matrix has to be computed. However, the convergence behavior is only linear (halving the error in each iteration). The choice of the starting value is much less problematic than with the Newton method.

As seen in Chap. 2, for structural or fluid mechanical problems one usually is faced with *coupled* systems of equations. As an example for such a coupled linear system we consider:

$$\underbrace{\begin{bmatrix} \mathbf{A}_{11} & \mathbf{A}_{12} \\ \mathbf{A}_{21} & \mathbf{A}_{22} \end{bmatrix}}_{\mathbf{A}} \underbrace{\begin{bmatrix} \phi_1 \\ \phi_2 \end{bmatrix}}_{\phi} = \underbrace{\begin{bmatrix} \mathbf{b}_1 \\ \mathbf{b}_2 \end{bmatrix}}_{\mathbf{b}}. \tag{7.12}$$

Such systems either can be solved *simultaneously* or *sequentially*. For a simultaneous solution the system is solved just in the form (7.12), e.g., with one of the solvers described in Sect. 7.1. For a sequential solution the system is solved within an iteration process successively for the different variables. For instance, for the system (7.12) in each iteration (starting value ϕ_2^0, $k = 0, 1, \ldots$) the following two steps have to be carried out:

(i)   Determine ϕ_1^{k+1} from $\mathbf{A}_{11}\phi_1^{k+1} = \mathbf{b}_1 - \mathbf{A}_{12}\phi_2^k$,

(ii) Determine ϕ_2^{k+1} from $\mathbf{A}_{22}\phi_2^{k+1} = \mathbf{b}_2 - \mathbf{A}_{21}\phi_1^{k+1}$.

In Fig. 7.4 the course of the iteration process with the corresponding computational steps is illustrated graphically.

For a simultaneous solution all coefficients of the system matrix and the right hand side have to be stored simultaneously. Also, auxiliary vectors, which

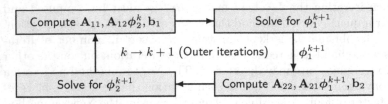

Fig. 7.4. Sequential solution of coupled equation systems

are needed for the corresponding solution method, are of the full size of the coupled system. For a sequential solution the coefficients of the individual subsystems can be stored in same arrays and the auxiliary vectors only have the size of the subsystems. A disadvantage of the sequential solution is the additional iteration process for coupling the subsystems. With a simultaneous solution (also in the case of an additional non-linearity), there is, as a result of better coupling of the unknowns, a faster convergence of the necessary outer iterations than with the sequential method. The combination of linearization and variable coupling usually leads to an improvement in the convergence.

In conclusion, we note that for non-linear and coupled systems the above considerations suggest the following two combinations of solution strategies:

- *Newton method with simultaneous solution:* high effort for the computation and inversion of the Jacobi matrix (reduction by quasi-Newton methods), high memory requirements, quadratic convergence behavior, starting value has to be "close" to the solution, Jacobian matrix is available (e.g., for stability investigations).
- *Successive iteration with sequential solution:* less computational effort per iteration, less memory requirements, decoupling of the individual equations is possible and can be combined with the linearization process, linear convergence behavior, less sensitive compared to "bad" starting values.

Depending on the actual problem, each variant has its advantages.

Exercises for Chap. 7

Exercise 7.1. Given is the linear equation system

$$\begin{bmatrix} 4 & -1 & 0 \\ -1 & 4 & -1 \\ 0 & -1 & 4 \end{bmatrix} \begin{bmatrix} \phi_1 \\ \phi_2 \\ \phi_3 \end{bmatrix} = \begin{bmatrix} 3 \\ 2 \\ 3 \end{bmatrix}.$$

(i) Determine the solution of the systems with the Gauß elimination method.
(ii) Determine the condition number of the system matrix and the maximum eigenvalues of the iteration matrices for the Jacobi, Gauß-Seidel, and SOR methods, where for the latter first the optimal relaxation parameter ω_{opt} is

to be determined. (iii) Carry out some iterations with the CG, Jacobi, Gauß-Seidel, and SOR methods (the latter with ω_{opt} from (ii)) each with the starting value $\phi^0 = 0$, and discuss the convergence properties of the methods taking into account the values determined in (ii).

Exercise 7.2. Consider the linear system from Exercise 7.1 as a coupled system, such that the first two and the third equation form a subsystem, i.e., $\phi_1 = (\phi_1, \phi_2)$ and $\phi_2 = \phi_3$. Carry out some iterations with the starting value $\phi_3^0 = 0$ acccording to the sequential solution approach defined in Sect. 7.2.

Exercise 7.3. Given is the matrix

$$\begin{bmatrix} 4 & -1 & 0 & 0 & -1 & 0 \\ -1 & 4 & -1 & 0 & 0 & -1 \\ 0 & -1 & 4 & -1 & 0 & 0 \\ 0 & 0 & -1 & 4 & 0 & 0 \\ -1 & 0 & 0 & 0 & 4 & -1 \\ 0 & -1 & 0 & 0 & -1 & 4 \end{bmatrix}$$

(i) Determine the complete LU decomposition of the matrix. (ii) Determine the ILU decomposition following the standard ILU approach and the SIP method with $\alpha = 1/2$.

Exercise 7.4. Given is the non-linear equation system

$$\phi_1 \phi_2^2 - 4\phi_2^2 + \phi_1 = 4 \,,$$
$$\phi_1^3 \phi_2 + 3\phi_2 - 2\phi_1^3 = 6 \,.$$

(i) Carry out some iterations with the Newton method with the starting values $(\phi_1^0, \phi_2^0) = (0,0)$, $(1,2)$, and $(9/5, 18/5)$. (ii) Transform the system suitably into the form (7.11) and carry out some iterations according to the successive iteration method with $\tilde{\mathbf{A}} = \mathbf{A}$, $\tilde{\mathbf{b}} = \mathbf{b}$, and the starting value $(\phi_1^0, \phi_2^0) = (0,0)$. (iii) Compare and discuss the results obtained in (i) and (ii).

8

Properties of Numerical Methods

In this chapter we summarize characteristic properties of numerical methods, which are important for the functionality and reliability of the corresponding methods as well as for the "proper" interpretation of the achieved results and, therefore, are most relevant for their practical application. The underlying basic mathematical concepts will be considered only as far as is necessary for the principal understanding. The corresponding properties with their implications for the computed solution will be exemplified by means of characteristic numerical examples.

8.1 Properties of Discretization Methods

When applying discretization methods to differential equations, for the unknown function, which we generally denote by $\phi = \phi(\mathbf{x}, t)$, at an arbitrary location $\mathbf{x}_P$ and an arbitrary time t_n the following three values have to be distinguished:

- the exact solution of the differential equation $\phi(\mathbf{x}_P, t_n)$,
- the exact solution of the discrete equation ϕ_P^n,
- the actually computed solution $\tilde{\phi}_P^n$.

In general, these values do not coincide because during the different steps of the discretization and solution processes errors are unavoidably made. The properties, which will be discussed in the following sections, mainly concern the relations between these values and the errors associated with this.

Besides the model errors, which we will exclude at this stage (see Chap. 2 for this), mainly two kinds of numerical errors occur when applying a numerical method:

- the *discretization error* $|\phi(\mathbf{x}_P, t_n) - \phi_P^n|$, i.e., the difference between the exact solution of the differential equation and the exact solution of the discrete equation,

- the *solution error* $|\tilde{\phi}_P^n - \phi_P^n|$, i.e., the difference between the exact solution of the discrete equation and the actually computed solution,

Most important for practice is the *total numerical error*

$$e_P^n = |\phi(\mathbf{x}_P, t_n) - \tilde{\phi}_P^n|,$$

i.e., the difference between the exact solution of the differential equation and the actual computed solution, which is composed of the discretization and solution errors. The solution error contains portions resulting from a possibly only approximative solution of the discrete equations. These portions are usually comparably small and can be controlled relatively easily by considering residuals. Therefore, we will not consider it further here.

For the relations between the different solutions and errors, which are illustrated schematically in Fig. 8.1, the concepts of consistency, stability, and convergence, in particular, play an important role. As an illustrative example problem we consider the one-dimensional unsteady transport equation

$$\rho \frac{\partial \phi}{\partial t} + \rho v \frac{\partial \phi}{\partial x} - \alpha \frac{\partial^2 \phi}{\partial x^2} = 0 \tag{8.1}$$

with constant values for α, ρ, and v for the problem domain $[0, L]$. With the boundary conditions $\phi(0) = \phi_0$ and $\phi(L) = \phi_L$ for $t \to \infty$ the problem has the analytical (steady) solution

$$\phi(x) = \phi_0 + \frac{e^{x\mathrm{Pe}/L} - 1}{e^{\mathrm{Pe}} - 1}(\phi_L - \phi_0)$$

with the *Peclet number*

$$\mathrm{Pe} = \frac{\rho v L}{\alpha},$$

which represents a measure for the ratio of convective and diffusive transport of ϕ.

8.1.1 Consistency

A discretization scheme is called *consistent* if the discretized equations for $\Delta x, \Delta t \to 0$ approach the original differential equations. Thus, the consistency defines a relation between the exact solutions of the differential and discrete equations (see Fig. 8.1). The consistency can be checked by an analysis of the *truncation error* τ_P^n, which is defined by the difference between the differential equation (interpreted by the Taylor series expansions of derivatives) and the discrete equation, if the exact variable values are inserted there. If the truncation error goes to zero with grid refinement, i.e.,

$$\lim_{\Delta x, \Delta t \to 0} \tau_P^n = 0,$$

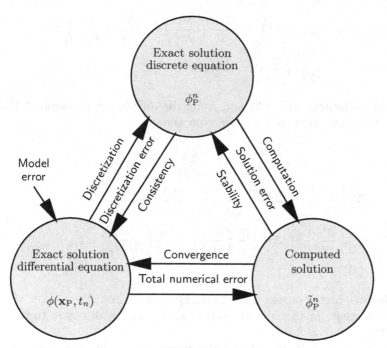

Fig. 8.1. Relation between solutions, errors, and properties

the method is consistent.

Let us consider as an example the discretization of problem (8.1) for an equidistant grid with grid spacing Δx and time step size Δt. Using a finite-volume method with CDS approximation for the spatial discretization and the explicit Euler method for the time discretization, the following discrete equation is obtained:

$$\rho\frac{\phi_P^{n+1} - \phi_P^n}{\Delta t} + \rho v\frac{\phi_E^n - \phi_W^n}{2\Delta x} - \alpha\frac{\phi_E^n + \phi_W^n - 2\phi_P^n}{\Delta x^2} = 0 \,. \tag{8.2}$$

To compare the differential equation (8.1) and the discrete equation (8.2) a local consideration at location x_P at time t_n is performed (the discrete solution is only defined locally). Evaluating the Taylor expansion around (x_P, t_n) of the exact solution at the points appearing in the discrete equation (8.1) yields:

$$\phi_P^{n+1} = \phi_P^n + \Delta t \left(\frac{\partial\phi}{\partial t}\right)_P^n + \frac{\Delta t^2}{2}\left(\frac{\partial^2\phi}{\partial t^2}\right)_P^n + O(\Delta t^2) \,,$$

$$\phi_E^n = \phi_P^n + \Delta x \left(\frac{\partial\phi}{\partial x}\right)_P^n + \frac{\Delta x^2}{2}\left(\frac{\partial^2\phi}{\partial x^2}\right)_P^n + \frac{\Delta x^3}{6}\left(\frac{\partial^3\phi}{\partial x^3}\right)_P^n$$

$$+ \frac{\Delta x^4}{24}\left(\frac{\partial^4\phi}{\partial x^4}\right)_P^n + O(\Delta x^5) \,,$$

$$\phi_W^n = \phi_P^n - \Delta x \left(\frac{\partial \phi}{\partial x}\right)_P^n + \frac{\Delta x^2}{2} \left(\frac{\partial^2 \phi}{\partial x^2}\right)_P^n - \frac{\Delta x^3}{6} \left(\frac{\partial^3 \phi}{\partial x^3}\right)_P^n$$
$$+ \frac{\Delta x^4}{24} \left(\frac{\partial^4 \phi}{\partial x^4}\right)_P^n - O(\Delta x^5),$$

Inserting these relations into (8.2) and using the differential equation (8.1), one gets for the truncation error τ_P^n the expression:

$$\tau_P^n = \rho \left(\frac{\partial \phi}{\partial t}\right)_P^n + \rho v \left(\frac{\partial \phi}{\partial x}\right)_P^n - \alpha \left(\frac{\partial^2 \phi}{\partial x^2}\right)_P^n +$$
$$\frac{\rho \Delta t}{2} \left(\frac{\partial^2 \phi}{\partial t^2}\right)_P^n + \frac{\rho v \Delta x^2}{6} \left(\frac{\partial^3 \phi}{\partial x^3}\right)_P^n - \frac{\alpha \Delta x^2}{12} \left(\frac{\partial^4 \phi}{\partial x^4}\right)_P^n + T_H =$$
$$= \Delta t \frac{\rho}{2} \left(\frac{\partial^2 \phi}{\partial t^2}\right)_P^n + \Delta x^2 \left[\frac{\rho v}{6} \left(\frac{\partial^3 \phi}{\partial x^3}\right)_P^n - \frac{\alpha}{12} \left(\frac{\partial^4 \phi}{\partial x^4}\right)_P^n\right] + T_H =$$
$$= O(\Delta t) + O(\Delta x^2),$$

where T_H denotes higher order terms. Thus, $\tau_P^n \to 0$ for $\Delta x, \Delta t \to 0$, i.e., the method is consistent. The consistency orders are 1 and 2 with regard to time and space, respectively.

For sufficiently small Δx and Δt a higher order of the truncation error also means a higher accuracy of the numerical solution. In general, the order of the truncation error gives information about how fast the errors decay when the grid spacing or time step size is refined. The absolute values of the solution error, which also depend from the solution itself, usually are not available.

To illustrate the influence of the consistency order on the numerical results we consider problem (8.1) in the steady case. Let the problem domain be the interval $[0, 1]$ and the boundary conditions given by $\phi(0) = 0$ and $\phi(1) = 1$. As problem parameters we take $\alpha = 1\,\text{kg}/(\text{ms})$, $\rho = 1\,\text{kg}/\text{m}^3$, and $v = 24\,\text{m/s}$. We consider the following finite-volume discretizations of 1st, 2nd, and 4th order:

- UDS1/CDS2: 1st order upwind differences for convective fluxes and 2nd order central differences for diffusive fluxes,
- CDS2/CDS2: 2nd order central differences for convective and diffusive fluxes,
- CDS4/CDS4: 4th order central differences for convective and diffusive fluxes.

For the 4th order method, in points x_1 und x_N adjacent to the boundaries the 2nd order CDS method is used. Figure 8.2 represents the dependence of the error

$$e_h = \frac{1}{N} \sum_{i=1}^{N} |\phi(x_i) - \tilde{\phi}_i|$$

on the grid spacing $\Delta x = 1/N$ for the different methods. One can observe the strongly varying decrease in error when the grid is refined according to the order of the methods (the higher the order, the more rapid). The slope of the corresponding straight line (for small enough Δx) corresponds to the order of the method.

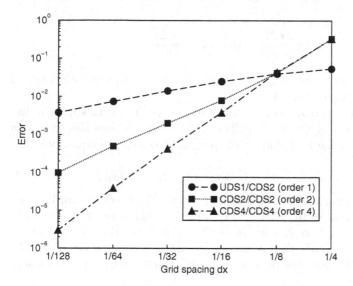

Fig. 8.2. Dependence of error on grid spacing for different discretization methods for one-dimensional transport equation

8.1.2 Stability

The concept of stability serves in the setting up of a relation between the actually computed solution and the exact solution of the discrete equation (see Fig. 8.1). A variety of different definitions of stability, which are useful for different purposes, exist in the literature. We will restrict ourselves here to a simple definition that is sufficient for our purpose. We call a discretization scheme *stable* if the solution error $|\tilde{\phi}_P^n - \phi_P^n|$ is bounded in the whole problem domain and for all time steps.

The general idea in proving the stability is to investigate how small perturbations (e.g., caused by round-off errors) influence the subsequent time steps. The important question here is, whether these perturbations are damped by the discretization scheme (then the method is stable) or not (then the method is unstable). Such a stability analysis can be performed by various methods, the most common being: the *von Neumann analysis*, the *matrix method*, and the *method of small pertubations (perturbation method)*. For general problems such investigations, if they can be done analytically at all, usually are rather difficult. We omit here an introduction of these methods (for this see, e.g., [12, 13]), and restrict ourselves to a heuristic consideration for our example problem (8.1) to illustrate the essential effects.

Two important characteristic numbers for stability considerations of transport problems are the *diffusion number* D and the *Courant number* C, which are defined as

$$D = \frac{\alpha \Delta t}{\rho \Delta x^2} \quad \text{and} \quad C = \frac{v \Delta t}{\Delta x}.$$

These numbers express the ratios of the time step size to the diffusive and convective transports, respectively.

Let us first consider an approximation of problem (8.1) with a spatial finite-volume discretization by the UDS1/CDS2 method and a time discretization by the explicit Euler method leading to the following discrete equation:

$$\phi_P^{n+1} = D\phi_E^n + (D+C)\phi_W^n + (1 - 2D - C)\phi_P^n. \tag{8.3}$$

A simple heuristic physical consideration of the problem requires an increase in ϕ_P^{n+1} if ϕ_P^n is increased. In general, such a uniform behavior of ϕ_P^n and ϕ_P^{n+1} is only guaranteed if all coefficients in (8.3) are positive. Since C and D are positive per definition, only the coefficient $(1 - 2D - C)$ can be negative. The requirement that this remains positive leads to a limitation for the time step size Δt:

$$\Delta t < \frac{\rho \Delta x^2}{2\alpha + \rho v \Delta x}. \tag{8.4}$$

Note that a corresponding stability analysis according to the von Neumann method (cf., e.g., [12]) leads to the same condition.

In particular, from the relation (8.4) for the two special cases of pure diffusion ($v = 0$) and pure convection ($\alpha = 0$) the time step limitations that follow are

$$\Delta t < \frac{\rho \Delta x^2}{2\alpha} \quad \text{and} \quad \Delta t < \frac{\Delta x}{v},$$

respectively. The latter is known in the literature as CFL condition (cf. Courant, Friedrichs, and Levy). Physically these conditions can be interpreted as follows: the selected time step size must be small enough, so that, due to the diffusive or convective transport, the information of the distribution of ϕ in one time step does not advance further than the next nodal point.

Chosing the CDS2/CDS2 scheme instead of the UDS1/CDS2 space discretization one gets the approximation:

$$\phi_P^{n+1} = (D - \frac{C}{2})\phi_E^n + (D + \frac{C}{2})\phi_W^n + (1 - 2D)\phi_P^n.$$

A von Neumann stability analysis for this scheme yields the time step size limitation (see, e.g., [12])

$$\Delta t < \min \left\{ \frac{2\alpha}{\rho v^2}, \frac{\rho \Delta x^2}{2\alpha} \right\}.$$

While in the case of pure diffusion the same time step limitation as for the UDS1/CDS2 scheme results (the discretization of the diffusive term has not changed), in the case of pure convection the CDS2/CDS2 method is *always* unstable regardless of the time step size. In Fig. 8.3 the relation between the time step limitation for stability and the grid spacing for the UDS1/CDS2 and CDS2/CDS2 methods (for a mixed convection-diffusion problem) is illustrated.

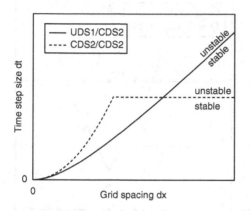

Fig. 8.3. Relation between time step limitation and grid spacing for convection-diffusion problem using the explicit Euler method with UDS1/CDS2 and CDS2/CDS2 space discretizations

Before we turn to the consequences of violating the time step limitation by means of an example, we first consider the case of an implicit time discretization. Using the implicit Euler method, we obtain with the UDS1/CDS2 scheme for problem (8.1) a discretization of the form

$$(1 + 2D + C)\phi_P^{n+1} = D\phi_E^{n+1} + (D + C)\phi_W^{n+1} + \phi_P^n.$$

All coefficients are positive, such that no problems with respect to a non-uniform change of ϕ are expected. A von Neumann analysis, which in this case requires an eigenvalue analysis of the corresponding coefficient matrix, also shows that the method is stable independent of the time step size for all values of D und C.

To illustrate the effects occurring in connection with the stability we consider our example problem (8.1) without convection ($v = 0$) with the boundary conditions $\phi_0 = \phi_1 = 0$ and a CDS approximation for the diffusive term. The exact (steady) solution $\phi = 0$ is perturbed at x_P and the perturbed solution is used as starting value for the computation with the explicit and implicit Euler methods for two different time step sizes. The problem parameters are chosen such that in one case the condition for the time step limitation (8.4) for the explicit Euler method is fulfilled ($D = 0.5$) and in the other case it is violated ($D = 1.0$). In Fig. 8.4 the corresponding course of the solution for some time steps for the different cases are indicated. One can observe damping behavior of the implicit method independent of the time step, while the explicit method damps the perturbation only if the time step limitation is fulfilled. If the time step is too large, the explicit method diverges.

In view of the above stability considerations we summarize again the most essential characteristic properties of explicit und implicit time discretization schemes with respect to their assets and drawbacks. Explicit methods limit the speed of spatial spreading of information. The time step must be adapted

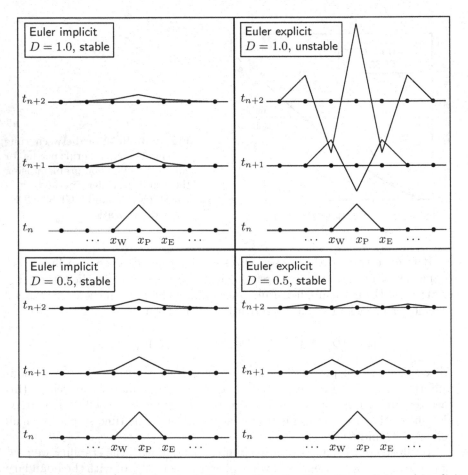

Fig. 8.4. Development of perturbation for discretization with explicit and implicit Euler methods for different time step sizes

to the spatial grid spacing to ensure numerical stability. The requirements on the time step size can be quite restrictive. The admissible time step size can be estimated by means of a stability analysis. The limitation in the time step size often constitutes a severe disadvantage (particularly when using fine spatial grids), such that explicit methods usually are less efficient. With implicit methods all variables of the new time level are coupled with each other. Thus, a change at an arbitrary spatial location immediately spreads over the full spatial domain. Therefore, there is no (or at least a much less restrictive) limitation of the admissible time step size. The time steps can be adapted optimally to the actual temporal course of the solution according to the desired accuracy. The increased numerical effort per time step, owing to the required resolution of the equation systems, is mostly more than compensated for by the possibility of using larger time steps.

8.1.3 Convergence

An essential requirement for a discretization method is that the actually computed solution approaches the exact solution when the spatial and temporal grids are refined:

$$\lim_{\Delta x, \Delta t \to 0} \tilde{\phi}_P^n = \phi(\mathbf{x}_P, t_n) \text{ for all } \mathbf{x}_P \text{ and } t_n \, .$$

This property is denoted as *convergence* of the method. The items consistency, stability, and convergence are closely related to each other. For linear problems the relation is provided by the fundamental *equivalence theorem of Lax*. Under certain assumptions about to the continuous problem, which we will not detail here (see, e.g., [12]), the theorem states:

> For a consistent discretization scheme, the stability is a necessary and sufficient condition for its convergence.

Based on the Lax theorem an analysis of a discretization scheme can be performed as follows:

- Analysis of consistency: one gets the truncation error and the order of the scheme.
- Analysis of stability: one gets information about the error behavior and the proper relation of the time step size to the spatial grid size.

This yields the information about the convergence of the method, which is the essential property.

For nonlinear problems, in general, the Lax theorem does not hold in this form. However, even in this case stability and consistency are essential prerequisites for a "reasonably" working method.

8.1.4 Conservativity

A discretization method is called *conservative* if the conservation properties of the differential equation, i.e., the balance of the underlying physical quantity, are also represented by the discrete equations independently from the choice of the numerical grid. If the discrete system is of the form

$$a_P \phi_P = \sum_c a_c \phi_c + b_P \,, \tag{8.5}$$

for a conservative method the relation

$$a_P = \sum_c a_c \tag{8.6}$$

has to be satisfied. However, not all methods satisfying (8.6) are necessarily conservative. As already pointed out elsewhere, the finite-volume method per definition is conservative because it directly works with the flux balances through the CV faces. Thus, the method automatically reflects the global conservation principle exactly.

For finite-element or also finite-difference methods conservativity is not ensured automatically. We illustrate the consequences for an example of a non-conservative finite-difference discretization. Consider the one-dimensional heat conduction equation

$$\frac{\partial}{\partial x} \left(\kappa \frac{\partial T}{\partial x} \right) = 0 \tag{8.7}$$

for the interval $[0, 1]$ with the boundary conditions $T(0) = 0$ and $T(1) = 1$. Applying the product rule, (8.7) can equivalently be written as

$$\frac{\partial \kappa}{\partial x} \frac{\partial T}{\partial x} + \kappa \frac{\partial^2 T}{\partial x^2} = 0 \,. \tag{8.8}$$

For the discretization we employ a grid with just one internal point, as shown in Fig. 8.5, such that only the temperature at x_2 has to be determined. The grid spacing is $\Delta x = x_2 - x_1 = x_3 - x_2$. Approximating the first derivatives in (8.8) with 1st order backward differences and the second derivative with central differences yields:

$$\frac{\kappa_2 - \kappa_1}{\Delta x} \frac{T_2 - T_1}{\Delta x} + \kappa_2 \frac{T_3 - 2T_2 + T_1}{\Delta x^2} = 0 \,.$$

Resolving this for T_2 and inserting the precribed values for T_1 und T_3 gives:

$$T_2 = \frac{\kappa_2}{\kappa_1 + \kappa_2} \,. \tag{8.9}$$

Now we consider the energy balance for the problem. This can be expressed by integrating (8.7) and applying the fundamental theorem of calculus as follows:

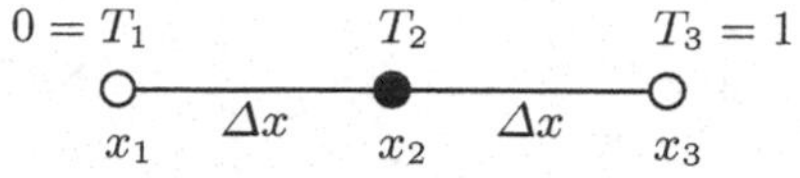

$$0 = T_1 \qquad T_2 \qquad T_3 = 1$$

Fig. 8.5. Grid for example of one-dimensional heat conduction

$$0 = \int_0^1 \frac{\partial}{\partial x}\left(\kappa \frac{\partial T}{\partial x}\right) dx = \kappa \frac{\partial T(1)}{\partial x} - \kappa \frac{\partial T(0)}{\partial x}. \tag{8.10}$$

Computing the right hand side of (8.10) using forward and backward differencing formulas at the boundary points and inserting the temperature according to (8.9) yields:

$$\kappa_3 \frac{T_3 - T_2}{\Delta x} - \kappa_1 \frac{T_2 - T_1}{\Delta x} = \frac{\kappa_1(\kappa_3 - \kappa_2)}{(\kappa_1 + \kappa_2)\Delta x}.$$

Thus, in general, i.e., if $\kappa_2 \neq \kappa_3$, this expression does not vanish, which means that the energy balance is not fulfilled.

Note that the conservativity does not directly relate to the global accuracy of a scheme. A conservative and a non-conservative scheme may have errors of the same size, they are just distributed differently over the problem domain.

8.1.5 Boundedness

From the conservation principles underlying continuum mechanical problems there result physical limits within which the solution for prescribed boundary conditions should be. These limits also should be met by a numerical solution. For example, a density always should be positive and a species concentration always should take values between 0% and 100%. This property of a discretization scheme is denoted as *boundedness*. Boundedness frequently is mixed up with the stability (in the sense described in Sect. 8.1.2). However, the boundedness concerns not the error development, but the accuracy of the discretization.

Let us consider the example problem (8.1) for the steady case with $\phi_0 < \phi_L$ for the boundary values. From the analytical solution it is easily seen that for ϕ the boundedness condition

$$\phi_0 \leq \phi \leq \phi_L \tag{8.11}$$

is satisfied in the problem domain $[0, L]$, i.e., in the interior of the problem domain the solution may not take smaller or larger values than on the boundary (this is also evident physically if (8.1), for instance, is interpreted as a heat transport equation). Now, to be physically meaningful the condition (8.11) should also be fulfilled by the numerical solution. One can show that for a discretization of the form (8.5), a sufficient condition for the boundedness is the validity of the inequality

$$|a_P| \geq \sum_c |a_c| \,.$$ (8.12)

For conservative methods, owing to the relation (8.6), this is fulfilled if and only if all non-zero coefficients a_c have the same sign.

In general, the adherence of the boundedness of a discretization method for transport problems poses problems if the convective fluxes are "too large" compared to the diffusive fluxes. The situation becomes worse with increasing order of the method. Among the finite-volume methods considered in Chap. 4 only the UDS method, which is only of 1st order, is unconditionally bounded. For all methods of higher order in the case of "too coarse" grids some coefficients may become negative, such that the inequality (8.12) is not fulfilled. An important quantity in this context is the *grid Peclet number* Pe_h, the discrete analogon to the Peclet number Pe defined in Sect. 8.1. For our example problem (8.1) the grid Peclet number is given by

$$Pe_h = \frac{\rho v \Delta x}{\alpha} \,.$$

For instance, from the coefficients a_c in (8.2) it follows that the CDS2/CDS2 method with equidistant grid for problem (8.1) is bounded if

$$Pe_h \leq 2 \quad \text{or} \quad \Delta x \leq \frac{2\alpha}{\rho v} \,.$$ (8.13)

For the QUICK method one obtains the condition $Pe_h \leq 8/3$. Thus, the requirement for the boundedness for higher order methods implicates a limitation of the admissible spatial grid spacing. This can be very restrictive in the case of a strong dominance of the convective transport (i.e., ρv is large compared to α).

If a solution is not bounded, this often shows up in the form of non-physical oscillations that can easily be identified. These show that the used grid is too coarse for the actual discretization scheme. In this case one can either choose a method of lower order with less restrictive boundedness requirements or refine the numerical grid. The question, what is preferable in a concrete case cannot be generally answered because this closely depends on the problem and the computing capacities available.

It should be noted that the condition (8.12) is sufficient, but not necessary, in order to obtain bounded solutions. As outlined in Sect. 7.1, the condition (8.12) also is of importance for the convergence of the iterative solution algorithms. With some iterative methods it can be difficult to get even a solution for high Peclet numbers.

To illustrate the dependence of the boundedness on the grid spacing, we consider problem (8.1) with the problem parameters $\alpha = \rho = L = 1$, $v = 24$ (each in the corresponding units) and the boundary conditions $\phi_0 = 0$ und $\phi_1 = 1$. We compute the solution for the two grid spacings

$\Delta x = 1/8$ and $\Delta x = 1/16$ with the UDS1/CDS2 and the CDS2/CDS2 methods (see Sect. 8.1.2). The corresponding grid Peclet numbers are $Pe_h = 3$ and $Pe_h = 3/2$, respectively. The criterion (8.13) is fulfilled in the second case, but not in the first. In Fig. 8.6 the solutions computed with the two methods for the two grid spacings are indicated together with the analytical solution. One can see that the CDS method gives physically wrong values for the coarse grid. Only for sufficiently small grid spacing (i.e., if $Pe_h < 2$) are physically meaningful results obtained. This effect does not occur if the UDS1 discretization is used. However, the results are relatively inaccurate, but they systematically approach the exact solution when the grid is refined.

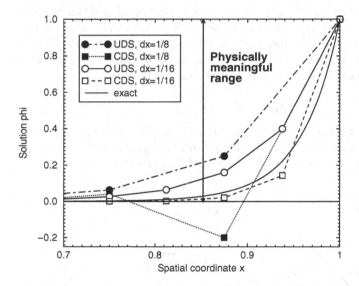

Fig. 8.6. Analytical solution and solutions with different discretization methods for $\Delta x = 1/8$ and $\Delta x = 1/16$ for one-dimensional transport equation

8.2 Estimation of Discretization Error

As outlined in the preceding sections, the discretization principally involves a discretization error. The size of this error mainly depends on

- the number and distribution of the nodal points,
- the discretization scheme employed.

For a concrete application the number and distribution of the nodal points must be chosen such that with the chosen discretization scheme the desired accuracy of the results can be achieved. Of course, it also should be possible to solve the resulting discrete equation system within a "reasonable" computing time with the computer resources available for the computation.

We now turn to the question of how the discretization error can be estimated – an issue most important for practical application. Since the exact solution of the differential equation is not known, the estimation must proceed in an approximative way based on the numerical results. This can be done by employing numerical solutions for several spatial and temporal grids whose grid spacings or time step sizes, respectively, are in some regular relation to each other. We will outline this procedure below and restrict ourselves to the spatial discretization (the temporal discretization can be handled completely analogously).

Let ϕ be a characteristic value of the exact solution of a given problem (e.g., the value at a certain grid point or some extremal value), h a measure for the grid size (e.g., the maximum grid point distance) of the numerical grid, and ϕ_h the approximative numerical solution on that grid. In general, for a method of order p one has (for a sufficiently fine grid):

$$\phi = \phi_h + Ch^p + O(h^{p+1}) \tag{8.14}$$

with a constant C that does not depend on h. Thus, the error $e_h = \phi - \phi_h$ is approximatively proportional to the p-th power of the grid size. The situation is illustrated in Fig. 8.7 for methods of 1st and 2nd order.

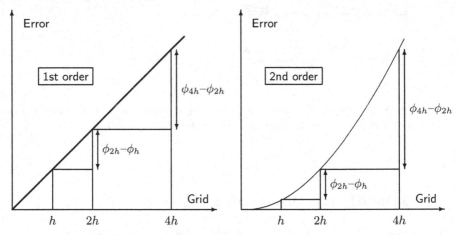

Fig. 8.7. Error versus grid size for discretization schemes of 1st and 2nd order

If the solutions for the grid sizes $4h$, $2h$, and h are known, these can be used for an error estimation. Inserting the three grid sizes into (8.14) and neglecting the terms of higher order one obtains three equations for the unknowns ϕ, C, and p. By resolving this for p, first the actual order of the scheme can be estimated:

$$p \approx \log \left(\frac{\phi_{2h} - \phi_{4h}}{\phi_h - \phi_{2h}} \right) / \log 2 . \tag{8.15}$$

This is necessary if one is not sure that the grid sizes employed are already in a range in which the asymptotic behavior, on which the relation (8.14) is based, is valid (i.e., if the higher order terms are "small enough"). Knowing p, the constant C results from (8.14) for h and $2h$ in

$$C \approx \frac{\phi_h - \phi_{2h}}{(2^p - 1)h^p} ,$$

such that for the error on the grid with grid size h (again from (8.14)) one gets the approximation

$$e_h \approx \frac{\phi_h - \phi_{2h}}{2^p - 1} .$$

With this, a *grid independent solution* can be estimated:

$$\phi \approx \phi_h + \frac{\phi_h - \phi_{2h}}{2^p - 1} . \tag{8.16}$$

This procedure is known as *Richardson extrapolation*.

In general, numerical simulations always should be done with at least one (better yet two, if there are doubts concerning the order of the method) systematic grid refinement (e.g., halving of grid spacing or time step size) in order to verify the quality of the solution with the above procedure.

Let us consider as a concrete example for the error estimation a bouyancy driven flow in a square cavity with a temperature gradient between the two side walls. The problem is computed for successively refined grids (10×10 CVs to 320×320 CVs) with a finite-volume method employing a 2nd order CDS discretization. The Nußelt numbers Nu_h (a measure for the convective heat transfer) that result from the computed temperatures for the different grids are given in Table 8.1 together with the corresponding differences $\mathrm{Nu}_{2h} - \mathrm{Nu}_h$ for two "subsequent" grids.

Table 8.1. Convergence of Nußelt numbers for successively refined grids

Grid	Nu_h	$\mathrm{Nu}_{2h} - \mathrm{Nu}_h$	p
10×10	8.461	−	−
20×20	10.598	-2.137	−
40×40	9.422	1.176	−
80×80	8.977	0.445	1.40
160×160	8.863	0.114	1.96
320×320	8.834	0.029	1.97

First the actual convergence order of the method is determined from the values for the three finest grids according to (8.15):

$$p \approx \log\left(\frac{\text{Nu}_{2h} - \text{Nu}_{4h}}{\text{Nu}_h - \text{Nu}_{2h}}\right) / \log 2 = \log\left(\frac{8.863 - 8.977}{8.834 - 8.863}\right) / \log 2 = 1.97 \,.$$

Table 8.1 also indicates the values for p when using three coarser grids. One can see that the method for a sufficiently fine grid – as expected for the CDS scheme – has an asymptotic convergence behavior of nearly 2nd order. Next, a grid independent solution is determined with the extrapolation formula (8.16) using the solutions for the two finest grids with 160×160 and 320×320 CVs:

$$\text{Nu} \approx \text{Nu}_h + \frac{\text{Nu}_h - \text{Nu}_{2h}}{2^p - 1} = 8.834 + \frac{8.834 - 8.863}{2^{1.97} - 1} = 8.824 \,.$$

In Fig. 8.8 the computed values for Nu and the grid independent solution (dashed line) are illustrated graphically.

By comparison with the grid independent solution the solution error on all grids can be determined. Representing the error depending on the grid size in a double-logarithmic diagram (see Fig. 8.9), for sufficiently fine grids one gets nearly a straight line with slope 2 corresponding to the order of the method. Further, one can observe that when using the coarsest grid one is not yet in the range of the asymptotic convergence. An extrapolation using this solution would lead to completely wrong results.

8.3 Influence of Numerical Grid

In Chap. 3 several properties of numerical grids were addressed that have different influences on the flexibility, discretization accuracy, and efficiency of

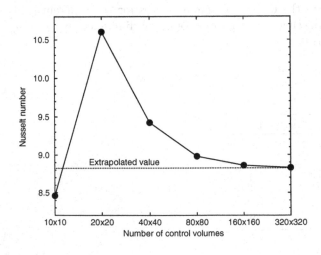

Fig. 8.8. Nußelt number depending on the number of CVs and extrapolated (grid independent) value

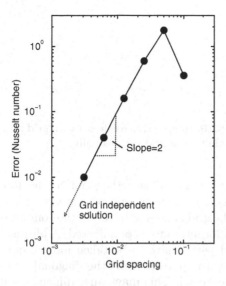

Fig. 8.9. Error for Nußelt number depending on grid spacing

a numerical scheme. Here we will summarize the most important grid properties most relevant to practical application involving the properties of the discretization and solution methods.

It should be stated again that, in general, compared to triangular or tetrahedral grids (for comparable discretizations) quadrilateral or hexahedral grids give more accurate results because portions of the error on opposite faces partially cancel each other. On the other hand the automatic generation of triangular or tetrahedral grids is simpler. Thus, for a concrete problem one should deliberate about which aspect is more important. The grid structure should also be included in these considerations – also with respect to an efficient solution of the resulting discrete equation systems. Structured triangular or tetrahedral grids usually do not make sense because quadrilateral or hexahedral grids can also be employed instead. In general, the error and efficieny aspects are more important for problems from fluid mechanics (in particular for turbulent flows), while for problems from structural mechanics mostly the geometrical flexibility plays a more important role.

Besides the global grid structure there are several local properties of the grids which are important for the efficiency and accuracy of a computation. In particular, these are the orthogonality of the grid lines, the expansion rate of adjacent grid cells, and the ratio of the side lengths of the grid cells. We will discuss these in connection with a finite-volume discretization for quadrilateral grids. Analogous considerations apply for finite-element discretizations and other types of cells.

The orthogonality of a grid is characterized by the intersection angle ψ between the grid lines (see Fig. 8.10). A grid is called *orthogonal* if all grid lines intersect at a right angle.

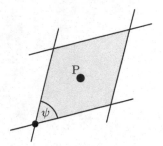

Fig. 8.10. Intersection angle between grid lines for definition of grid orthogonality

If the connecting line of the points P and E is orthogonal to the face S_e, then only the derivative in this direction has to be approximated (see also Fig. 4.14). As outlined in Sect. 4.5, if the grid is non-orthogonal, the computation of the diffusive fluxes becomes significantly more complicated. Additional neighboring relations between the grid points have to be taken into account and by the appearance of coefficients with opposite signs the diagonal dominance of the system matrix can be weakened. This may cause difficulties in the convergence of the solvers. The same applies if the additional neighboring values are treated explicitly in the way described in Sect. 7.1.4. Therefore, an attempt should always be made to keep the numerical grid as orthogonal as possible (as far as this is possible with respect to the problem geometry).

As we have already seen in Sect. 4.4, the truncation error of a discretization scheme also depends on the expansion ratio

$$\xi_e = \frac{x_E - x_e}{x_e - x_P}$$

of the grid (see Fig. 8.11). In the case of a central difference discretization of the first derivative, for instance, for a one-dimensional problem the truncation error at the point x_e becomes:

$$\tau_e = \left(\frac{\partial \phi}{\partial x}\right)_e - \frac{\phi_E - \phi_P}{x_E - x_P} =$$
$$= \frac{(1 - \xi_e)\Delta x}{2}\left(\frac{\partial^2 \phi}{\partial x^2}\right)_e + \frac{(1 - \xi_e + \xi_e^2)\Delta x^2}{6}\left(\frac{\partial^3 \phi}{\partial x^3}\right)_e + O(\Delta x^3).$$

where $\Delta x = x_e - x_P$. The leading error term, which with respect to the grid spacing is only of 1st order, only vanishes if $\xi_e = 1$. The more the expansion rate deviates from 1 (non-equidistance), the larger this portion of the error becomes – with a simultaneous decrease in the order of the scheme. Corresponding considerations apply for all spatial directions as well as for a discretization of the time interval in the case of time-dependent problems.

In order not to deteriorate the accuracy of the discretization too much, when generating the grid (in space and time) it should be paid attention that the grid expansion ratio should not be allowed to become too large (e.g.,

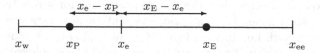

Fig. 8.11. Definition of expansion ratio between two grid cells

between 0.5 and 2), at least in areas with strong variations of the unknown variables in direction of the corresponding coordinate.

A further important quantity, which influences the condition number of the discrete equation system (and therefore the efficiency of the solution algorithms, see Sect. 7.1), is the ratio λ_P between the length and height of a control volume. This is called the *grid aspect ratio*. For an orthogonal CV, λ_P is defined by (see Fig. 8.12)

$$\lambda_P = \frac{\Delta x}{\Delta y},$$

where Δx and Δy are the length and height of the CV, respectively. For non-orthogonal CVs a corresponding quantity can be defined, for instance, by the ratio of the minimum of the face lengths δS_e and δS_w to the maximum of the face lengths δS_n und δS_s.

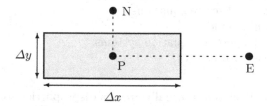

Fig. 8.12. Definition of aspect ratio of grid cells

The values for λ_P particularly influence the size of the contributions of the discretization of the diffusive parts in the off-diagonals of the system matrix. As an example, let us consider the diffusive term in (8.1) for an equidistant rectangular grid. Using a central difference approximation one obtains for the east and north coefficients the expressions

$$a_E = \alpha \frac{\Delta y}{\Delta x} = \frac{\alpha}{\lambda_P} \quad \text{and} \quad a_N = \alpha \frac{\Delta x}{\Delta y} = \alpha \lambda_P.$$

Thus, the ratio between a_N and a_E amounts to λ_P^2. If λ_P strongly deviates from 1 (in this case one speaks of *anisotropic grids*), it, in particular, negatively influences the eigenvalue distribution of the system matrix, which – as outlined in Sect. 7.1 – determines the convergence rate of the most common iterative solution algorithms. The larger λ_P is, the slower the convergence of the iterative methods. It should be noted, however, that there are also specially designed iterative solution methods (e.g., special variants of the

ILU method), which possess acceptable convergence properties for strongly anisotropic grids. Thus, when generating the grid, moderate aspect ratios of the grid cells should be ensured, i.e., in the range $0.1 \leq \lambda_P \leq 10$, or, if this is not possible, a corresponding special linear system solver should be employed.

In practical applications, it is usually not possible to satisfy all the above grid properties in the whole problem domain simultaneously in an optimal way. This makes it necessary to find an adequate compromise.

8.4 Cost Effectiveness

Besides the accuracy, also the cost effectiveness of numerical computations, i.e., the *costs* that have to be *paid* to obtain a numerical solution with a certain accuracy, for practical applications is a very important aspect that always should be taken into account when employing numerical methods. As is apparent from the considerations in the preceding sections, the accuracy and cost effectiveness of a method depends on a variety of different factors:

- the extent of detail in the geometry modeling,
- the structure of the grid and the shapes of the cells,
- the mathematical model underlying the computation,
- the number and distribution of grid cells and time steps,
- the number of coefficients in the discrete equations,
- the order of the discretization scheme,
- the solution algorithm for the algebraic equation systems,
- the stopping criteria for iteration processes,
- the available computer, . . .

Since issues of accuracy and cost effectiveness usually are in a disproportionate relation in close interaction with each other, it is necessary to find here a reasonable compromise with respect to the concrete requirements of the actual problem. For example, approximations of higher order are generally more accurate than lower order methods, but sometimes they are more "costly". This is because, for instance, the iterative solution of the resulting equation systems on the available computer system is much more time consuming, so that the use of a lower order method with a larger number of nodes might be advantageous. Here, practical experience has shown that for a large number of applications 2nd order methods represent a reasonable compromise.

Exercises for Chap. 8

Exercise 8.1. The differential equation $\phi' = \cos \phi$ for the function $\phi = \phi(t)$ is discretized with an implicit time discretization according to

$$\frac{2\phi^{n+1} + a\phi^n + b\phi^{n-1}}{\Delta t} = \cos \phi^n .$$

For which real parameters a and b is the method consistent? What is the leading term of the truncation error in this case?

Exercise 8.2. For the one-dimensional convection equation (ρ and v constant, no source term) the discretization scheme

$$\frac{(1+\xi)\phi^{n+1} - (1+2\xi)\phi^n + \xi\phi^{n-1}}{\Delta t} = \theta\mathcal{L}(\phi^{n+1}) + (1-\theta)\mathcal{L}(\phi^n)$$

with a central differencing approximation $\mathcal{L}(\phi)$ for the convective term and two real parameters ξ and θ is given. (i) Determine the truncation error of the scheme. (ii) Discuss the consistency order of the method depending on ξ and θ.

Exercise 8.3. A discretization of an unsteady two-dimensional problem yields for the unknown function $\phi = \phi(x, y, t)$ the discrete equation

$$\phi_P^{n+1} = (1+\alpha)\phi_P^n + \alpha\phi_E^n + (1-\alpha^2)\phi_W^n + \phi_S^n + (4-\alpha^2)\phi_N^n$$

with a real parameter $\alpha > 0$. Discuss the stability of the scheme depending on α.

Exercise 8.4. Given is the steady one-dimensional convection-diffusion equation (ρ, v, and α constant, $v > 0$, no source term). (i) Formulate a finite-volume discretization using the flux-blending scheme of Sect. 4.3.3 (with the UDS and CDS methods) for the convective term. (ii) Check the validity of condition (8.6) for the conservativity of the method. (iii) Determine a condition for the admissible grid spacing ensuring the boundedness of the method.

Exercise 8.5. Consider the unsteady one-dimensional convection-diffusion equation with the spatial discretization of Exercise 8.4. Formulate the explicit Euler method and determine a condition on the time step size for the stability of the resulting method.

Exercise 8.6. Given is the steady one-dimensional convection-diffusion equation (ρ, v, and α constant, no source term) on the interval $[0, 1]$ with boundary conditions $\phi(0) = 0$ and $\phi(1) = 1$. Let the problem be discretized according to the UDS1/CDS2 and CDS2/CDS2 finite-volume methods with equidistant grid spacing Δx. (i) Consider the ansatz $\phi_i = C_1 + C_2 b^i$ for the discrete solution ϕ_i in the node $x_i = (i-1)/N$ ($i = 0, 1, \ldots, N+1$) and determine for both discretization methods the constants C_1, C_2, and b, such that ϕ_i solves the discrete equation exactly and ϕ_0 and ϕ_{N+1} fulfill the boundary conditions. (ii) Compare the result of (i) with the analytical solution of the differential equation and discuss the behavior of the discrete solution for $\alpha \to 0$ and $\Delta x \to 0$.

Exercise 8.7. A finite-element discretization results in the discrete equation system

$$\begin{bmatrix} -1+\alpha & 1-\alpha \\ 5\alpha & 1-2\alpha \end{bmatrix} \begin{bmatrix} \phi_1 \\ \phi_2 \end{bmatrix} = \begin{bmatrix} 18\alpha \\ -\alpha \end{bmatrix}.$$

For which values of the real parameter α is the scheme in any case bounded?

Exercise 8.8. For the function $\phi = \phi(x, y)$ the integral $I = \int_{S_e} \phi \, dS$ over the face S_e of the square control volume $[1, 3]^2$ has to be computed. (i) Determine for the approximation $I \approx \phi(3, \alpha) \Delta y$ the leading term of the truncation error and the order (with respect to the length Δy of S_e) depending on the real parameter $\alpha \in [1, 3]$. (ii) Compute I for the function $\phi(x, y) = x^3 y^4$ directly (analytically) and with the approximation defined in (i) with $\alpha = 2$. Compare the results.

Exercise 8.9. A finite-volume discretization results in the discrete equation

$$\alpha \phi_P = 2\phi_E + \phi_W + \beta \phi_S - 2\phi_N \,.$$

For which combinations of the two real parameters α and β the method is (i) definitely not conservative? (ii) in any case bounded?

Exercise 8.10. A finite-element computation on equidistant grids with the grid spacings $4h$, $2h$, and h gives the solutions $\phi_{4h} = 20$, $\phi_{2h} = 260$, and $\phi_h = 275$. (i) What is the order of the method? (ii) What is the grid independent solution?

Exercise 8.11. Discretize the steady two-dimensional diffusion equation (α constant, Dirichlet boundary conditions) with the 2nd order CDS method for the finite-volume grids shown in Fig. 8.13. Determine in each case the condition number of the system matrix and the spectral radius of the iteration matrix for the Gauß-Seidel method and compare the corresponding values.

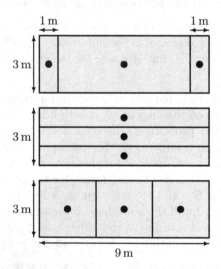

Fig. 8.13. Numerical grids for Exercise 8.11

9

Finite-Element Methods in Structural Mechanics

The investigation of deformations and stresses in solids belongs to the most frequent tasks in engineering applications. In practice nowadays the numerical study of such problems involves almost exclusively finite-element methods. Due to the great importance of these methods, in this chapter we will address in more detail the particularities and the practical treatment of corresponding problems. In particular, the important concept of isoparametric finite elements will be considered. We will do this exemplarily by means of linear two-dimensional problems for a 4-node quadrilateral element. However, the formulations employed allow in a very simple way an understanding of the necessary modifications if other material laws, other strain-stress relations, and/or other types of elements are used. The considerations simultaneously serve as an example of the application of the finite-element method to *systems* of partial differential equations.

9.1 Structure of Equation System

As an example we consider the equations of linear elasticity theory for the plane stress state (see Sect. 2.4.3). To save indices we denote the two spatial coordinates by x and y and the two unknown displacements by u and v (see Fig. 9.1). Furthermore, since different index ranges occur, for clarity all occuring summations will be given explicitly (i.e., no Einstein summation convention).

The underlying linear strain-displacement relations are

$$\varepsilon_{11} = \frac{\partial u}{\partial x}, \quad \varepsilon_{22} = \frac{\partial v}{\partial y}, \quad \text{and} \quad \varepsilon_{12} = \frac{1}{2}\left(\frac{\partial u}{\partial y} + \frac{\partial v}{\partial x}\right), \tag{9.1}$$

and the (linear) elastic material law for the plane stress state will be used in the form

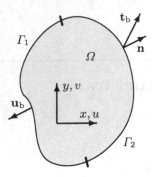

Fig. 9.1. Disk in plane stress state with notations

$$
\begin{bmatrix} T_{11} \\ T_{22} \\ T_{12} \end{bmatrix} = \underbrace{\frac{E}{1-\nu^2} \begin{bmatrix} 1 & \nu & 0 \\ \nu & 1 & 0 \\ 0 & 0 & 1-\nu \end{bmatrix}}_{\mathbf{C}} \underbrace{\begin{bmatrix} \varepsilon_{11} \\ \varepsilon_{22} \\ \varepsilon_{12} \end{bmatrix}}_{\boldsymbol{\varepsilon}}.
\tag{9.2}
$$

Here, exploiting the symmetry properties, we summarize the relevant components of the strain and stress tensors into the vectors $\boldsymbol{\varepsilon} = (\varepsilon_1, \varepsilon_2, \varepsilon_3)$ and $\mathbf{T} = (T_1, T_2, T_3)$. As boundary conditions at the boundary part Γ_1 the displacements

$$u = u_{\mathrm{b}} \quad \text{and} \quad v = v_{\mathrm{b}}$$

and on the boundary part Γ_2 the stresses

$$T_1 n_1 + T_3 n_2 = t_{\mathrm{b}1} \quad \text{and} \quad T_3 n_1 + T_2 n_2 = t_{\mathrm{b}2}$$

are prescribed (see Fig. 9.1).

As a basis for the finite-element approximation the weak formulation of the problem (see Sect. 2.4.1) is employed. We denote the test functions (virtual displacements) by φ_1 and φ_2 and define

$$\psi_1 = \frac{\partial \varphi_1}{\partial x}, \quad \psi_2 = \frac{\partial \varphi_2}{\partial y}, \quad \text{and} \quad \psi_3 = \frac{1}{2} \left(\frac{\partial \varphi_1}{\partial y} + \frac{\partial \varphi_2}{\partial x} \right).$$

The weak form of the equilibrium condition (momentum conservation) then can be formulated as follows:

Find (u, v) with $(u, v) = (u_{\mathrm{b}}, v_{\mathrm{b}})$ on Γ_1, such that

$$
\sum_{k,j=1}^{3} \int_{\Omega} C_{kj} \varepsilon_k \psi_j \, \mathrm{d}\Omega = \sum_{j=1}^{2} \left(\int_{\Omega} \rho f_j \varphi_j \, \mathrm{d}\Omega + \int_{\Gamma_2} t_{\mathrm{b}j} \varphi_j \, \mathrm{d}\Gamma \right)
\tag{9.3}
$$

for all test functions (φ_1, φ_2) with $\varphi_1 = \varphi_2 = 0$ on Γ_1.

Employing the problem formulation in this form, further considerations are largely independent of the special choices of the material law and the strain-stress relation. Only the correspondingly modified definitions for the material matrix $\mathbf{C}$ and the strain tensor ε have to be taken into account. In this way also an extension to nonlinear material laws (e.g., plasticity) or large deformations (e.g., for rubberlike materials, cp. Sect. 2.4.5) is quite straightforward (see, e.g., [2]).

9.2 Finite-Element Discretization

A practically important element class for structural mechanics applications are the *isoparametric elements*, which we will introduce by means of an example. The basic idea of the isoparametric concepts is to employ the same (isoparametric) mapping to represent the displacements as well as the geometry with local coordinates (ξ, η) in a reference unit area. The mapping to the unit area (triangle or square) is accomplished by a variable transformation, which corresponds to the ansatz for the unknown function.

As an example, we consider an isoparametric quadrilatral 4-node element, which also is frequently used in practice because it usually provides a good compromise between accuracy requirements and computational effort. However, it should be noted that the considerations are largely independent from the element employed (triangles or quadrilaterals, ansatz functions). The considered element can be seen as a generalization of the bilinear parallelogram element, which was introduced in Sect. 5.6.4. The procedure for the assembling of the discrete equations is analogous to a large extent.

A coordinate transformation of a general quadrilateral Q_i to the unit square Q_0 (see Fig. 9.2) is given by

$$x = \sum_{j=1}^{4} N_j^e(\xi, \eta) x_j \quad \text{and} \quad y = \sum_{j=1}^{4} N_j^e(\xi, \eta) y_j \, , \tag{9.4}$$

where $P_j = (x_j, y_j)$ are the vertices of the quadrilateral (here and in the following we omit for simplicity the index i on the element quantities). The bilinear isoparametric ansatz functions

$$N_1^e(\xi, \eta) = (1 - \xi)(1 - \eta) \, , \; N_2^e(\xi, \eta) = \xi(1 - \eta) \, ,$$
$$N_3^e(\xi, \eta) = \xi\eta \, , \qquad\qquad N_4^e(\xi, \eta) = (1 - \xi)\eta$$

correspond to the local shape functions, which were already used for the bilinear parallelogram element. For the displacements one has the local shape functions representation

$$u(\xi, \eta) = \sum_{j=1}^{4} N_j^e(\xi, \eta) u_j \quad \text{and} \quad v(\xi, \eta) = \sum_{j=1}^{4} N_j^e(\xi, \eta) v_j \tag{9.5}$$

with the displacements u_j and v_j at the vertices of the quadrilateral as nodal variables. By considering the relations (9.4) and (9.5) the principal idea of the

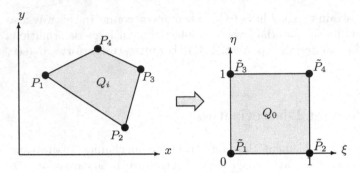

Fig. 9.2. Transformation of arbitrary quadrilateral to unit square

isoparametric concept becomes apparent, i.e., for the coordinate transformation and the displacements the same shape functions are employed.

According to the elementwise approach to assemble the discrete equation system described in Sect. 5.3, we next determine the element stiffness matrix and the element load vector. As a basis we employ a weak form of the equilibrium condition within an element with the test functions $(N_j^e, 0)$ and $(0, N_j^e)$ for $j = 1, \ldots, 4$. In order to allow a compact notation, it is helpful to introduce the nodal displacement vector as

$$\phi = [u_1, v_1, u_2, v_2, u_3, v_3, u_4, v_4]^T$$

and write the test functions in the following matrix form:

$$\mathbf{N} = \begin{bmatrix} N_1^e & 0 & N_2^e & 0 & N_3^e & 0 & N_4^e & 0 \\ 0 & N_1^e & 0 & N_2^e & 0 & N_3^e & 0 & N_4^e \end{bmatrix}.$$

As analogon to $\psi = (\psi_1, \psi_2, \psi_3)$ within the element we further define the matrix

$$\mathbf{A} = \begin{bmatrix} \dfrac{\partial N_1^e}{\partial x} & 0 & \dfrac{\partial N_2^e}{\partial x} & 0 & \dfrac{\partial N_3^e}{\partial x} & 0 & \dfrac{\partial N_4^e}{\partial x} & 0 \\[2mm] 0 & \dfrac{\partial N_1^e}{\partial y} & 0 & \dfrac{\partial N_2^e}{\partial y} & 0 & \dfrac{\partial N_3^e}{\partial y} & 0 & \dfrac{\partial N_4^e}{\partial y} \\[2mm] \dfrac{1}{2}\dfrac{\partial N_1^e}{\partial y} & \dfrac{1}{2}\dfrac{\partial N_1^e}{\partial x} & \dfrac{1}{2}\dfrac{\partial N_2^e}{\partial y} & \dfrac{1}{2}\dfrac{\partial N_2^e}{\partial x} & \dfrac{1}{2}\dfrac{\partial N_3^e}{\partial y} & \dfrac{1}{2}\dfrac{\partial N_3^e}{\partial x} & \dfrac{1}{2}\dfrac{\partial N_4^e}{\partial y} & \dfrac{1}{2}\dfrac{\partial N_4^e}{\partial x} \end{bmatrix}.$$

The equilibrium relation corresponding to (9.3) for the element Q_i now reads

$$\sum_{k,l=1}^{3} \int_{Q_i} C_{kl} \varepsilon_k A_{lj} \, d\Omega = \sum_{k=1}^{2} \left(\int_{Q_i} \rho f_k N_{kj} \, d\Omega + \int_{\Gamma_{2i}} t_{bk} N_{kj} \, d\Gamma \right) \qquad (9.6)$$

for all $j = 1, \ldots, 8$. Γ_{2i} denotes the edges of the element Q_i located on the boundary part Γ_2 with a stress boundary condition. If there are no such edges the corresponding term is just zero.

Inserting the expressions (9.5) into the strain-stress relations (9.1) we get the strains ε_i dependent on the ansatz functions:

$$\varepsilon_1 = \sum_{j=1}^{4} \frac{\partial N_j^e}{\partial x} u_j \,, \quad \varepsilon_2 = \sum_{j=1}^{4} \frac{\partial N_j^e}{\partial y} v_j \,, \quad \varepsilon_3 = \frac{1}{2} \sum_{j=1}^{4} \left(\frac{\partial N_j^e}{\partial y} u_j + \frac{\partial N_j^e}{\partial x} v_j \right).$$

With the matrix $\mathbf{A}$ these relations can be written in compact form as

$$\varepsilon_j = \sum_{k=1}^{8} A_{jk} \phi_k \quad \text{for } j = 1, 2, 3 \,. \tag{9.7}$$

Inserting this into the weak element formulation (9.6) we finally obtain:

$$\sum_{k=1}^{8} \phi_k \sum_{n,l=1}^{3} \int_{Q_i} C_{nl} A_{nj} A_{lk} \, \mathrm{d}\Omega = \sum_{k=1}^{2} \left(\int_{Q_i} \rho f_k N_{kj} \, \mathrm{d}\Omega + \int_{\Gamma_{2i}} t_{bk} N_{kj} \, \mathrm{d}\Gamma \right)$$

for $j = 1, \ldots, 8$. For the components of the element stiffness matrix $\mathbf{S}^i$ and the element load vector $\mathbf{b}^i$ we thus have the following expressions:

$$S_{jk}^i = \sum_{n,l=1}^{3} \int_{Q_i} C_{nl} A_{nj} A_{lk} \, \mathrm{d}\Omega \,, \tag{9.8}$$

$$b_j^i = \sum_{k=1}^{2} \left(\int_{Q_i} \rho f_k N_{kj} \, \mathrm{d}\Omega + \int_{\Gamma_{2i}} t_{bk} N_{kj} \, \mathrm{d}\Gamma \right) \tag{9.9}$$

for $k, j = 1, \ldots, 8$.

In the above formulas, for the computation of the element contributions the derivatives of the shape functions with respect to x and y appear, which cannot be computed directly because the shape functions are given as functions depending on ξ and η. The relation between the derivatives in the two coordinate systems is obtained by employing the chain rule as:

$$\begin{bmatrix} \dfrac{\partial N_k^e}{\partial x} \\[2mm] \dfrac{\partial N_k^e}{\partial y} \end{bmatrix} = \frac{1}{\det(\mathbf{J})} \underbrace{\begin{bmatrix} \dfrac{\partial y}{\partial \eta} & -\dfrac{\partial y}{\partial \xi} \\[2mm] -\dfrac{\partial x}{\partial \eta} & \dfrac{\partial x}{\partial \xi} \end{bmatrix}}_{\mathbf{J}^{-1}} \begin{bmatrix} \dfrac{\partial N_k^e}{\partial \xi} \\[2mm] \dfrac{\partial N_k^e}{\partial \eta} \end{bmatrix}$$

with the Jacobi matrix $\mathbf{J}$. The derivatives of x and y with respect to ξ and η can be expressed by using the transformation rules (9.4) by derivatives of the shape function with respect to ξ and η as well. In this way one obtains for the derivatives of the shape functions with respect to x and y:

$$\frac{\partial N_k^e}{\partial x} = \frac{1}{\det(\mathbf{J})} \sum_{j=1}^{4} \left(\frac{\partial N_j^e}{\partial \eta} \frac{\partial N_k^e}{\partial \xi} - \frac{\partial N_j^e}{\partial \xi} \frac{\partial N_k^e}{\partial \eta} \right) y_j ,$$

$$\frac{\partial N_k^e}{\partial y} = \frac{1}{\det(\mathbf{J})} \sum_{j=1}^{4} \left(\frac{\partial N_j^e}{\partial \xi} \frac{\partial N_k^e}{\partial \eta} - \frac{\partial N_j^e}{\partial \eta} \frac{\partial N_k^e}{\partial \xi} \right) x_j ,$$

(9.10)

where

$$\det(\mathbf{J}) = \left(\sum_{j=1}^{4} \frac{\partial N_j^e}{\partial \xi} x_j \right) \left(\sum_{j=1}^{4} \frac{\partial N_j^e}{\partial \eta} y_j \right) - \left(\sum_{j=1}^{4} \frac{\partial N_j^e}{\partial \xi} y_j \right) \left(\sum_{j=1}^{4} \frac{\partial N_j^e}{\partial \eta} x_j \right) .$$

Thus, all quantities required for the computations of the element contributions are computable directly from the shape functions and the coordinates of the nodal variables.

For the unification of the computation over the elements the integrals are transformed to the unit square Q_0. For example, for the element stiffness matrix one gets:

$$S_{jk}^i = \sum_{n,l=1}^{3} \int_{Q_0} C_{nl} A_{nj}(\xi, \eta) A_{lk}(\xi, \eta) \det(\mathbf{J}(\xi, \eta)) \, d\xi d\eta .$$

(9.11)

The computation of the element contributions – different from the bilinear parallelogram element – in general can no longer be performed exactly because due to the factor $1/\det(\mathbf{J})$ in the relations (9.10) rational functions appear in the matrix $\mathbf{A}$ (transformed in the coordinates ξ und η). Thus, numerical integration is required, for which Gauß quadrature should be advantageously employed (see Sect. 5.7). Here, the order of the numerical integration formula has to be compatible with the order of the finite-element ansatz. We will not address the issue in detail (see, e.g., [2]), and mention only that for the considered quadrilateral 4-node element a second-order Gauß quadrature is sufficient. For instance, the contributions to the element stiffness matrix are computed with this according to

$$\mathbf{S}_{jk}^i = \frac{1}{4} \sum_{n,l=1}^{3} \sum_{p=1}^{4} C_{nl} A_{nj}(\xi_p, \eta_p) A_{lk}(\xi_p, \eta_p) \det(\mathbf{J}(\xi_p, \eta_p))$$

(9.12)

with the nodal points $(\xi_p, \eta_p) = (3 \pm \sqrt{3}/6, 3 \pm \sqrt{3}/6)$ (cf. Table 5.12). The computation of the element load vector can be performed in a similar way (see, e.g., [2]).

We again mention that corresponding expressions for the element contributions of other element types can be derived in a fully analogous way.

Having computed the element stiffness matrices and element load vectors for all elements, the assembling of the global stiffness matrix and the global load vector can be done according to the procedure described in Sect. 5.6.3. We will illustrate this in the next section by means of an example.

In practical applications in most cases one is directly interested not in the displacements, but in the resulting stresses (T_1, T_2, T_3). For the considered quadrilateral 4-node element these are not continuous at the element interfaces and advantageously are determined from (9.2) by using the representation (9.7) in the centers of the elements (there one gets the most accurate values):

$$T_j = \sum_{k=1}^{8} \sum_{n=1}^{3} C_{nj} A_{nk} \left(\frac{1}{2}, \frac{1}{2} \right) \phi_k . \tag{9.13}$$

A further quantity relevant in practice is the *strain energy*

$$\Pi = \frac{1}{2} \sum_{n,j=1}^{3} \int_\Omega C_{nj} \varepsilon_n \varepsilon_j \, d\Omega , \tag{9.14}$$

which characterizes the work performed for the deformation. With (9.7) the strain energy is obtained according to

$$\Pi = \frac{1}{2} \sum_{i} \sum_{k,j=1}^{8} \sum_{n,l=1}^{3} \sum_{p=1}^{4} C_{nl} A_{nj}(\xi_p, \eta_p) A_{lk}(\xi_p, \eta_p) \phi_k \phi_j \det(\mathbf{J}(\xi_p, \eta_p)) ,$$

where the first summation has to be carried out over all elements Q_i.

9.3 Examples of Applications

As a simple example for the approach outlined in the previous section we first consider an L-shaped device under pressure load, which is clamped at one end. The problem configuration with all corresponding data is indicated in Fig. 9.3. At the bottom boundary the displacements $u = v = 0$ and at the top boundary the stress components $t_{b1} = 0$ und $t_{b2} = -2 \cdot 10^{-4} \, \text{N/m}^2$ are prescribed. All other boundaries are free, i.e., there the stress boundary condition $t_{b1} = t_{b2} = 0$ applies.

For the solution of the problem we use a discretization with only two quadrilatreal 4-node elements. The elements and the numbering of the nodal variables (u_j, v_j) for $j = 1, \ldots, 6$ are indicated in Fig. 9.4. After the computation of the two element stiffness matrices and element load vectors (using the formulas derived in the preceding section), the assembling of the global stiffness matrix and load vector $\mathbf{S}$ and $\mathbf{b}$, respectively, can be done in the usual way (see Sect. 5.6.3). With the assignment of the nodal variables given by the coincidence matrix in Table 9.1, $\mathbf{S}$ and $\mathbf{b}$ get the following structure:

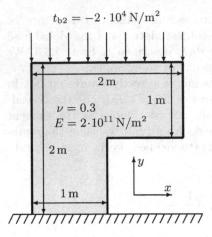

Fig. 9.3. L-shaped device under pressure load

$$
\mathbf{S} =
\begin{bmatrix}
* & * & * & * & * & * & * & * & 0 & 0 & 0 & 0 \\
* & * & * & * & * & * & * & * & 0 & 0 & 0 & 0 \\
* & * & * & * & * & * & * & * & 0 & 0 & 0 & 0 \\
* & * & * & * & * & * & * & * & 0 & 0 & 0 & 0 \\
* & * & * & * & * & * & * & * & * & * & * & * \\
* & * & * & * & * & * & * & * & * & * & * & * \\
* & * & * & * & * & * & * & * & * & * & * & * \\
* & * & * & * & * & * & * & * & * & * & * & * \\
0 & 0 & 0 & 0 & * & * & * & * & * & * & * & * \\
0 & 0 & 0 & 0 & * & * & * & * & * & * & * & * \\
0 & 0 & 0 & 0 & * & * & * & * & * & * & * & * \\
0 & 0 & 0 & 0 & * & * & * & * & * & * & * & *
\end{bmatrix}
\quad \text{and} \quad
\mathbf{b} =
\begin{bmatrix}
0 \\ 0 \\ 0 \\ 0 \\ 0 \\ * \\ 0 \\ 0 \\ 0 \\ * \\ 0 \\ 0
\end{bmatrix},
$$

where "*" denotes a non-zero entry.

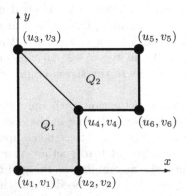

Fig. 9.4. Discretization of L-shaped device with two quadrilateral 4-node elements

Table 9.1. Assignment of nodal values and elements (coincidence matrix) for L-shaped device

Element	Local nodal variable							
	ϕ_1	ϕ_2	ϕ_3	ϕ_4	ϕ_5	ϕ_6	ϕ_7	ϕ_8
1	u_1	v_1	u_2	v_2	u_4	v_4	u_3	v_3
2	u_3	v_3	u_4	v_4	u_6	v_6	u_5	v_5

Next the geometric boundary conditions have to be taken into account, i.e., the equation system has to be modified to ensure that the nodal variables u_1, v_1, u_2, and v_2 get the value zero. According to the procedure outlined in Sect. 5.6.3 this finally leads to a discrete equation system of the form:

$$
\begin{bmatrix}
1 & 0 & 0 & 0 & 0 & 0 & 0 & 0 & 0 & 0 & 0 & 0 \\
0 & 1 & 0 & 0 & 0 & 0 & 0 & 0 & 0 & 0 & 0 & 0 \\
0 & 0 & 1 & 0 & 0 & 0 & 0 & 0 & 0 & 0 & 0 & 0 \\
0 & 0 & 0 & 1 & 0 & 0 & 0 & 0 & 0 & 0 & 0 & 0 \\
0 & 0 & 0 & 0 & * & * & * & * & * & * & * & * \\
0 & 0 & 0 & 0 & * & * & * & * & * & * & * & * \\
0 & 0 & 0 & 0 & * & * & * & * & * & * & * & * \\
0 & 0 & 0 & 0 & * & * & * & * & * & * & * & * \\
0 & 0 & 0 & 0 & * & * & * & * & * & * & * & * \\
0 & 0 & 0 & 0 & * & * & * & * & * & * & * & * \\
0 & 0 & 0 & 0 & * & * & * & * & * & * & * & * \\
0 & 0 & 0 & 0 & * & * & * & * & * & * & * & *
\end{bmatrix}
\begin{bmatrix}
u_1 \\ v_1 \\ u_2 \\ v_2 \\ u_3 \\ v_3 \\ u_4 \\ v_4 \\ u_5 \\ v_5 \\ u_6 \\ v_6
\end{bmatrix}
=
\begin{bmatrix}
0 \\ 0 \\ 0 \\ 0 \\ 0 \\ * \\ 0 \\ 0 \\ 0 \\ * \\ 0 \\ 0
\end{bmatrix}.
$$

Figure 9.5 shows the computed deformed state (in 10^5-fold magnification) as well as the (linearly interpolated) distribution of the absolute values of the displacement vector. In Table 9.2 the maximum displacement $u_{\max}$, the maximum stress $T_{\max}$, and the strain energy Π are given. Besides the values obtained with the two quadrilateral 4-node elements, also the corresponding results when using two quadrilateral 8-node elements and four triangular 3-node elements are given. The associated element and node distributions are indicated in Fig. 9.6. For an appraisal of the different results in Table 9.2, the values of a reference solution also are given, which has been obtained with a very fine grid, as well as the corresponding relative errors.

Several conclusions, which also have some universal validity, can be drawn from the results. One can observe that – independent from the element type – the error is smallest for the strain energy and largest for the stresses. Generally, owing to the rather low number of elements, of course, the errors are relatively large. Comparing the results for the 4-node and 8-node quadrilateral elements, one can observe the gain in accuracy when using a polynomial ansatz of higher order (biquadratic instead of bilinear). The results for the 4-node element – with the same number of nodal variables (i.e., with comparable computational effort) – are significantly more accurate than for the triangular 3-node element.

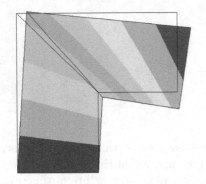

Fig. 9.5. Deformation and distribution of absolute value of displacement vector for L-shaped device

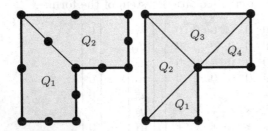

Fig. 9.6. Discretization for L-shaped device with quadrilateral 8-node elements (left) and triangular 3-node elements (right)

Table 9.2. Comparison of reference and numerical solutions with different elements for L-shaped device

Element type	Displacement [μm]		Energy [Ncm]		Stress [N/mm²]	
	u_{max}	Error	Π	Error	T_{max}	Error
4-node quadrilateral	2,61	35%	1,75	28%	0,12	50%
8-node quadrilateral	2,96	26%	1,90	22%	0,16	33%
3-node triangle	1,07	73%	0,94	61%	0,06	75%
Reference solution	4,02	–	2,43	–	0,24	–

To illustrate the convergence behavior of the 4-node quadrilateral elements when increasing the number of elements, Fig. 9.7 gives in a double-logarithmic diagram the relative error of the maximum displacement u_{max} for computations with increasing number of elements. One can observe the systematic reduction of the error with increasing element number, corresponding to a quadratic convergence order of the displacements. As outlined in Sect. 8.2, this behavior can be exploited for an error estimation.

As a second, more complex example we consider the determination of stresses in a unilaterally clamped disk with three holes under tensile load. The configuration is sketched in Fig. 9.8 together with the corresponding problem parameters. A typical question for this problem is, for instance, the determination of the location and value of the maximum occurring stresses.

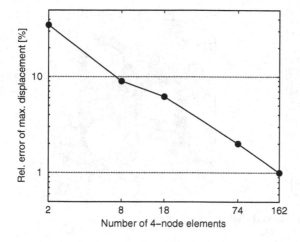

Fig. 9.7. Relative error of maximum displacement depending on number of elements for L-shaped device

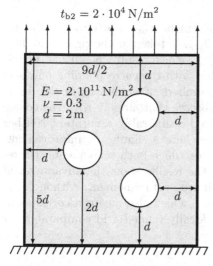

Fig. 9.8. Pierced disk under tensile load

For the discretization again the isoparametric quadrilateral 4-node element is employed. We consider three different numerical grids for the subdivision into elements: a largely uniform grid with 898 elements, a uniformly refined grid with 8,499 elements, and a locally refined grid with 3,685 elements. The coarser uniform and the locally refined grids are shown in Fig. 9.9. The generation of the latter involved first an estimation for the local discretization error from a computation with the coarse uniform grid by using the stress gradients as an error indicator, which then served as criterion for the local refinement at locations where large gradients (i.e., errors) occur. More details about adaptive refinement techniques are given in Sect. 12.1.

In Fig. 9.10 the deformed state (10^6-fold enlarged) computed with the locally refined grid is shown together with the distribution of the normal stress

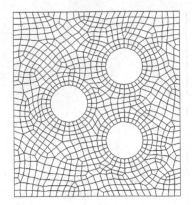

Fig. 9.9. Numerical grids for pierced disk

T_2 in the y-direction. The maximum displacement u_{max} is achieved at the upper boundary, while the maximum stress T_{max} is taken at the right boundary of the left hole. The values for u_{max} and T_{max}, which result when using the different grids, are given in Table 9.3 together with the corresponding numbers of nodal variables corresponding to the numbers of unknowns in the linear equation systems to be solved. One can observe that already with the coarse uniform grid the displacements are captured comparably accurately. Neither the uniform nor the local refinement gives here a siginficant improvement. Differences show up for the stress values, for which both refinements still result in noticeable changes. In particular, the results show the advantages of a local grid refinement compared to a uniform grid refinement. Although the number of nodal variables is much smaller – which also means a correspondingly shorter computing time – with the locally refined grid comparable (or even better) values can be achieved.

Table 9.3. Numerical solutions with different element subdivisions for pierced disk

Grid	Elements	Nodes	u_{max} [µm]	T_{max} [Ncm^2]
uniform	898	2 008	1,57	9,98
uniform	8 499	17 316	1,58	9,81
locally refined	3 685	7 720	1,58	9,76

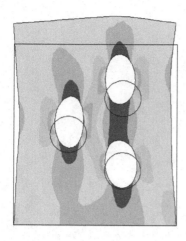

Fig. 9.10. Computed deformations and stresses for pierced disk

Exercises for Chap. 9

Exercise 9.1 Compute the element stiffness matrix and the element load vector for the plane stress state for a quadrilateral 4-node element with the vertices $P_1 = (0,0)$, $P_2 = (2,0)$, $P_3 = (1,1)$, and $P_4 = (0,1)$.

Exercise 9.2 Derive an expression corresponding to (9.8) for the element stiffness matrix for the triangular 3-node element. Compute the matrix for the element 1 of the triangulation shown in Fig. 9.6

Exercise 9.3 Determine the isoparametric shape functions $N_j = N_j(\xi, \eta)$ $(j = 1, \ldots, 8)$ for the quadrilateral 8-node element.

Exercise 9.4 Determine the shape functions and the element stiffness matrix for the three-dimensional prism element with triangular base surface.

Finite-Volume Methods for Incompressible Flows

In this chapter we will specially address the application of finite-volume methods for the numerical computation of flows of incompressible Newtonian fluids. This subject matter is of particular importance because most flows in practical applications are of this type and nearly all commercial codes that are available for such problems are based on finite-volume discretizations. Special emphasis will be given to the coupling of velocity and pressure which constitutes a major problem in the incompressible case.

10.1 Structure of Equation System

The conservation equations for the description of incompressible flows for Newtonian fluids have already been presented in Sect. 2.5.1. Here, we restrict ourselves to the two-dimensional case, for which the equations can be written as follows:

$$\frac{\partial(\rho u)}{\partial t} + \frac{\partial}{\partial x}\left[\rho u u - 2\mu\frac{\partial u}{\partial x}\right] + \frac{\partial}{\partial y}\left[\rho v u - \mu\left(\frac{\partial u}{\partial y} + \frac{\partial v}{\partial x}\right)\right] + \frac{\partial p}{\partial x} = \rho f_u\,, \quad (10.1)$$

$$\frac{\partial(\rho v)}{\partial t} + \frac{\partial}{\partial x}\left[\rho u v - \mu\left(\frac{\partial u}{\partial y} + \frac{\partial v}{\partial x}\right)\right] + \frac{\partial}{\partial y}\left[\rho v v - 2\mu\frac{\partial v}{\partial y}\right] + \frac{\partial p}{\partial y} = \rho f_v\,, \quad (10.2)$$

$$\frac{\partial u}{\partial x} + \frac{\partial v}{\partial y} = 0\,, \quad (10.3)$$

$$\frac{\partial(\rho\phi)}{\partial t} + \frac{\partial}{\partial x}\left(\rho u\phi - \alpha\frac{\partial\phi}{\partial x}\right) + \frac{\partial}{\partial y}\left(\rho v\phi - \alpha\frac{\partial\phi}{\partial y}\right) = \rho f_\phi\,. \quad (10.4)$$

The unknowns in the equation system are: the two Cartesian velocity components u and v, the pressure p, and the scalar quantity ϕ, which denotes some transport quantity that – depending on the specific application – additionally has to be determined (e.g., temperature, concentration, or turbulence quantities). The density ρ, the dynamic viscosity μ, the diffusion coefficient α, as well as the source terms f_u, f_v, and f_ϕ are prescribed (maybe also depending on the unknowns).

The equation system (10.1)-(10.4) has to be completed by boundary conditions and, in the unsteady case, by initial conditions for the velocity components and the scalar quantity (no boundary and initial conditions for the pressure!). The types of boundary conditions have already been discussed in Sect. 2.5.1 (see also Sect. 10.4).

In general, the equation system (10.1)-(10.4) has to be considered as a coupled system and, therefore, also has to be solved correspondingly. Summarizing the unknowns in the vector $\psi = (u, v, p, \phi)$, the structure of the system can be represented as follows:

$$\begin{bmatrix} A_{11}(\psi) & A_{12}(\psi) & A_{13} & A_{14}(\psi) \\ A_{21}(\psi) & A_{22}(\psi) & A_{23} & A_{24}(\psi) \\ A_{31}(\psi) & A_{32}(\psi) & 0 & 0 \\ A_{41}(\psi) & A_{42}(\psi) & 0 & A_{44}(\psi) \end{bmatrix} \begin{bmatrix} u \\ v \\ p \\ \phi \end{bmatrix} = \begin{bmatrix} b_1(\psi) \\ b_2(\psi) \\ 0 \\ b_4(\psi) \end{bmatrix}, \qquad (10.5)$$

where $A_{11}, \ldots, A_{44}$ and $b_1, \ldots, b_4$ are the operators defined according to (10.1)-(10.4). Looking at (10.5), the special difficulty when computing incompressible flows becomes apparent, i.e., the lack of a "reasonable" equation for the pressure, which is expressed by the zero element on the main diagonal of the system matrix. How to deal with this problem will be discussed in greater detail later.

If the material parameters in the mass and momentum equations do not depend on the scalar quantity ϕ, first the equations (10.1)-(10.3) can be solved for u, v, and p, independently of the scalar equation (10.4). Afterwards ϕ can be determined from the latter independently of (10.1)-(10.3) with the velocities determined before. However, in the general case all material parameters are dependent on all variables, such that the full system has to be solved simultaneously.

10.2 Finite-Volume Discretization

For the finite-volume discretization of the system (10.1)-(10.4) we apply the techniques introduced in Chap. 4. The starting point is a subdivision of the flow domain into control volumes (CVs), where we again restrict ourselves to quadrilaterals. For coupled systems it is basically possible to define different CVs and nodal value locations for different variables. For simple geometries, which allow the usage of Cartesian grids, incompressible flow computations in the past frequently were carried out with a staggered arrangement of the variables, where different CVs and nodal value locations are used for the velocity components and the pressure. The corresponding arrangement of the variables and CVs is indicated in Fig. 10.1. The u- and v-equations are discretized with respect to the u- and v-CVs, and the continuity and scalar equations with respect to the scalar CVs. The major reason for this procedure is to avoid an oscillating pressure field. However, for complex geometries the staggered variable arrangement appears to be disadvantageous. We will return to these issues in Sect. 10.3.2, after we have dealt a bit closely with the reasons for the pressure oscillations.

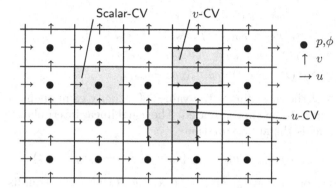

Fig. 10.1. Staggered arrangement of variables and CVs

In the following we assume the usual (non-staggered) cell-oriented variable arrangement and use Cartesian velocity components – also on non-Cartesisan grids (see Fig. 10.2). With grid-oriented velocity components the balance equations would lose their conservativity, since the conservation of a vector only ensures the conservation of its components if they have a fixed direction. In addition, the momentum equations would be significantly more complex, such that their discretization, in particular for three-dimensional flows, would become more involved.

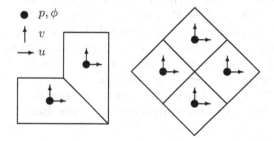

Fig. 10.2. Non-staggered variable arrangement on non-Cartesian grids with Cartesian velocity components

We consider a general quadrilateral CV with the notations introduced in Sect. 4.1 (see Fig. 4.5). The finite-volume discretization of the scalar equation (10.4) has already been described in detail in Chap. 4, so we will not separately consider this in the following. The balance equation for a CV resulting from the discretization of the continuity equation (10.3) can be formulated with the mass fluxes through the CV faces as follows:

$$\frac{\dot{m}_e}{\rho_e} + \frac{\dot{m}_w}{\rho_w} + \frac{\dot{m}_n}{\rho_n} + \frac{\dot{m}_s}{\rho_s} = 0 \,. \tag{10.6}$$

Using the midpoint rule the following approximations result for the mass fluxes:

$$\dot{m}_e = \rho_e u_e (y_{ne} - y_{se}) - \rho_e v_e (x_{ne} - x_{se}) \,,$$
$$\dot{m}_w = \rho_w u_w (y_{sw} - y_{nw}) - \rho_w v_w (x_{sw} - x_{nw}) \,,$$
$$\dot{m}_n = \rho_n v_n (x_{ne} - x_{nw}) - \rho_n u_n (y_{ne} - y_{nw}) \,,$$
$$\dot{m}_s = \rho_s v_s (x_{sw} - x_{se}) - \rho_s u_s (y_{sw} - y_{se}) \,.$$

For the discretization of the mass fluxes the values of velocity components u and v at the CV faces have to be approximated. The usual linear interpolation for $\dot{m}_e$, for instance, yields the approximation

$$\dot{m}_e = \rho_e \left(\gamma_{e1} u_E + \gamma_{e2} u_P \right) (y_{ne} - y_{se}) - \rho_e \left(\gamma_{e1} v_E + \gamma_{e2} v_P \right) (x_{ne} - x_{se})$$

with suitable interpolation factors γ_{e1} and γ_{e2} (see, e.g., (4.11)). For equidistant Cartesian grids with the grid spacings Δx and Δy we obtain for (10.6) with this approach the approximation

$$(u_E - u_W)\Delta y + (v_N - v_S)\Delta x = 0 \,.$$

As we will see later (see Sect. 10.3.2), the above seemingly obvious linear interpolation rule will turn out to be unusable, since it leads to (non-physical) oscillations in the numerical solution scheme. Before giving a detailed explanation of this effect, we first deal with the discretization of the momentum equations.

The discretization of the convective fluxes in the momentum equations can be carried out in a manner analogous to that for the general transport equation (see Sect. 4.5), if one first assumes that the mass flux is known. This way a linearization of the convective term is achieved. Formally, this corresponds to a Picard iteration, as introduced in Sect. 7.2 for the solution of non-linear equations. We will see later how this can be concretely realized within an iteration process involving the continuity equation.

Using the midpoint rule for the approximation of the surface integrals, for instance for the convective flux F_e^C in the u-equation through the face S_e, yields

$$F_e^C = \int_{S_e} \rho (u n_1 + v n_2) u \, dS_e \approx \underbrace{\left[\rho_e u_e (y_{ne} - y_{se}) - \rho_e v_e (x_{ne} - x_{se}) \right]}_{\dot{m}_e} u_e \,.$$

For the approximation of u_e we use the flux blending technique according to (4.8) introduced in Sect. 4.3.3 with a combination of the UDS and CDS methods:

$$\dot{m}_e u_e \approx \dot{m}_e u_e^{\text{UDS}} + \underbrace{\beta(\dot{m}_e u_e^{\text{CDS}} - \dot{m}_e u_e^{\text{UDS}})}_{b_\beta^{u,e}}, \tag{10.7}$$

where

$$\dot{m}_e u_e^{\text{UDS}} = \max\{\dot{m}_e, 0\} u_{\text{P}} + \min\{\dot{m}_e, 0\} u_{\text{E}} = \begin{cases} \dot{m}_e u_{\text{P}} & \text{if } \dot{m}_e > 0 \\ \dot{m}_e u_{\text{E}} & \text{if } \dot{m}_e < 0 \end{cases}$$

and

$$u_e^{\text{CDS}} = \gamma_{e1} u_{\text{E}} + \gamma_{e2} u_{\text{P}} .$$

Using the coefficient $b_\beta^{u,e}$ defined in (10.7) we have altogether for the convective flux the following approximation:

$$F_e^{\text{C}} \approx \max\{\dot{m}_e, 0\} u_{\text{P}} + \min\{\dot{m}_e, 0\} u_{\text{E}} + b_\beta^{u,e} .$$

To shorten the notation we summarize the b_β-terms for all CV faces in the u- and v-equations each into a single term:

$$b_\beta^u = b_\beta^{u,e} + b_\beta^{u,w} + b_\beta^{u,n} + b_\beta^{u,s} \quad \text{and} \quad b_\beta^v = b_\beta^{v,e} + b_\beta^{v,w} + b_\beta^{v,n} + b_\beta^{v,s}.$$

In the iteration procedure to be described in the next section these terms will be treated "explicitly", i.e., the corresponding variable values are computed with known values according to the technique described at the end of Sect. 7.1.4. In principle the blending factors for the u- and v-equations can be different. However, due to missing decision criteria the same value is usually chosen.

Also the discretization of the diffusive fluxes in the momentum equations can be done largely analogous to the procedure described in Sect. 4.5. One should note that the momentum equations (10.1) and (10.2) involve additional diffusive terms compared to the general scalar transport equation (10.4). For instance, for the diffusive flux in the u-equation through the CV face S_c one has

$$F_c^{\text{D}} = -\int_{S_c} \mu \left[2 \frac{\partial u}{\partial x} n_1 + \left(\frac{\partial u}{\partial y} + \frac{\partial v}{\partial x} \right) n_2 \right] dS_c . \tag{10.8}$$

The additional terms

$$-\int_{S_c} \mu \left(\frac{\partial u}{\partial x} n_1 + \frac{\partial v}{\partial x} n_2 \right) dS_c$$

only vanish (due to the continuity equation) if μ is constant over the flow domain.

The approximation of the diffusive flux (10.8), e.g., for the face S_e, gives

$$F_e^{\mathrm{D}} \approx D_e^1(u_{\mathrm{E}} - u_{\mathrm{P}}) + \underbrace{D_e^2(u_{\mathrm{ne}} - u_{\mathrm{se}}) + D_e^3(v_{\mathrm{E}} - v_{\mathrm{P}}) + D_e^4(v_{\mathrm{ne}} - v_{\mathrm{se}})}_{b_{\mathrm{D}}^{u,\mathrm{e}}} \qquad (10.9)$$

with

$$D_e^1 = \frac{\mu_e \left[2(y_{\mathrm{ne}} - y_{\mathrm{se}})^2 + (x_{\mathrm{ne}} - x_{\mathrm{se}})^2\right]}{(x_{\mathrm{ne}} - x_{\mathrm{se}})(y_{\mathrm{E}} - y_{\mathrm{P}}) - (y_{\mathrm{ne}} - y_{\mathrm{se}})(x_{\mathrm{E}} - x_{\mathrm{P}})},$$

$$D_e^2 = \frac{\mu_e \left[2(y_{\mathrm{E}} - y_{\mathrm{P}})(y_{\mathrm{ne}} - y_{\mathrm{se}}) + (x_{\mathrm{E}} - x_{\mathrm{P}})(x_{\mathrm{ne}} - x_{\mathrm{se}})\right]}{(y_{\mathrm{ne}} - y_{\mathrm{se}})(x_{\mathrm{E}} - x_{\mathrm{P}}) - (x_{\mathrm{ne}} - x_{\mathrm{se}})(y_{\mathrm{E}} - y_{\mathrm{P}})},$$

$$D_e^3 = \frac{\mu_e(x_{\mathrm{ne}} - x_{\mathrm{se}})(y_{\mathrm{ne}} - y_{\mathrm{se}})}{(y_{\mathrm{ne}} - y_{\mathrm{se}})(x_{\mathrm{E}} - x_{\mathrm{P}}) - (x_{\mathrm{ne}} - x_{\mathrm{se}})(y_{\mathrm{E}} - y_{\mathrm{P}})},$$

$$D_e^4 = \frac{\mu_e(x_{\mathrm{ne}} - x_{\mathrm{se}})(y_{\mathrm{E}} - y_{\mathrm{P}})}{(x_{\mathrm{ne}} - x_{\mathrm{se}})(y_{\mathrm{E}} - y_{\mathrm{P}}) - (y_{\mathrm{ne}} - y_{\mathrm{se}})(x_{\mathrm{E}} - x_{\mathrm{P}})}.$$

The values of the velocity components at the vertices of the CV can be approximated as described in Sect. 4.5 by linear interpolation from the four neighboring nodes (cf. Fig. 4.15). Analogous expressions result for the other three CV faces and for the v-equation.

The term denoted in (10.9) by $b_{\mathrm{D}}^{u,\mathrm{e}}$, as well as the corresponding terms for the other CV faces and in the v-equation, also can be treated "explicitly" within an iteration procedure (as the terms b_{β}^u und b_{β}^v), i.e., they can be interpreted as *diffusive flux sources* at the corresponding CV face. Again, for a compact notation we summarize all the corresponding terms:

$$b_{\mathrm{D}}^u = b_{\mathrm{D}}^{u,\mathrm{e}} + b_{\mathrm{D}}^{u,\mathrm{w}} + b_{\mathrm{D}}^{u,\mathrm{n}} + b_{\mathrm{D}}^{u,\mathrm{s}} \quad \text{and} \quad b_{\mathrm{D}}^v = b_{\mathrm{D}}^{v,\mathrm{e}} + b_{\mathrm{D}}^{v,\mathrm{w}} + b_{\mathrm{D}}^{v,\mathrm{n}} + b_{\mathrm{D}}^{v,\mathrm{s}}.$$

The source terms ρf_u and ρf_v in the momentum equations can be approximated, for instance by using the midpoint rule for volume integrals, as outlined in Chap. 4:

$$\int_V \rho f_u \, \mathrm{d}V \approx (\rho f_u)_{\mathrm{P}} \, \delta V = b_f^u \quad \text{and} \quad \int_V \rho f_v \, \mathrm{d}V \approx (\rho f_v)_{\mathrm{P}} \, \delta V = b_f^v.$$

The pressure terms in the momentum equations, due to the relation

$$\int_V \frac{\partial p}{\partial x_i} \, \mathrm{d}V = \int_S p n_i \, \mathrm{d}S \qquad (10.10)$$

resulting from the Gauß integral theorem, can be approximated either as volume or surface integrals. In the case of a Cartesian grid both methods produce the same approximation. In the general case only the approximation via the surface integral is strictly conservative. Applying the midpoint rule to the surface integral, for instance, the pressure term in the u-equation can be approximated by

$$\int_S pn_1 \, dS \approx \sum_c (pn_1)_c \, \delta S_c = p_e(y_{ne} - y_{se}) - p_w(y_{nw} - y_{sw}) + b_p^u$$

with

$$b_p^u = p_s(y_{se} - y_{sw}) - p_n(y_{ne} - y_{nw}).$$

The expression for the pressure term in the v-equation follows analogously.

To express the pressure terms by values in the CV centers, the values at the CV faces have to interpolated. With linear interpolation one obtains, e.g., for p_e, the approximation

$$p_e = \gamma_{e1} p_E + \gamma_{e2} p_P,$$

where γ_{e1} and γ_{e2} are the corresponding interpolation factors.

For unsteady flows the time discretization of the momentum equations, can basically be carried out with all of the methods introduced in Chap. 6 for this purpose. With the implicit Euler method, for instance, one obtains for the time derivative term in the u-equation the approximation:

$$\int_V \frac{\partial(\rho u)}{\partial t} \, dV \approx \frac{(\rho u)_P^{n+1} - (\rho u)_P^n}{\Delta t_n} \delta V =: a_t^u u_P^{n+1} - b_t^u$$

with

$$a_t^u = \frac{\rho_P^{n+1} \delta V}{\Delta t_n} \quad \text{and} \quad b_t^u = \frac{\rho_P^n \delta V}{\Delta t_n} u_P^n,$$

where we have assumed that the size of the CV does not change with time. The values with index n are known from the previous time step.

To summarize, the discretization of the momentum equations for each CV results in algebraic equations of the form:

$$a_P^u u_P = \sum_c a_c^u u_c + b^u - p_e(y_{ne} - y_{se}) + p_w(y_{nw} - y_{sw}), \qquad (10.11)$$

$$a_P^v v_P = \sum_c a_c^v v_c + b^v - p_n(x_{ne} - x_{nw}) + p_s(x_{se} - x_{sw}). \qquad (10.12)$$

Using the approximations outlined exemplarily above, the coefficients, for instance, for the u-equation read:

$$a_{\mathrm{E}}^u = -D_{\mathrm{e}}^1 - \min\{\dot{m}_{\mathrm{e}}, 0\}, \quad a_{\mathrm{W}}^u = -D_{\mathrm{w}}^1 - \min\{\dot{m}_{\mathrm{w}}, 0\},$$

$$a_{\mathrm{N}}^u = -D_{\mathrm{n}}^1 - \min\{\dot{m}_{\mathrm{n}}, 0\}, \quad a_{\mathrm{S}}^u = -D_{\mathrm{s}}^1 - \min\{\dot{m}_{\mathrm{s}}, 0\},$$

$$a_{\mathrm{P}}^u = a_{\mathrm{E}}^u + a_{\mathrm{W}}^u + a_{\mathrm{N}}^u + a_{\mathrm{S}}^u + a_t^u + \dot{m}_{\mathrm{e}} + \dot{m}_{\mathrm{w}} + \dot{m}_{\mathrm{n}} + \dot{m}_{\mathrm{s}},$$

$$b^u = b_f^u - b_{\mathrm{D}}^u - b_\beta^u - b_p^u + b_t^u.$$

The coefficients for the v-equation are given analogously. If $\beta = 0$ (pure UDS method), the term b_β^u vanishes. For time dependent problems the values for u and v in (10.11) and (10.12) have to be interpreted as values at the "new" time level t_{n+1} (whereas the coefficients and source terms only involve values at the "old" time level t_n). For steady problems the terms a_t^u and b_t^u vanish, and the sum of the mass fluxes in the coefficient a_{P}^u is zero due to the continuity equation (the latter also applies in the case of a spatially constant density). If the grid is Cartesian, the term b_p^u vanishes.

For CVs located at the boundary of the problem domain, the coefficients have to be suitably modified. Before turning to this topic, however, we will first discuss possible solution methods for the coupled discrete equation system (10.6), (10.11), and (10.12).

10.3 Solution Algorithms

As already mentioned, the computation of the pressure constitutes a particular problem for incompressible flows. The pressure only appears in the momentum equations, but not in the continuity equation, which in a sense would be available for this purpose. There are several techniques for dealing with this problem. One possibility of involving the continuity equation in the computation of the pressure is offered by what is called *pressure-correction methods*, which can be derived in different variants and are mostly used in actual flow simulation programs. We will give some important examples in the next section.

Alternatively, there are *artificial compressibility methods*, which are based on the addition of a time derivative of the pressure in the continuity equation:

$$\frac{1}{\rho \beta_{\mathrm{c}}} \frac{\partial p}{\partial t} + \frac{\partial u}{\partial x} + \frac{\partial v}{\partial y} = 0$$

with an arbitrary parameter $\beta_{\mathrm{c}} > 0$, with which the portion of the artificial compressibility can be controlled. The "proper" choice of this parameter (possibly also adaptively) is crucial for the efficiency of the method. However, obvious criteria for this are not available. The solution techniques employed together with artificial compressibility are derived from schemes developed for compressible flows (which usually fail in the borderline cases of incompressibility). Although systematic comparisons are missing, the author believes that these methods, which are rarely used in actual flow simulation codes, are less efficient than pressure-correction methods. Therefore, these methods will not be considered further here (details can be found, e.g., in [12]).

It should be mentioned that pressure-correction methods, which originally were developed for incompressible flows, also can be generalized for the computation of compressible flows (see, e.g., [8]).

10.3.1 Pressure-Correction Methods

The major problem when solving the coupled discrete equation system consists in the simultaneous fulfillment of the momentum *and* continuity equations (the coupling of the scalar equation usually does not pose problems). The general idea of a pressure-correction method to achieve this is to first compute preliminary velocity components from the momentum equations and then to correct this together with the pressure, such that the continuity equation is fulfilled. This proceeding is integrated into an iterative solution process, at the end of which both the momentum and continuity equations are approximately fulfilled. We will consider some examples of such iteration procedures, which at the same time involve a linearization of the equations.

In order to describe the basic ideas and to concentrate on the special features of variable coupling, we will first restrict ourselves to the case of an equidistant Cartesian grid. Although it will prove to be unsuitable, we will describe the principal ideas by using a central difference approximation for the pressure terms and the continuity equation. In this way we will also point out the problems that arise. Afterwards we will discuss how to modify the method in order to circumvent these problems.

First we introduce an iteration process

$$\{u^k, v^k, p^k, \phi^k\} \to \{u^{k+1}, v^{k+1}, p^{k+1}, \phi^{k+1}\},$$

which is based on the assumption that all matrix coefficients and the source terms in the momentum equations are already known. The iteration procedure is defined as follows for each CV:

$$a_{\mathrm{P}}^{u,k} u_{\mathrm{P}}^{k+1} - \sum_c a_c^{u,k} u_c^{k+1} + \frac{\Delta y}{2}(p_{\mathrm{E}}^{k+1} - p_{\mathrm{W}}^{k+1}) = b^{u,k}, \qquad (10.13)$$

$$a_{\mathrm{P}}^{v,k} v_{\mathrm{P}}^{k+1} - \sum_c a_c^{v,k} v_c^{k+1} + \frac{\Delta x}{2}(p_{\mathrm{N}}^{k+1} - p_{\mathrm{S}}^{k+1}) = b^{v,k}, \qquad (10.14)$$

$$(u_{\mathrm{E}}^{k+1} - u_{\mathrm{W}}^{k+1})\Delta y + (v_{\mathrm{N}}^{k+1} - v_{\mathrm{S}}^{k+1})\Delta x = 0, \qquad (10.15)$$

$$a_{\mathrm{P}}^{\phi,k} \phi_{\mathrm{P}}^{k+1} - \sum_c a_c^{\phi,k} \phi_c^{k+1} = b^{\phi,k}. \qquad (10.16)$$

The task now is to compute the values for the $(k+1)$-th iteration from these equations (all quantities of the k-th iteration are assumed to be already computed). In principle the equation system (10.13)-(10.16) could be solved directly with respect to the unknowns u^{k+1}, v^{k+1}, p^{k+1}, and ϕ^{k+1} in the above form. However, since the pressure does not appear in (10.15), the system is very ill-conditioned (and also rather large), such that a direct solution would mean a relatively high computational effort. It turns out that a successive solution procedure, which allows for a decoupled computation of u^{k+1}, v^{k+1}, p^{k+1}, and ϕ^{k+1}, is more appropriate.

In the first step we consider the discrete momentum equations (10.13) and (10.14) with an estimated (known) pressure field p^*. This, for instance, can be the pressure field p^k from the k-th iteration or also simply $p^* = 0$. In the latter case the resulting methods are also known as *fractional-step methods* (or *projection methods*). We obtain the two linear equation systems (considered over all CVs)

$$a_{\mathrm{P}}^{u,k} u_{\mathrm{P}}^* - \sum_c a_c^{u,k} u_c^* = b^{u,k} - \frac{\Delta y}{2}(p_{\mathrm{E}}^* - p_{\mathrm{W}}^*), \tag{10.17}$$

$$a_{\mathrm{P}}^{v,k} v_{\mathrm{P}}^* - \sum_c a_c^{v,k} v_c^* = b^{v,k} - \frac{\Delta x}{2}(p_{\mathrm{N}}^* - p_{\mathrm{S}}^*), \tag{10.18}$$

which can be solved numerically with respect to the (provisional) velocity components u^* and v^* (e.g., with one of the solvers described in Sect. 7.1). The velocity components u^* and v^* determined this way do not fulfill the continuity equation (this has not yet been taken into account). Thus, setting up a mass balance with these velocities, i.e., inserting u^* and v^* into the discrete continuity equation (10.15), yields a mass source b_{m}:

$$(u_{\mathrm{E}}^* - u_{\mathrm{W}}^*)\Delta y + (v_{\mathrm{N}}^* - v_{\mathrm{S}}^*)\Delta x = -b_{\mathrm{m}}. \tag{10.19}$$

In the next step the velocity components v^{k+1} and v^{k+1} that actually have to be determined as well as the corresponding pressure p^{k+1} are searched, such that the continuity equation is fulfilled. For the derivation of the corresponding equations we first introduce the corrections

$$u' = u^{k+1} - u^* , \quad v' = v^{k+1} - v^* , \quad p' = p^{k+1} - p^* .$$

By respective subtraction of (10.17), (10.18), and (10.19) from (10.13), (10.14), and (10.15) one gets the relations

$$a_{\mathrm{P}}^{u,k} u_{\mathrm{P}}' + \sum_c a_c^{u,k} u_c' = -\frac{\Delta y}{2}(p_{\mathrm{E}}' - p_{\mathrm{W}}'), \tag{10.20}$$

$$a_{\mathrm{P}}^{v,k} v_{\mathrm{P}}' + \sum_c a_c^{v,k} v_c' = -\frac{\Delta x}{2}(p_{\mathrm{N}}' - p_{\mathrm{S}}'), \tag{10.21}$$

$$(u_{\mathrm{E}}' - u_{\mathrm{W}}')\Delta y + (v_{\mathrm{N}}' - v_{\mathrm{S}}')\Delta x = b_{\mathrm{m}}. \tag{10.22}$$

A characteristic approach for pressure-correction methods is that now the sum terms in the relations (10.20) and (10.21), which still contain the unknown velocity corrections in the neighboring points of P, are suitably approximated. There are different possibilities for this, the simplest of which is to simply neglect these terms:

$$\sum_c a_c^{u,k} u_c' \approx 0 \quad \text{and} \quad \sum_c a_c^{v,k} v_c' \approx 0.$$

This approach yields the *SIMPLE method (Semi-Implicit Method for Pressure-Linked Equations)* proposed by Patankar und Spalding in 1972. We will first continue with this assumption and discuss alternative approaches later.

Solving for u_P' and v_P' after neglecting the sum terms in (10.20) and (10.21) gives the relations:

$$u_P' = -\frac{\Delta y}{2a_P^{u,k}}(p_E' - p_W'), \tag{10.23}$$

$$v_P' = -\frac{\Delta x}{2a_P^{v,k}}(p_N' - p_S'). \tag{10.24}$$

Inserting these values into the continuity equation (10.22) yields

$$\left[-\frac{\Delta y}{2a_{P,E}^{u,k}}(p_{EE}' - p_P') + \frac{\Delta y}{2a_{P,W}^{u,k}}(p_P' - p_{WW}')\right]\Delta y$$

$$+ \left[-\frac{\Delta x}{2a_{P,N}^{v,k}}(p_{NN}' - p_P') + \frac{\Delta x}{2a_{P,S}^{u,k}}(p_P' - p_{SS}')\right]\Delta x = b_m, \tag{10.25}$$

where, for instance, $a_{P,E}^{u,k}$ denotes the central coefficient in the u-equation for the CV around the point E. Summarizing the terms suitably, the following equation for the pressure correction p' results from (10.25):

$$a_P^{p,k}p_P' = a_{EE}^{p,k}p_{EE}' + a_{WW}^{p,k}p_{WW}' + a_{NN}^{p,k}p_{NN}' + a_{SS}^{p,k}p_{SS}' + b_m \tag{10.26}$$

with

$$a_{EE}^{p,k} = \frac{\Delta y^2}{2a_{P,E}^{u,k}}, \quad a_{WW}^{p,k} = \frac{\Delta y^2}{2a_{P,W}^{u,k}}, \quad a_{NN}^{p,k} = \frac{\Delta x^2}{2a_{P,N}^{v,k}}, \quad a_{SS}^{p,k} = \frac{\Delta x^2}{2a_{P,S}^{v,k}}$$

and

$$a_P^{p,k} = a_{EE}^{p,k} + a_{WW}^{p,k} + a_{NN}^{p,k} + a_{SS}^{p,k}.$$

Considering (10.26) for all CVs, one has a linear equation system from which the pressure correction p' can be determined. If this is known, the velocity corrections u' and v' can be computed from (10.23) and (10.24). With these corrections the searched quantities u^{k+1}, v^{k+1}, and p^{k+1} can finally be determined.

The last step is the determination of ϕ^{k+1} from the equation:

$$a_P^{\phi,k+1}\phi_P^{k+1} - \sum_c a_c^{\phi,k+1}\phi_c^{k+1} = b^{\phi,k+1}. \tag{10.27}$$

The index $k+1$ on the coefficients indicates that for their computation the "new" velocity components can already be used. If the coefficients also depend on ϕ, the values of ϕ from the k-th iteration are used.

This completes one iteration of the pressure-correction method and the next one starts again with the solution of the two discrete momentum equations (10.17) and (10.18) with respect to the provisional velocity components. The course of the overall procedure is illustrated schematically in Fig. 10.3. Within the scheme one generally can distinguish between *inner iterations* and *outer iterations*.

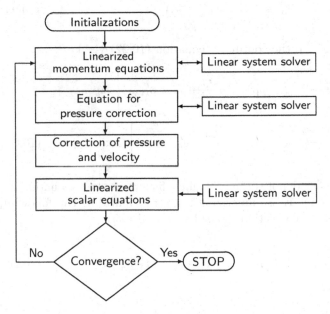

Fig. 10.3. Schematic representation of pressure-correction method

The inner iterations involve the repeated implementation of a solution algorithm for the linear equation systems for the different variables u, v, p', and ϕ. During these iterations the coefficients and source terms of the corresponding linear system remain constant and only the variable values change. Of course, when using a direct solver these iterations do not apply. The methods described in Sect. 7.1 are examples of iteration schemes that can be employed, although different methods can be used for different variables (which can make sense since the pressure-correction equation, for instance, is usually "harder" to solve than the others).

The outer iterations denote the repetition of the cycle, in which the coupled discrete equation system for *all* variables is solved up to a prescribed accuracy (for each time step in the unsteady case). After an outer iteration the coefficients and source terms are usually updated. Due to the non-linearity of the momentum equations and the coupling between velocity and pressure, such an outer iteration process is always necessary when solving the incompressible Navier-Stokes equations (as well as when using other solution algorithms). As convergence criterion for the global procedure, for instance, it can be required that the sum of the absolute residuals for all equations – normalized with suitable norming factors – becomes smaller than a prescribed bound.

10.3.2 Pressure-Velocity Coupling

We turn next to a problem mentioned earlier, which arises when for the described method a central differencing approximation for the mass flux computation in the continuity equation is used. Equation (10.26) corresponds to an algebraic relation for the pressure correction, which one would obtain when discretizing a diffusion equation with a central finite-difference scheme of second order, but with a "double" grid spacing $2\Delta x$ and $2\Delta y$ (see Fig. 10.4). This means that the pressure at point P is not directly linked with its nearest neighbor points (E, W, N, and S). Thus, four discrete solutions exist (all correctly fulfilling the equations), which are completely independent from each other. Consequently, oscillatory solutions may occur when applying the scheme in the given form. For a problem actually having a constant pressure distribution as a correct solution, an alternating solution as illustrated in Fig. 10.5 can result with the scheme.

We will clarify the above issue in another way by means of a one-dimensional example. We will consider a problem for an equidistant grid that has the alternating pressure distribution shown in Fig. 10.6 as correct solu-

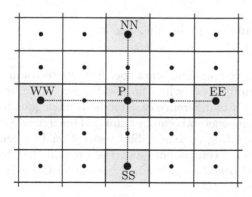

Fig. 10.4. Neighboring relations for pressure corrections with central difference approximation

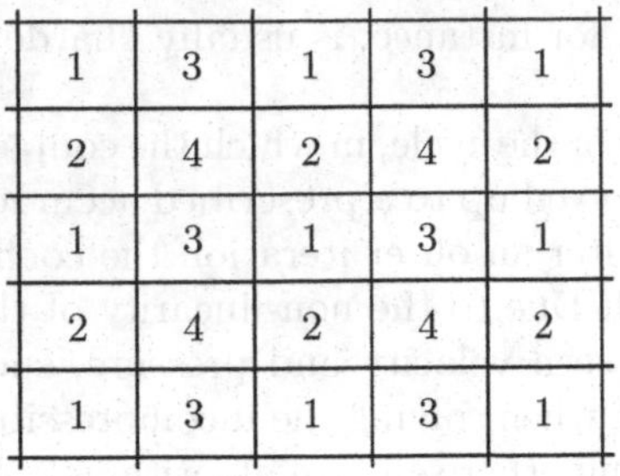

1	3	1	3	1
2	4	2	4	2
1	3	1	3	1
2	4	2	4	2
1	3	1	3	1

Fig. 10.5. Decoupled solutions for two-dimensional problem with constant pressure distribution

tion. For the pressure gradient in the momentum equations one obtains with linear interpolation:

$$p_e - p_w = \frac{1}{2}(p_E + p_P) - \frac{1}{2}(p_P + p_W) = 0 \,.$$

Thus, this pressure distribution produces no contribution to the source term in the (discrete) momentum equation in the CV center, i.e., there is no resulting pressure force there (which is physically correct). If there are no further source terms in the momentum equations, this yields the velocity field $u = 0$. Using this for the computation of the velocities at the CV faces these also are zero. However, this does not represent the correct physical situation, since by the given pressure distribution there is a pressure gradient, i.e., a resulting pressure force, at the CV faces, so that a velocity different from zero should result.

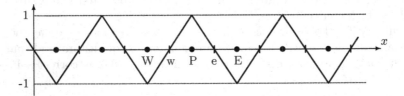

Fig. 10.6. Alternating pressure distribution for one-dimensional problem

It should be noted that the depicted problem is not specific to finite-volume methods, but also occurs in a similar way when finite-difference or finite-element methods are used. In the finite-difference case methods analogous to those decribed below for finite-volume methods can be employed to circumvent the problem. In a finite-element setting a compatibility condition between the ansatz functions for the velocity components and the pressure has to be ensured. This is known as *LBB condition* (after Ladyzhenskaya-Babuska-Brezzi) or *inf-sup condition* (see, e.g., [16]).

We now turn to the question of how to avoid the depicted decoupling of the pressure within the iterative solution procedure. Several approaches have been developed for this since the mid-1960s. The two most relevant techniques are:

- use of a staggered grid,
- selective interpolation of the mass fluxes.

The possibility of using staggered grids, as proposed by Harlow and Welch in 1965, was already mentioned in Sect. 10.2 (see Fig. 10.1). For a Cartesian grid in this case the variable values necessary for the computation of the pressure gradients in the momentum equations and of the mass fluxes through the CV faces in the continuity equation are available exactly at the locations where they are needed. The result is a pressure-correction equation, which again corresponds to a central difference discretization of a diffusion equation, but now with the "normal" grid spacings Δx and Δy so that no oscillations due to a decoupling arise. However, the advantages of the staggered grid largely vanish as soon as the grid is non-Cartesian and no grid oriented velocity components are used. The situation is pointed out in Fig. 10.7. For instance, in the case of a redirection of grid lines by $90°$, the velocity components located at the CV faces contribute nothing to the mass flux. When using grid-oriented velocities the advantages of the staggered grid could largely be maintained, since only the velocity component located at the corresponding CV face contributes to the mass flux (see Fig. 10.7). However, as already mentioned, in this case the momentum equations become much more complex and lose their conservative form. Last but not least, staggered grids for complex geometries also are difficult to manage with respect to the data structures. In particular, if multigrid algorithms (see Sect. 12.2) are used for the solution of the equation systems, it is advantageous if all variables are stored at the same location and only *one* grid is used.

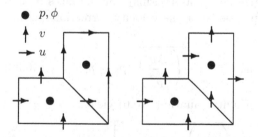

Fig. 10.7. Staggered variable arrangement with Cartesian (left) and grid-oriented (right) velocity components

The possibility of avoiding the decoupling of the pressure also on a non-staggered grid offers the technique known as selective interpolation, which first was proposed by Rhie and Chow (1983). Here, the velocity components required for the computation of the mass fluxes through the CV faces are determined by a special interpolation method that ensures that the velocity components at the CV faces only depend from pressures in the directly neighboring CV centers (e.g., P and E for the face S_e).

The discretized momentum equations (10.13) and (10.14) can serve as a starting point for a selective interpolation. Solving for u_P, for instance, the discrete u-equation (still without pressure interpolation) reads:

$$u_P = \frac{\sum_c a_c^u u_c + b^u}{a_P^u} - \frac{\Delta x \Delta y}{a_P^u}\left(\frac{\partial p}{\partial x}\right)_P . \qquad (10.28)$$

For the determination of u_e all terms on the right hand side of this equation are linearly interpolated except for the pressure gradient, which is approximated by a central difference with the corresponding values in the points P and E:

$$u_e = \overline{\left(\frac{\sum_c a_c^u u_c + b^u}{a_P^u}\right)}_e - \overline{\left(\frac{\Delta y}{a_P^u}\right)}_e (p_E - p_P) . \qquad (10.29)$$

The overbar denotes a linear interpolation from neighboring CV centers. For the considered face S_e, for instance, these are the points P and E, i.e., we have

$$\overline{\left(\frac{\Delta y}{a_P^u}\right)}_e = \gamma_{e1}\left(\frac{\Delta y}{a_P^u}\right)_E + \gamma_{e2}\left(\frac{\Delta y}{a_P^u}\right)_P \qquad (10.30)$$

with $\gamma_{e1} = \gamma_{e2} = 1/2$ for an equidistant grid. For the value v_n at the face S_n one correspondingly obtains:

$$v_n = \overline{\left(\frac{\sum_c a_c^v v_c + b^v}{a_P^v}\right)}_n - \overline{\left(\frac{\Delta x}{a_P^v}\right)}_n (p_N - p_P) , \qquad (10.31)$$

where the interpolation denoted by the overbar now has to be carried out with respect to the points P and N. The equations (10.29) and (10.31) can be interpreted as approximated momentum equations for the corresponding points on the CV faces.

According to the methodology described in Sect. 10.3.1 for the derivation of the SIMPLE method (i.e., neglection of sum terms), for the faces S_e and S_n, for example, the following expressions for the velocity corrections result:

$$u_e' = -\overline{\left(\frac{\Delta y}{a_P^{u,k}}\right)}_e (p_E' - p_P') \quad \text{and} \quad v_n' = -\overline{\left(\frac{\Delta x}{a_P^{v,k}}\right)}_n (p_N' - p_P') . \qquad (10.32)$$

Inserting these values into the continuity equation (10.6) yields:

$$\left[-\overline{\left(\frac{\Delta y}{a_P^{u,k}}\right)}_e (p_E' - p_P') + \overline{\left(\frac{\Delta y}{a_P^{u,k}}\right)}_w (p_P' - p_W')\right] \Delta y$$

$$+ \left[-\overline{\left(\frac{\Delta x}{a_P^{v,k}}\right)}_n (p_N' - p_P') + \overline{\left(\frac{\Delta x}{a_P^{v,k}}\right)}_s (p_P' - p_S')\right] \Delta x = b_m . \qquad (10.33)$$

The algebraic equation for the pressure correction gets the form

$$a_P^{p,k} p_P' = a_E^{p,k} p_E' + a_W^{p,k} p_W' + a_N^{p,k} p_N' + a_S^{p,k} p_S' + b_m \qquad (10.34)$$

with the coefficients

$$a_E^{p,k} = \overline{\left(\frac{\Delta y^2}{a_P^{u,k}}\right)}_e, \quad a_W^{p,k} = \overline{\left(\frac{\Delta y^2}{a_P^{u,k}}\right)}_w, \quad a_N^{p,k} = \overline{\left(\frac{\Delta x^2}{a_P^{v,k}}\right)}_n, \quad a_S^{p,k} = \overline{\left(\frac{\Delta x^2}{a_P^{v,k}}\right)}_s$$

and

$$a_P^{p,k} = a_E^{p,k} + a_W^{p,k} + a_N^{p,k} + a_S^{p,k} .$$

Therefore, one now obtains a discrete Poisson equation with "normal" grid spacings Δx and Δy (see Fig. 10.8). Considering once again the one-dimensional example of the alternating pressure distribution from Sect. 10.3.2 (see Fig. 10.6), one can see that the velocity computed according to (10.32) is no longer zero because now the pressure gradient is computed from the values p_P and p_E. Thus, a force acts on the face S_e, which is physically correct for the given pressure distribution. Thus no oscillatory effects occur.

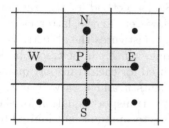

Fig. 10.8. Neighboring relations for pressure corrections with selective interpolation

It should be noted that only the values of the velocities at the CV faces fulfill the continuity equation. For the nodal values, in general, it is not possible to guarantee this (it is also not necessary). The nodal values can be corrected analogous to the relations (10.32). The mass fluxes are also corrected this way so that they are available in the next outer iteration for the computation of the convective fluxes at the CV faces.

For non-Cartesian grids, the process of setting up the pressure-correction equation is basically the same as in the Cartesian case, except that the corresponding expressions become a bit more complicated. So, the interpolation according to (10.30) may require the incorporation of additional nodal values, e.g., the six neighboring values P, E, N, S, NE, and SE (see Fig. 10.9). For further details we refer to [8].

10.3.3 Under-Relaxation

For steady problems (or in the unsteady case for large time step sizes) a pressure-correction method in the described form will not simply converge.

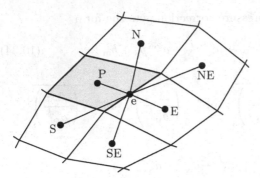

Fig. 10.9. Computation of values in CV face midpoints for non-Cartesian grids for selective interpolation

This is due to the coupling within the equation system and the fact that strong variation of one variable may immoderately influence the others causing the iteration process to diverge (in practice this is often the case). To obtain a converging scheme an *under-relaxation* may help. It can be introduced in different variants into an iterative solution procedure. Generally, the objective of an under-relaxation is to reduce the change of a variable from one iteration to the other. The principle approach is the same as that employed for the derivation of the SOR-method from the Gauß-Seidel method for the solution of linear equation systems (see Sect. 7.1.2). However, in the latter case a *stronger* change in the variable (over-relaxation) constitutes the objective.

We first describe an under-relaxation technique, which can be traced back to Patankar (1980) and which generally can be used for each transport quantity (in our case these are the velocity components u and v as well as the scalar quantity ϕ). As an example, we consider the transport equation for the scalar quantity ϕ. Note that the application of this technique is not limited to finite-volume methods, but can be used in an analogous way for finite-element or finite-difference methods as well.

The starting point is the algebraic equation resulting from the discretization of the continuous problem:

$$a_P^\phi \phi_P = \sum_c a_c^\phi \phi_c + b^\phi. \qquad (10.35)$$

Let an iteration process for the computation of ϕ_P^{k+1} be defined from already known values ϕ_P^k (e.g., the pressure-correction method as described in the previous section or an iterative linear system solver). The "new" value ϕ_P^{k+1} now is not computed directly with the given iteration rule, but by a linear combination with a certain portion of the value from the k-th iteration:

$$\phi_P^{k+1} = \alpha_\phi \frac{\sum_c a_c^\phi \phi_c^{k+1} + b^\phi}{a_P^\phi} + (1 - \alpha_\phi)\phi_P^k \qquad (10.36)$$

with the *under-relaxation parameter* $0 < \alpha_\phi \leq 1$. Equation (10.36) can again be put into the form (10.35), if the coefficients a_P^ϕ and b^ϕ are modified as follows:

$$\underbrace{\frac{a_P^\phi}{\alpha_\phi}}_{\tilde{a}_P^\phi} \phi_P^{n+1} = \sum_c a_c^\phi \phi_c^{n+1} + \underbrace{b^\phi + (1 - \alpha_\phi)\frac{a_P^\phi}{\alpha_\phi}\phi_P^k}_{\tilde{b}^\phi} .$$

There is a close relationship between this under-relaxation technique and methods that solve steady problems via the solution of unsteady equations (pseudo-time stepping, see Sect. 6.1). Discretizing the unsteady equation corresponding to (10.35) with the implicit Euler method, for instance, one gets:

$$\underbrace{\left(a_P^\phi + \frac{\rho_P \delta V}{\Delta t_n}\right)}_{\tilde{a}_P^\phi} \phi_P^{n+1} = \sum_c a_c^\phi \phi_c^{n+1} + \underbrace{\left(b^\phi + \frac{\rho_P \delta V}{\Delta t_n} \phi_P^n\right)}_{\tilde{b}^\phi},$$

where n denotes the time step and Δt_n the time step size. With an under-relaxation as described above, the coefficient a_P^ϕ is enlarged by the division with $\alpha_\phi < 1$, whereas this occurs with the pseudo-time stepping by the addition of the term $\rho_P \delta V / \Delta t_n$. The following relations between Δt_n and α_ϕ can easily be derived:

$$\Delta t_n = \frac{\rho_P \alpha_\phi \delta V}{a_P^\phi (1 - \alpha_\phi)} \quad \text{or} \quad \alpha_\phi = \frac{a_P^\phi \Delta t_n}{a_P^\phi \Delta t_n + \rho_P \delta V} .$$

Thus, a value of α_ϕ constant for all CVs corresponds to a time step size Δt_n varying form CV to CV. Conversely, one time step with Δt_n can be interpreted as an under-relaxation with α_ϕ varying from CV to CV.

Depending on the choice of the approximation of the sum terms in (10.20) and (10.21), it might be necessary to also introduce an under-relaxation for the pressure in order to ensure the convergence of a pressure-correction method. The "new" pressure p^{k+1} is corrected only with a certain portion of the full pressure correction p':

$$p^{k+1} = p^* + \alpha_p p' ,$$

where $0 < \alpha_p \leq 1$. This under-relaxation is necessary if in the derivation of the pressure-correction equation strong simplifications are made, e.g., the simple neglection of the sum terms in the SIMPLE method or the neglection of terms due to grid non-orthogonality.

It should be pointed out that neither of the under-relaxation techniques has an influence on the finally computed solution. In other words, no matter how the parameters are chosen, in the case of convergence one always gets the same solution (only the "approach" to this is different).

It can be shown by a simple analysis (see [8]) that for the SIMPLE method with

$$\alpha_p = 1 - \alpha_u \,, \tag{10.37}$$

where α_u is the under-relaxation parameter for both momentum equations, a "good" convergence rate can be achieved. The question that remains is how the value for α_u should be chosen. In general, this is a difficult issue because the corresponding optimal value strongly depends on the underlying problem. A methodology that would allow the determination of the optimum values for the different under-relaxation parameters in an adaptive and automatic way is not yet available. On the other hand, however (in particular for steady problems) the "right" choice of the under-relaxation parameter is essential for the efficiency of a pressure-correction method. Frequently, only by a well-directed change of these parameters is it possible to get a solution at all. Therefore, under-relaxation plays an essential role for practical application and we will discuss the interactions of the parameters in more detail.

We will study the typical convergence behavior of the SIMPLE method depending on the under-relaxation parameters for a concrete example problem. As problem configuration we consider the flow around a circular cylinder in a channel, which is already known from Sect. 6.4. As inflow condition a steady parabolic velocity profile is prescribed, corresponding to a Reynolds number of Re = 30 (based to the cylinder diameter). The problem in this case has a steady solution with two characteristic vortices behind the cylinder. Figure 10.10 displays the corresponding streamlines for the "interesting" cut-out of the problem domain (the asymmetry of the vortices is due to the slightly asymmetrical problem geometry, cf. Fig. 6.11). The numerical solution of the problem is computed with different relaxation parameters for the velocity components and the pressure for the grid shown in Fig. 10.11 with 1 536 CVs.

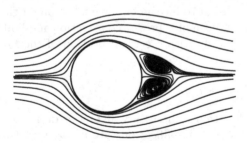

Fig. 10.10. Streamlines for steady flow around circular cylinder

In Fig. 10.12 the number of required pressure-correction iterations depending on the relaxation parameters is shown. The results allow for some characteristic conclusions that also apply to other problems:

• The optimal values for α_u and α_p mutually depend on each other.

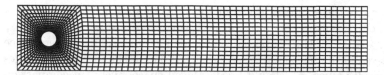

Fig. 10.11. Block-structured grid for computation of flow around circular cylinder

- Slightly exceeding the optimal value for α_u leads to the divergence of the iteration. With an undershooting of this value the iterations still converge, but the rate of convergence decreases relatively strongly.
- The larger α_p, the "narrower" the "opportune" range for α_u.
- If α_u is close to the optimal value, α_p has a relatively strong influence on the rate of convergence. If, in this case, the optimal value for α_p is slightly exceeded the iterations diverge. An undershooting of this value leads to only a slight deterioration in the convergence rate (except for extremely small values).
- If α_u is distinctly smaller than the optimal value, one obtains for arbitrary α_p (except for extremely small values) nearly the same convergence rate (also for $\alpha_p = 1$).

For the SIMPLE method typically "good" values for α_u are in the range of 0.6 to 0.8 and according to (10.37) correspondingly for α_p in the range of 0.2 to 0.4. For a concrete application one should first try to compute a solution

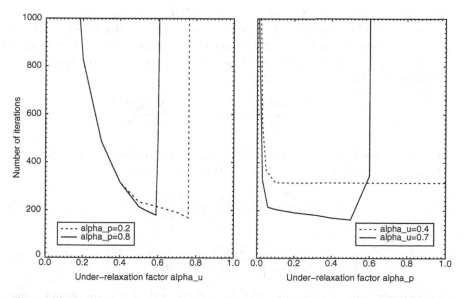

Fig. 10.12. Number of required pressure-correction iterations depending on under-relaxation parameters for velocity and pressure

with values out of these ranges. If the method diverges with these values, it is advisable to check with a computation with very small values for α_u and α_p (e.g., $\alpha_u = \alpha_p = 0.1$) if the convergence problems really are due to the choice of the under-relaxation factors. If this is the case, i.e., if the method converges with these small values, α_u should be decreased successively (in comparison to the initially diverging computation), until a converging method results. Hereby, α_p should be adapted according to (10.37).

Usually, the under-relaxation parameters α_u, α_p, and α_ϕ are chosen in advance and are kept at these values for all iterations. In principle, a dynamic adaptive control during the computation would be possible. However, suitable reliable criteria for such an adaption are hard to find.

10.3.4 Pressure-Correction Variants

Besides the described SIMPLE method there are a variety of further variants of pressure-correction methods for the pressure-velocity coupling. We will briefly discuss two of these methods, which are implemented in many actual programs. Again we concentrate on a non-staggered variable arrangement in connection with the selective interpolation technique described in Sect. 10.3.2.

As outlined in Sect. 10.3.1, the SIMPLE method results by the neglection of the sum terms

$$\sum_c a_c^{u,k} u_c' \quad \text{and} \quad \sum_c a_c^{v,k} v_c'$$

in (10.20) and (10.21). The idea of the *SIMPLEC method* (the "C" stands for "consistent"), proposed by Van Doormal and Raithby (1984), is to approximate these terms by velocity values in neighboring points. It is assumed here that u_P' and v_P' represent mean values of the corresponding values in the neighboring CVs (see Fig. 10.13):

$$u_P' \approx \frac{\sum_c a_c^{u,k} u_c'}{\sum_c a_c^{u,k}} \quad \text{and} \quad v_P' \approx \frac{\sum_c a_c^{v,k} v_c'}{\sum_c a_c^{v,k}}.$$

Inserting these expressions into (10.28) (and the corresponding equation for v), results in the following relations between u', v', and p':

$$u_P' = -\frac{\Delta x \Delta y}{a_P^{u,k} - \sum_c a_c^{u,k}} \left(\frac{\partial p'}{\partial x}\right)_P, \tag{10.38}$$

$$v_P' = -\frac{\Delta x \Delta y}{a_P^{v,k} - \sum_c a_c^{v,k}} \left(\frac{\partial p'}{\partial y}\right)_P. \tag{10.39}$$

Analogously, one obtains by insertion of the velocity components resulting from the selective interpolation into the continuity equation a pressure-correction equation with the following coefficients:

$$a_E^{p,k} = \overline{\left(\frac{\Delta y^2}{a_P^{u,k} - \sum_c a_c^{u,k}} \right)}_e , \quad a_W^{p,k} = \overline{\left(\frac{\Delta y^2}{a_P^{u,k} - \sum_c a_c^{u,k}} \right)}_w ,$$

$$a_N^{p,k} = \overline{\left(\frac{\Delta x^2}{a_P^{v,k} - \sum_c a_c^{v,k}} \right)}_n , \quad a_S^{p,k} = \overline{\left(\frac{\Delta x^2}{a_P^{v,k} - \sum_c a_c^{v,k}} \right)}_s ,$$

$$a_P^{p,k} = a_E^{p,k} + a_W^{p,k} + a_N^{p,k} + a_S^{p,k} .$$

After the pressure-correction equation is solved for p', the velocity corrections can be computed according to (10.38) and (10.39).

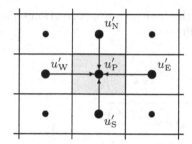

Fig. 10.13. Approximation of velocity corrections in SIMPLEC method

For the SIMPLEC method an under-relaxation of the pressure is not necessary (i.e., $\alpha_p = 1$ can be used). The velocities, however, in general also have to be under-relaxed in this case. If $b^{u,k}$ and $b^{v,k}$ vanish, the SIMPLEC and SIMPLE methods become identical, if in the latter the under-relaxation factor for the pressure is chosen according to

$$\alpha_p = 1 - \alpha_u .$$

For problems, for which $b^{u,k}$ and $b^{v,k}$ have a big influence, the SIMPLEC method usually is more efficient than the SIMPLE method (at comparable computational effort per iteration). For non-orthogonal grids there are also some disadvantages, which, however, we will not discuss further here.

A further variant of the SIMPLE method is the *PISO method* proposed by Issa (1986). With this the first correction step is identical to that for the SIMPLE method, such that the same pressure and velocity corrections p', u', and v' are obtained first. The idea of the PISO method now is to compensate the simplifications of the SIMPLE method when deriving the pressure-correction equation by further correction steps. For this, starting from (10.20) and (10.21), further corrections u'', v'', and p'' are searched. In a way analogous to the first corrections, one gets the relations:

$$u_e'' = \overline{\left(\frac{\sum_c a_c^{u,k} u_c'}{a_P^{u,k}} \right)}_e - \overline{\left(\frac{\Delta y}{a_P^{u,k}} \right)}_e (p_E'' - p_P'') , \qquad (10.40)$$

$$v_n'' = \left(\frac{\overline{\sum_c a_c^{v,k} v_c'}}{a_P^{v,k}}\right)_n - \left(\overline{\frac{\Delta x}{a_P^{v,k}}}\right)_n (p_N'' - p_P'') . \tag{10.41}$$

From the continuity equation one gets:

$$\frac{\dot{m}_e''}{\rho_e} + \frac{\dot{m}_w''}{\rho_w} + \frac{\dot{m}_n''}{\rho_n} + \frac{\dot{m}_s''}{\rho_s} = 0 . \tag{10.42}$$

Inserting the expressions (10.40) and (10.41) for u'' and v'' into (10.42), a second pressure-correction equation of the form

$$a_P^{p,k} p_P'' = \sum_c a_c^{p,k} p_c'' + \tilde{b}_m \tag{10.43}$$

results. The coefficients are the same as in the first pressure-correction equation and only the source term is defined differently:

$$\tilde{b}_m = \left(\overline{\Delta y \frac{\sum_c a_c^{u,k} u_c'}{a_P^{u,k}}}\right)_e - \left(\overline{\Delta y \frac{\sum_c a_c^{u,k} u_c'}{a_P^{u,k}}}\right)_w +$$
$$\left(\overline{\Delta x \frac{\sum_c a_c^{v,k} v_c'}{a_P^{v,k}}}\right)_n - \left(\overline{\Delta x \frac{\sum_c a_c^{v,k} v_c'}{a_P^{v,k}}}\right)_s .$$

The fact that the coefficients of the first and second pressure-correction equations are the same can be exploited when using, for instance, an ILU method (see Sect. 7.1.3) for the solution of the systems, since the decomposition has to be computed only once.

Having solved the system (10.43), the velocity corrections u'' and v'' are computed according to (10.40) and (10.41). Afterwards one obtains u^{k+1}, v^{k+1}, and p^{k+1} by:

$$u^{k+1} = u^{**} + u'' , \quad v^{k+1} = v^{**} + v'' , \quad p^{k+1} = p^{**} + p'' ,$$

where u^{**}, v^{**}, and p^{**} denote the values obtained after the first correction.

Basically, further corrections are possible in a similar way in order to put the approximations of each of the first terms in the right hand sides of (10.40) and (10.41) closer to the "right" values (i.e., same velocities on left and right hand sides). However, rarely more than two corrections are applied because it is not worthwile to fulfill the linearized momentum equations exactly and since the coefficients have to be newly computed anyway (due to the non-linearity and the coupling of u and v). The PISO method also does not need an under-relaxation of the pressure (but for the velocity components).

The assets and drawbacks of the PISO method compared to the SIMPLE or SIMPLEC methods are

- The number of outer iterations usually is lower.

- The effort per iteration is higher, since one (or more) additional pressure-correction equations have to be solved.
- To compute the source term $\tilde{b}^p$ for the second pressure-correction equation, the coefficients from the two momentum equations as well as the values of u' and v' must be available. This increases the memory requirements.

Which method is best for a certain problem strongly depends on the problem. However, the differences are usually insignificant.

10.4 Treatment of Boundary Conditions

The general proceedings for the integration of different boundary conditions into a finite-volume method have already been described in Sect. 4.7. Since these can also be employed for corresponding boundary conditions for flow problems, we will address here only particularities that arise for the flow boundary types (see Sect. 2.5.1). The conditions for the scalar quantity ϕ as well as for the two velocity components in the case of an inflow boundary do not need to be discussed since there no particularities arise.

As an example we consider a general quadrilateral CV whose south face S_s is located at the boundary of the problem domain (see Fig. 10.14). We decompose the velocity vector $\mathbf{v} = (u, v)$ at the boundary into normal and tangential components v_n and v_t, respectively:

$$\mathbf{v_s} = v_n \mathbf{n} + v_t \mathbf{t},$$

where $\mathbf{n} = (n_1, n_2)$ and $\mathbf{t} = (t_1, t_2)$ are the unit vectors normal and tangential to the wall, respectively (see Fig. 10.14). We have $\mathbf{t} = (-n_2, n_1)$, such that v_n and v_t are related to the Cartesian velocity components u and v by

$$v_n = un_1 + vn_2 \quad \text{and} \quad v_t = vn_1 - un_2.$$

Let us consider first the boundary conditions at an impermeable wall. Due to the no-slip conditions the velocity there is equal to the (prescribed) wall velocity: $(u, v) = (u_b, v_b)$. Since there can be no flow through an impermeable

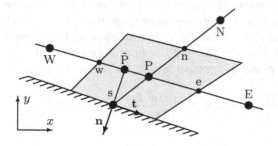

Fig. 10.14. Quadrilateral CV at south boundary with notations

wall, the convective fluxes for all variables are zero there. This can easily be taken into account by setting the convective flux through the corresponding CV face to zero.

The treatment of the diffusive fluxes in the momentum equations deserves special attention. Since the tangential velocity v_t along a wall is constant, its derivative vanishes in the tangential direction:

$$\frac{\partial v_t}{\partial x} t_1 + \frac{\partial v_t}{\partial y} t_2 = 0. \tag{10.44}$$

Writing the continuity equation (10.3) in terms of the tangential and normal components yields:

$$\frac{\partial v_n}{\partial x} n_1 + \frac{\partial v_n}{\partial y} n_2 + \frac{\partial v_t}{\partial x} t_1 + \frac{\partial v_t}{\partial y} t_2 = 0. \tag{10.45}$$

From the relations (10.44) and (10.45) it follows that at the wall, besides $(u, v) = (u_b, v_b)$, the normal derivative of v_n also must vanish:

$$\frac{\partial v_n}{\partial x} n_1 + \frac{\partial v_n}{\partial y} n_2 = 0. \tag{10.46}$$

Physically this means that the normal stress at the wall is zero, and that the exchange of momentum is transmitted only by the shear stress, i.e., the *wall shear stress* (see Fig. 10.15).

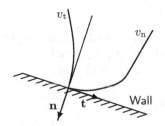

Fig. 10.15. Course of tangential and normal velocity at wall boundary

Condition (10.46) will not be satisfied automatically by a discrete solution and, therefore, (in addition to the Dirichlet wall condition) should be considered directly when approximating the diffusive fluxes in the momentum equations. This can be done by using the correspondingly modified diffusive flux as a basis for the approximation. Otherwise, since v_n does not vanish in P, a value different from zero would result for the approximations of the normal derivative of v_n. The modified diffusive flux for the boundary face S_s in the u-momentum equation with a possible approximation, for instance, reads:

$$-\int\limits_{S_s} \mu \left(\frac{\partial v_t}{\partial x} n_1 + \frac{\partial v_t}{\partial y} n_2 \right) t_1 \, \mathrm{d}S_s \approx \mu_s \frac{v_{t,\tilde{P}} - v_{t,s}}{|\mathbf{x}_{\tilde{P}} - \mathbf{x}_s|} t_1 \delta S_s,$$

where $v_{t,s} = v_b n_1 - u_b n_2$ is determined by the prescribed wall velocity and the point $\tilde{P}$ is defined according to Fig. 10.14. If the non-orthogonality of the grid is not too severe, $v_{t,\tilde{P}}$ simply can be approximated by $v_{t,P}$. Otherwise it is necessary to carry out an interpolation involving further neighboring points (depending on the location of $\tilde{P}$). One obtains a corresponding relation (with t_2 instead of t_1) for the v-momentum equation. For the actual implementation, v_t can be expressed again by the Cartesian velocity components.

At a symmetry boundary one has the conditions

$$\frac{\partial v_t}{\partial x} n_1 + \frac{\partial v_t}{\partial y} n_2 = 0 \quad \text{and} \quad v_n = 0 \,.$$

Since $v_n = 0$, the convective flux through the boundary face is zero in this case also. For the diffusive flux, since in general

$$\frac{\partial v_n}{\partial x} n_1 + \frac{\partial v_n}{\partial y} n_2 \neq 0 \,,$$

compared to a wall one has a reversed situation: the shear stress is zero and the exchange of momentum is transmitted only by the normal stress (see Fig. 10.16). Again this can be considered directly by a modification of the corresponding diffusive flux. For the u-equation and the boundary face S_s, for instance, one has

$$-\int_{S_s} \mu \left(\frac{\partial v_n}{\partial x} n_1 + \frac{\partial v_n}{\partial y} n_2 \right) n_1 \, dS_s \approx \mu_s \frac{v_{n,\tilde{P}}}{|\mathbf{x}_{\tilde{P}} - \mathbf{x}_s|} n_1 \delta S_s \,,$$

where for the approximation of the normal derivative the boundary condition $v_n = 0$ has been used. Also v_n can be expressed again by the Cartesian velocity components. The values of the velocity components in the symmetry boundary points can be determined by suitable extrapolation from inner points.

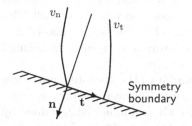

Fig. 10.16. Course of tangential and normal velocities at symmetry boundary

An outflow boundary constitutes a special problem for flow computations. Here, usually no exact conditions are known and have to be prescribed "artificially" in some suitable way. Therefore, in general an outflow boundary should be located sufficiently far away from the part of the flow domain in which the processes relevant for the problem take place. In this way it becomes possible to make certain assumptions about the courses of the variables at the outflow boundary without influencing the solution in the "interesting" parts of the domain. A usual assumption for such a boundary then is that the normal derivatives of both velocity components vanish:

$$\frac{\partial u}{\partial x} n_1 + \frac{\partial u}{\partial y} n_2 = 0 \quad \text{and} \quad \frac{\partial v}{\partial x} n_1 + \frac{\partial v}{\partial y} n_2 = 0.$$

These conditions can be realized, for instance, for the face S_s by setting the coefficients $a_S^{u,k}$ and $a_S^{v,k}$ to zero. The boundary values u_s and v_s can be determined by extrapolation from inner values, where a subsequent correction of these values ensures that the sum of the outflowing mass fluxes equals the inflowing ones.

A special case with respect to the boundary conditions is the pressure-correction equation. If the values of the normal velocity components are pre-scribed, as is the case for all boundary types discussed above (also at an outflow boundary, via the above correction), this component needs no correc-tion at the boundary. This must be taken into account when assembling the pressure-correction equation by setting the corresponding term for the correc-tions in the mass conservation equation to zero. This corresponds to a zero normal derivative for p' at the boundary. As an example, we will explain this briefly for the face S_s for the Cartesian case. Due to $v'_s = 0$ with

$$v'_s = -\left(\frac{\Delta y}{a_P^{u,k}}\right)_s (p'_s - p'_P),$$

we have $p'_s = p'_P$ for the pressure correction at the boundary. The pressure-correction equation for the boundary CV reads:

$$\underbrace{\left(a_E^{p,k} + a_N^{p,k} + a_W^{p,k}\right)}_{a_P^{p,k}} p'_P = a_E^{p,k} p'_E + a_N^{p,k} p'_N + a_W^{p,k} p'_W + b_m,$$

i.e., the coefficient $a_S^{p,k}$ vanishes. In order for a solution for the p'-equation system theoretically to exist, the sum of the mass sources b_m over all CVs must be zero. This condition is fulfilled if the sum of the outflowing mass fluxes equals the inflowing mass fluxes. This has to be ensured by the aforementioned correction of the velocity components at the outflow.

The use of non-staggered grids for the assembly of the discrete momentum equations requires pressure values at the boundary. However, the pressure can-not be prescribed at the boundary since it already is uniquely determined by the differential equation and the velocity boundary conditions. The required pressure values can simply be extrapolated (e.g., linearly) from inner values. There is also the possibility of prescribing the pressure instead of the veloci-ties. For the corresponding modifications in the pressure-correction equation we refer to [8].

10.5 Example of Application

For illustration we will retrace the course of a pressure-correction procedure for a simple example, for which the individual steps can be carried out "manually". For this we consider a two-dimensional channel flow as sketched in Fig. 10.17. The problem is described by the system (10.1)-(10.3) (without the time derivative term) with the boundary conditions

$$u = 0, \quad v = 0 \quad \text{for } y = 0 \text{ and } y = H,$$

$$\frac{\partial u}{\partial x} = 0, \quad \frac{\partial v}{\partial x} = 0 \quad \text{for } x = L,$$

$$u(y) = \frac{4u_{\max}}{H^2}\left(Hy - y^2\right), \quad v = 0 \quad \text{for } x = 0.$$

Let the problem data be given as follows:

$$\rho = 142 \,\text{kg/m}^3, \quad \mu = 2 \,\text{kg/ms}, \quad u_{\max} = 3 \,\text{m/s}, \quad L = 4 \,\text{m}, \quad H = 1 \,\text{m}.$$

The problem possesses the analytical solution

$$u = \frac{4u_{\max}}{H^2}\left(Hy - y^2\right), \quad v = 0, \quad p = -\frac{8\mu u_{\max}}{H^2}x + C$$

with an arbitrary constant C.

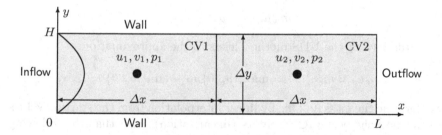

Fig. 10.17. Two-dimensional channel flow with discretization by two CVs

To clarify the principles of the pressure-correction procedure we can restrict ourselves to the following two equations (mass and u-momentum conservation, setting all terms with v to zero):

$$\frac{\partial u}{\partial x} = 0 \quad \text{and} \quad \rho\frac{\partial(uu)}{\partial x} - \mu\left(\frac{\partial^2 u}{\partial x^2} + \frac{\partial^2 u}{\partial y^2}\right) = -\frac{\partial p}{\partial x}.$$

For the discretization of the problem domain we use two CVs as shown in Fig. 10.17 with $\Delta x = L/2$ and $\Delta y = H$. As pressure-correction method we use the SIMPLE method with selective interpolation. Note that the latter for the considered example would not be necessary since both CVs are located directly at the boundary. However, we will use it to exemplify the corresponding interpolations. As starting values for the iteration process we choose

$$p_1^0 = p_2^0 = 0 \quad \text{and} \quad u_1^0 = u_2^0 = 1 \,.$$

By integration and application of the Gauß integral theorem the u-momentum equation yields the relation

$$\rho \sum_c \int_S \rho u u n_1 \, \mathrm{d}S - \mu \sum_c \int_S \left(\frac{\partial u}{\partial x} n_1 + \frac{\partial u}{\partial y} n_2 \right) \mathrm{d}S = - \sum_c \int_S p n_1 \mathrm{d}S \,.$$

With the midpoint rule the following general approximation results:

$$\dot{m}_e u_e + \dot{m}_w u_w - \mu \Delta y \left[\left(\frac{\partial u}{\partial x} \right)_e - \left(\frac{\partial u}{\partial x} \right)_w \right]$$

$$-\mu \Delta x \left[\left(\frac{\partial u}{\partial y} \right)_n - \left(\frac{\partial u}{\partial y} \right)_s \right] = -(p_e - p_w) \Delta y \,. \tag{10.47}$$

The convective fluxes through the faces S_n and S_s vanish due to the boundary condition $v = 0$ at the channel walls.

Let us consider first the CV1. For the convective flux through the face S_w we obtain from the inflow condition

$$\dot{m}_w u_w = -\rho u_{\max}^2 \Delta y$$

and for the face S_e the UDS method leads to the approximation:

$$\dot{m}_e u_e \approx \dot{m}_e u_e^{\mathrm{UDS}} = \max\{\dot{m}_e, 0\} u_P + \min\{\dot{m}_e, 0\} u_E \,.$$

We determine the mass flux $\dot{m}_e$ by linear interpolation from the starting values for u in the neighboring CV centers (linearization), such that the following approximation results (with $\dot{m}_e^0 > 0$):

$$\dot{m}_e u_e \approx \dot{m}_e^0 u_e \approx \frac{\rho \Delta y}{2} \left(u_1^0 + u_2^0 \right) u_P \,.$$

The discretization of the diffusive fluxes and the pressure by central differences results in:

$$\left(\frac{\partial u}{\partial x} \right)_e \approx \frac{u_E - u_P}{\Delta x} \,, \quad \left(\frac{\partial u}{\partial x} \right)_w \approx \frac{u_P - u_w}{\Delta x/2} \,,$$

$$\left(\frac{\partial u}{\partial y} \right)_n \approx \frac{u_n - u_P}{\Delta y/2} \,, \quad \left(\frac{\partial u}{\partial y} \right)_s \approx \frac{u_P - u_s}{\Delta y/2}$$

and

$$(p_e - p_w)\Delta y \approx \left[\frac{1}{2}(p_P + p_E) - \frac{1}{2}(3p_P - p_E)\right]\Delta y = (p_E - p_P)\Delta y \,,$$

where for the determination of the boundary pressure p_w a linear extrapolation from the values p_P and p_E was employed. Inserting the above approximations in (10.47) results with $u_P = u_1$, $u_E = u_2$, $p_P = u_1$, and $p_E = u_2$ in the equation:

$$\left[3\mu\frac{\Delta y}{\Delta x} + 4\mu\frac{\Delta x}{\Delta y} + \frac{\rho\Delta y}{2}\left(u_1^0 + u_2^0\right)\right]u_1 - \mu\frac{\Delta y}{\Delta x}u_2 = \\ \left(\rho u_{\max}\Delta y + 2\mu\frac{\Delta y}{\Delta x}\right)u_{\max} - (p_2 - p_1)\Delta y \,. \tag{10.48}$$

Writing (10.48) as usual in the form (the index 1 refers to the CV, the index u is omitted)

$$a_P^1 u_1 - a_E^1 u_2 = b^1 - (p_2 - p_1)\Delta y \tag{10.49}$$

and by inserting the corresponding numbers we get the following values for the coefficients:

$$a_P^1 = 3\mu\frac{\Delta y}{\Delta x} + 4\mu\frac{\Delta x}{\Delta y} + \frac{\rho\Delta y}{2}\left(u_1^0 + u_2^0\right) = 161 \,,$$

$$a_E^1 = \mu\frac{\Delta y}{\Delta x} = 1 \,,$$

$$b^1 = \left(\rho u_{\max}\Delta y + 2\mu\frac{\Delta y}{\Delta x}\right)u_{\max} = 1284 \,.$$

For the CV2 from the discretized u-momentum equation we obtain in a similar way first the expression

$$\left(\mu\frac{\Delta y}{\Delta x} + 4\mu\frac{\Delta x}{\Delta y} + \max\{\dot{m}_w, 0\} + \dot{m}_e\right)u_2 \\ - \left(\mu\frac{\Delta y}{\Delta x} - \min\{\dot{m}_w, 0\}\right)u_1 = -(p_2 - p_1)\Delta y \,.$$

At the face S_e the outflow boundary condition $\partial u/\partial x = 0$ was used as follows: from a backward difference approximation at x_e follows $u_e = u_P = u_2$, such that the convective flux through S_e becomes $\dot{m}_e u_2$, and the diffusive flux directly from the boundary condition becomes zero. Since the mass flux $\dot{m}_w$ of the CV2 must be equal to the negative of the mass flow through the face S_e of the CV1, we employ the approximation

$$\dot{m}_w \approx \dot{m}_w^0 = -\frac{\rho\Delta y}{2}\left(u_1^0 + u_2^0\right) \,,$$

and $\dot{m}_e$ is approximated from the staring value u_2^0 according to

$$\dot{m}_e^0 \approx \rho \Delta y u_2^0 \,.$$

Of course, for the special situation of the example problem it could be exploited that $\dot{m}_e$ is the outflow mass flux, which should equal the inflow mass flux $\rho u_{\max}^2 \Delta y$. However, because this is not possible in general, we will also not do it here. Written in the form

$$-a_W^2 u_1 + a_P^2 u_2 = b^2 - (p_2 - p_1)\Delta y \qquad (10.50)$$

the coefficients of the equation for CV2 become:

$$a_P^2 = \mu \frac{\Delta y}{\Delta x} + 4\mu \frac{\Delta x}{\Delta y} + \rho \Delta y u_2^0 = 159 \,,$$

$$a_W^2 = \mu \frac{\Delta y}{\Delta x} + \frac{\rho \Delta y}{2} \left(u_1^0 + u_2^0\right) = 143 \,,$$

$$b^2 = 0 \,.$$

For the first step of the SIMPLE method, i.e., for the determination of the preliminary velocities u_1^* and u_2^*, the two equations (10.49) and (10.50) are available. Introducing an under-relaxation (with under-relaxation factor α_u) as described in Sect. 10.3.3 yields the modified system

$$\frac{a_P^1}{\alpha_u} u_1^* - a_E^1 u_2^* = b^1 - (p_2^* - p_1^*)\Delta y + \frac{a_P^1(1 - \alpha_u)}{\alpha_u} u_1^0 \,, \qquad (10.51)$$

$$-a_W^2 u_1^* + \frac{a_P^2}{\alpha_u} u_2^* = b^2 - (p_2^* - p_1^*)\Delta y + \frac{a_P^2(1 - \alpha_u)}{\alpha_u} u_2^0 \,. \qquad (10.52)$$

Taking $\alpha_u = 1/2$ and using the starting values with $p_1^* = p_1^0$ and $p_2^* = p_2^0$ gives the following two equations for the determination of u_1^* and u_2^*:

$$322u_1^* - u_2^* = 1445 \quad \text{and} \quad -143u_1^* + 318u_2^* = 159 \,.$$

The resolution of this systems yields:

$$u_1^* \approx 4.4954 \quad \text{and} \quad u_2^* \approx 2.5215 \,.$$

Next the velocity at the outflow boundary u_{out} (i.e., u_e for the CV2) is determined. This is needed for setting up the pressure-correction equation. From the discretization of the outflow boundary condition with a backward difference, which has been used in the discrete momentum equation for the CV2, it follows that $u_{\text{out}} = u_2^*$. One can see that due to $u_{\max} \neq u_{\text{out}}$ with this value the global conservativity of the method is not ensured:

$$\dot{m}_{\mathrm{in}} = \rho u_{\mathrm{max}} \Delta y \neq \rho u_{\mathrm{out}} \Delta y = \dot{m}_{\mathrm{out}} \,.$$

If one uses this value for u_{out}, the pressure-correction equation, which we set up afterwards, would not be solvable. Thus, according to the requirement of global conservativity we set

$$u_{\mathrm{out}} = u_{\mathrm{max}} = 3 \,,$$

such that the condition $\dot{m}_{\mathrm{in}} = \dot{m}_{\mathrm{out}}$ is fulfilled.

In the next step the pressure and velocity corrections are determined by taking into account the mass conservation equation. The general approximation of the latter with the midpoint rule yields:

$$\int_V \frac{\partial u}{\partial x} \, dV = \sum_c \int_{S_c} u n_1 \, dS_c \approx (u_{\mathrm{e}} - u_{\mathrm{w}}) \Delta y = 0 \,. \tag{10.53}$$

With the selective interpolation introduced in Sect. 10.3.2 we obtain after division by Δy for the velocity at the common face of the two CVs:

$$u_{\mathrm{e},1} = u_{\mathrm{w},2} = \frac{1}{2} \left(\frac{a_{\mathrm{E}}^1 u_2 + b^1}{a_{\mathrm{P}}^1} + \frac{a_{\mathrm{W}}^2 u_1 + b^2}{a_{\mathrm{P}}^2} \right) - \frac{1}{2} \left(\frac{\Delta y}{a_{\mathrm{P}}^1} + \frac{\Delta y}{a_{\mathrm{P}}^2} \right) (p_2 - p_1) \,.$$

With this, using $u_{\mathrm{w},1} = u_{\mathrm{max}}$ and $u_{\mathrm{e},2} = u_{\mathrm{out}}$, the following discrete continuity equations result for the two CVs:

$$\frac{1}{2} \left(\frac{a_{\mathrm{E}}^1 u_2 + b^1}{a_{\mathrm{P}}^1} + \frac{a_{\mathrm{W}}^2 u_1 + b^2}{a_{\mathrm{P}}^2} \right) - \frac{1}{2} \left(\frac{\Delta y}{a_{\mathrm{P}}^1} + \frac{\Delta y}{a_{\mathrm{P}}^2} \right) (p_2 - p_1) - u_{\mathrm{max}} = 0 \,,$$

$$u_{\mathrm{out}} - \frac{1}{2} \left(\frac{a_{\mathrm{E}}^1 u_2 + b^1}{a_{\mathrm{P}}^1} + \frac{a_{\mathrm{W}}^2 u_1 + b^2}{a_{\mathrm{P}}^2} \right) + \frac{1}{2} \left(\frac{\Delta y}{a_{\mathrm{P}}^1} + \frac{\Delta y}{a_{\mathrm{P}}^2} \right) (p_2 - p_1) = 0 \,.$$

By inserting the preliminary velocity and pressure values the mass sources for the two CVs become:

$$b_{\mathrm{m}}^1 = u_{\mathrm{max}} - \frac{1}{2} \left(\frac{a_{\mathrm{E}}^1 u_2^* + b^1}{a_{\mathrm{P}}^1} + \frac{a_{\mathrm{W}}^2 u_1^* + b^2}{a_{\mathrm{P}}^2} \right) - \frac{1}{2} \left(\frac{\Delta y}{a_{\mathrm{P}}^1} + \frac{\Delta y}{a_{\mathrm{P}}^2} \right) (p_2^* - p_1^*) \,,$$

$$b_{\mathrm{m}}^2 = \frac{1}{2} \left(\frac{a_{\mathrm{E}}^1 u_2^* + b^1}{a_{\mathrm{P}}^1} + \frac{a_{\mathrm{W}}^2 u_1^* + b^2}{a_{\mathrm{P}}^2} \right) + \frac{1}{2} \left(\frac{\Delta y}{a_{\mathrm{P}}^1} + \frac{\Delta y}{a_{\mathrm{P}}^2} \right) (p_2^* - p_1^*) - u_{\mathrm{out}} \,.$$

Inserting the numbers gives:

$$b_m^1 = -b_m^2 \approx -3.0169 \,.$$

Subtraction from the corresponding "exact" equations (see Sect. 10.3.1, Eqs. (10.20)-(10.22)) and involving the characteristic approximations for the SIMPLE method

$$a_E^1 u_2' \approx 0 \quad \text{and} \quad a_W^2 u_1' \approx 0$$

gives for the corrections $u' = u^1 - u^*$ and $p' = p^1 - p^*$ (the index 1 denotes the value to be computed from the first SIMPLE iteration) at the faces of the two CVs:

$$u_{e,1}' = u_{w,2}' = \frac{1}{2}\left(\frac{\Delta y}{a_P^1} + \frac{\Delta y}{a_P^2}\right)(p_1' - p_2') \quad \text{and} \quad u_{w,1}' = u_{e,2}' = 0. \tag{10.54}$$

Inserting these values into the continuity equation for the corrections, which in general form is given by

$$u_e' - u_w' = b_m, \tag{10.55}$$

leads to the following equation system for the pressure corrections:

$$\frac{1}{2}\left(\frac{\Delta y}{a_P^2} + \frac{\Delta y}{a_P^1}\right)p_1' - \frac{1}{2}\left(\frac{\Delta y}{a_P^2} + \frac{\Delta y}{a_P^1}\right)p_2' = b_m^1, \tag{10.56}$$

$$-\frac{1}{2}\left(\frac{\Delta y}{a_P^1} + \frac{\Delta y}{a_P^2}\right)p_1' + \frac{1}{2}\left(\frac{\Delta y}{a_P^1} + \frac{\Delta y}{a_P^2}\right)p_2' = b_m^2. \tag{10.57}$$

It is obvious that these two equations are linearly dependent, which actually must be the case since the pressure is uniquely determined only up to an additive constant. In order for the system to be solvable, the condition

$$b_m^1 + b_m^2 = 0$$

must be fulfilled, which in our example is the case (if we had not adapted the outflow velocity, this condition would not be fulfilled!). The pressure can thus be arbitrarily prescribed in one CV and the value in the other can be computed relative to this. We set $p_1' = 0$ and obtain from (10.56) with the concrete numerical values:

$$p_2' = \frac{-2\, b_m^1 a_P^1 a_P^2}{\Delta y \alpha_u \left(a_P^1 + a_P^2\right)} + p_1' \approx 482.69.$$

With this, the velocity corrections at the CV faces from (10.54) become

$$u_{e,1}' = u_{w,2}' \approx 3.0169.$$

Of course, both corrections must be equal, since the face S_e of CV1 is identical to the face S_w of CV2.

For the correction of the velocity values in the CV centers relations that are analogous to (10.54) are used:

$$u_1^1 = u_1^* + \frac{\Delta y}{a_P^1}(p_2' - p_1') \approx 7.4935\,,$$

$$u_2^1 = u_2^* + \frac{\Delta y}{a_P^2}(p_2' - p_1') \approx 5.5573\,.$$

Choosing the relaxation factor $\alpha_p = 1/2$ for the pressure under-relaxation, we obtain for the corrected pressure:

$$p_2^1 = p_2^* + \alpha_p p_2' \approx 241.345\,.$$

This completes the first SIMPLE iteration. The second iteration starts with the solution of the system

$$\frac{a_P^1}{\alpha_u}u_1^* - a_E^1 u_2^* = b^1 - (p_2^1 - p_1^1)\Delta y + \frac{a_P^1(1 - \alpha_u)}{\alpha_u}u_1^1,$$

$$-a_W^2 u_1^* + \frac{a_P^2}{\alpha_u}u_2^* = b^2 - (p_2^1 - p_1^1)\Delta y + \frac{a_P^2(1 - \alpha_u)}{\alpha_u}u_2^1$$

with respect to u_1^* and u_2^*. What then proceeds is completely analogous to the first iteration. In Fig. 10.18 the development of the absolute relative errors with respect to the exact values $u_1 = u_2 = 3$ and $p_2 = -48$ are given in the course of further SIMPLE iterations.

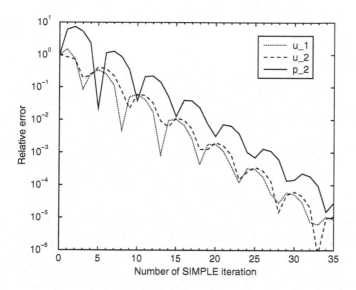

Fig. 10.18. Convergence behavior for velocity and pressure for computation of channel flow example with SIMPLE method

Exercises for Chap. 10

Exercise 10.1. Consider the example from Sect. 10.5 with the given finite-volume discretization. (i) Carry out one iteration with the SIMPLEC method and with the PISO method (with two pressure corrections). (ii) Formulate the discrete equations as a coupled equation system (without pressure-correction method). Linearize the system with the Newton method and with successive iteration and carry out one iteration in each case. (iii) Compare the corresponding results.

Exercise 10.2. The flow of a fluid with constant density ρ in a nozzle with length L with cross section $A = A(x)$ under certain assumptions can be described by the one-dimensional equations (mass and momentum balance)

$$(Au)' = 0 \quad \text{and} \quad \rho(Au^2)' + Ap' = 0$$

for $0 \leq x \leq L$. At the inflow $x = 0$ the velocity u_0 is prescribed. Formulate the SIMPLE method with staggered grid and with non-staggered grid with and without selective interpolation. Use in each case a second-order finite-volume method with three equidistant CVs.

Computation of Turbulent Flows

Flow processes in practical applications are in most cases turbulent. Although the Navier-Stokes equations introduced in Sect. 2.5 are valid for turbulent flows as well – as we will see in the following section – due to the enormous computational effort that would be related to this, it usually is not possible to compute the flows directly on the basis of these equations. Therefore, it is necessary to introduce special modeling techniques to achieve numerical results for turbulent flows. In this section we will consider this subject in an introductory way. In particular, we will address statistical turbulence models, the usage of which mostly constitutes the only way to compute practically relevant turbulent flows with "reasonable" computational effort. Again we restrict ourselves to the incompressible case.

11.1 Characterization of Computational Methods

A distinguishing feature of turbulent flows is chaotic fluid motion, which is characterized by irregular, highly frequent spatial and temporal fluctuations of the flow quantities. Therefore, turbulent flows are basically always unsteady and three-dimensional. As an example, Fig. 11.1 shows the transition of a laminar into a turbulent flow.

In order to be able to fully resolve the turbulent structures numerically, very fine discretizations in space and time are required:

- the spatial step size has to be smaller than the smallest turbulent eddies,
- the time step size has to be smaller than the shortest turbulent fluctuations.

Following the pioneering work of Kolmogorov (1942) by dimensional analysis for the smallest spatial and temporal scales the expressions $l_k = (\nu^3/\epsilon)^{1/4}$ and $t_k = (\nu/\epsilon)^{1/2}$ can be derived, respectively, with the *dynamic viscosity* ν and the dissipation rate of turbulent kinetic energy ϵ. The latter is related to large scale quantities by $\epsilon \sim \bar{v}/L$, where $\bar{v}$ and L are characteristic values of the

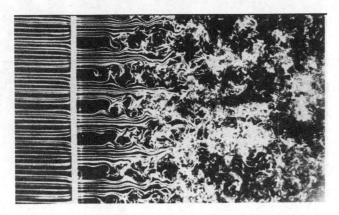

Fig. 11.1. Example of a turbulent flow (from [27])

velocity and the length of the underlying problem. Involving the Reynolds number $Re = \bar{v}L/\nu$, it follows that $l_k \sim Re^{3/4}$ and $t_k \sim Re^{1/2}$. Thus, the larger the Reynolds number, the smaller the occuring scales and, therefore, the finer the spatial and temporal resolution has to be. Since the relations are strongly over proportional, the numerical effort tremendously increases with the Reynolds number. In Table 11.1 the resulting asymptotic dependence of the memory requirements and the computing effort from the Reynolds number are given for the cases of free and near-wall turbulence.

Table 11.1. Asymptotic memory requirement and computing effort for computing turbulent flows

	Free turbulence	Near-wall turbulence
Memory	$\sim Re^{2.25}$	$\sim Re^{2.625}$
Computing time	$\sim Re^{3}$	$\sim Re^{3.5}$

We will explain the above issue by means of a simple example and quantify the corresponding computational effort. For this, let us consider a turbulent channel flow with Reynolds number $Re = \rho \bar{v} H/\mu = 10^6$ (see Fig. 11.2). Subsequently, characteristic quantities related to the numerical solution of the model equations given in Sect. 2.5 are given:

- the size of the smallest eddies is around 0.2 mm,
- the resolution of the eddies requires 10^{14} grid points,
- to get meaningful mean values around 10^4 time steps are necessary,
- solving the equations requires about 500 Flop per grid point and time step,
- the total number of computing operations is about $5 \cdot 10^{20}$ Flop,

- on a high-performance computer with 10^{10} Flops the total computing time for the simulation would be around 1600 years.

One can observe that even for this rather simple example the possibility of such a direct computation, also denoted as *direct numerical simulation (DNS)*, is out of reach.

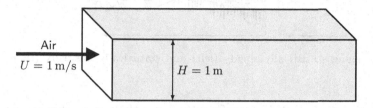

Fig. 11.2. Turbulent channel flow with Re $= 10^6$

A DNS nowadays can only be carried out for (geometrically) simpler turbulent flows with Reynolds numbers up to around Re $= 20\,000$ spending months of computing time on the fastest supercomputers. For a practical application it is necessary to employ alternative approaches for computing turbulent flows. For this, one can generally distinguish between two approaches:

- the *large eddy simulation (LES)*,
- the simulation with *statistical turbulence models*.

We will address the basic ideas of both approaches in the following, where we start with the latter as being the most relevant for engineering practice today.

11.2 Statistical Turbulence Modeling

When using a statistical turbulence model, (temporally) averaged flow equations are solved with respect to mean values of the flow quantities. *All* turbulence effects are taken into account by a suitable modeling. The starting point is an averaging process, where each flow variable, which we generally denote by ϕ, is expressed by a mean value $\overline{\phi}$ and a fluctuation ϕ'. The mean value can either be *statistically steady* or *statistically unsteady* (see Fig. 11.3).

In the statistically steady case one has

$$\phi(\mathbf{x}, t) = \overline{\phi}(\mathbf{x}) + \phi'(\mathbf{x}, t), \tag{11.1}$$

where the mean value can be defined by

$$\overline{\phi}(\mathbf{x}) = \lim_{T \to \infty} \frac{1}{T} \int_{t_0}^{t_0 + T} \phi(\mathbf{x}, t)\, dt$$

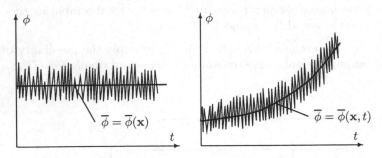

Fig. 11.3. Averaging statistically steady (left) and statistically unsteady (right) flows

with the averaging time T. If T is sufficiently large, the mean value $\overline{\phi}$ does not depend on the point of time t_0 at which the averaging is started. For statistically unsteady processes also the mean value is time-dependent, i.e., in (11.1) $\overline{\phi}(\mathbf{x})$ has to be replaced by $\overline{\phi}(\mathbf{x}, t)$ and the mean value must be defined by *ensemble averaging*:

$$\overline{\phi}(\mathbf{x}, t) = \lim_{N \to \infty} \frac{1}{N} \sum_{n=1}^{N} \phi(\mathbf{x}, t) \, .$$

N can be interpreted as the number of imaginary experiments (each under the same conditions), which are necessary to obtain mean values that are independent of the fluctuations, but time-dependent.

Inserting the expressions (11.1) for all variables into the corresponding balance equations for mass and momentum (2.70) and (2.71) and then averaging the equations results in the following averaged equations, which are denoted as *Reynolds averaged Navier-Stokes (RANS)* equations (or *Reynolds equations*):

$$\frac{\partial \overline{v}_i}{\partial x_i} = 0 \, , \quad (11.2)$$

$$\frac{\partial (\rho \overline{v}_i)}{\partial t} + \frac{\partial}{\partial x_j} \left[\rho \overline{v}_i \, \overline{v}_j + \rho \overline{v'_i v'_j} - \mu \left(\frac{\partial \overline{v}_i}{\partial x_j} + \frac{\partial \overline{v}_j}{\partial x_i} \right) \right] + \frac{\partial \overline{p}}{\partial x_i} = \rho f_i \, . \quad (11.3)$$

The energy equation (2.72) or other scalar equations can be handled in a fully analogous way.

On the one hand the averaging simplifies the equations, i.e., the mean values are either time-independent or the time dependence can be resolved with a "justifiable" number of time steps, but on the other hand some new unknowns arise, i.e., the averaged products of the fluctuations $\overline{v'_i v'_j}$, which represent a measure of the statistical dependence (correlation) of the corresponding quantities. If these terms vanish, the quantities would be statistically independent. However, this is normally not the case. The terms $\rho \overline{v'_i v'_j}$ are denoted as *Reynolds stresses*. In order to be able to solve the equation system (11.2) and (11.3) with respect to the mean values (in practice this is usually the only information required), the system has to be closed by employing suitable approximations for the correlations. This task is known as *turbulence modeling* and the corresponding models are denoted as *RANS models*. Numerous such models exist, the most important of which are:

- algebraic models (zero-equation models),
- one- and two-equation models,
- Reynolds stress models.

Algebraic models model the Reynolds stresses just by algebraic expressions. With one- and two-equation models one or two additional differential equations for suitable turbulence quantities (e.g., turbulent kinetic energy, dissipation rate, ...) are formulated. In the case of Reynolds stress models transport equations are formulated directly for the Reynolds stresses, which in the three-dimensional case leads to 7 additional differential equations: 6 for the components of the (symmetric) Reynolds stress tensor and one, for instance, for the dissipation rate.

In order to exemplify the special features which arise with respect to numerical issues when using a statistical turbulence model, we consider as an example the very popular k-ε model. This model belongs to the class of two-equation models, which frequently represent a reasonable compromise between the physical modeling quality and the numerical effort to solve the corresponding equation systems.

11.2.1 The k-ε Turbulence Model

The basis of the k-ε model, which was developed at the end of the 1960s by Spalding and Launder, is the assumption of the validity of the following relation for the Reynolds stresses:

$$\rho \overline{v_i' v_j'} = -\mu_t \left(\frac{\partial \overline{v}_i}{\partial x_j} + \frac{\partial \overline{v}_j}{\partial x_i} \right) + \frac{2}{3} \rho \, \delta_{ij} k \,, \qquad (11.4)$$

which is also known as *Boussinesq approximation*. μ_t denotes the *turbulent viscosity* (or also *eddy viscosity*), which, contrary to the dynamic viscosity μ, is not a material parameter but a variable depending on the flow variables. The assumption of the existence of such a quantity is known as *eddy viscosity hypotheses*. k is the *turbulent kinetic energy*, which is defined by

$$k = \frac{1}{2} \overline{v_i' v_i'} \,.$$

The relation (11.4), which relates the fluctuations to the mean values, has a strong similarity with the constitutive law for the Cauchy stress tensor in the case of a Newtonian fluid (see (2.64)).

With the relation (11.4) the problem of closing the system (11.2) and (11.3) is not yet solved, since μ_t and k also are unknowns. A further assumption of the k-ε model relates μ_t to k and a further physically interpretable quantity, i.e., the *dissipation rate* of the turbulent kinetic energy ε:

$$\mu_t = C_\mu \rho \frac{k^2}{\varepsilon} . \tag{11.5}$$

C_μ is an empirical constant and the dissipation rate ε is defined by:

$$\varepsilon = \frac{\mu}{\rho} \overline{\frac{\partial v_i'}{\partial x_j} \frac{\partial v_i'}{\partial x_j}} .$$

The relation (11.5) is based on the assumption that the rates of production and dissipation of turbulence are in equilibrium. In this case one has the relation

$$\varepsilon \approx \frac{k^{3/2}}{l} \tag{11.6}$$

with the *turbulent length scale* l, which represents a measure of the size of the eddies in the turbulent flow. Together with the relation

$$\mu_t = C_\mu \rho l \sqrt{2k} ,$$

which results from similarity considerations, this yields (11.5).

The task that remains is to set up suitable equations from which k and ε can be computed. By further model assumptions, which will not be given in further detail here (see, e.g., [17]), for both quantities transport equations can be derived, which possess the same form as a general scalar transport equation (only the diffusion coefficients and source terms are specific):

$$\frac{\partial(\rho k)}{\partial t} + \frac{\partial}{\partial x_j} \left[\rho \overline{v}_j k - \left(\mu + \frac{\mu_t}{\sigma_k} \right) \frac{\partial k}{\partial x_j} \right] = G - \rho \varepsilon , \tag{11.7}$$

$$\frac{\partial(\rho \varepsilon)}{\partial t} + \frac{\partial}{\partial x_j} \left[\rho \overline{v}_j \varepsilon - \left(\mu + \frac{\mu_t}{\sigma_\varepsilon} \right) \frac{\partial \varepsilon}{\partial x_j} \right] = C_{\varepsilon 1} G \frac{\varepsilon}{k} - C_{\varepsilon 2} \rho \frac{\varepsilon^2}{k} . \tag{11.8}$$

Here, σ_k, σ_ε, $C_{\varepsilon 1}$, and $C_{\varepsilon 2}$ are further empirical constants. The standard values for the constants involved in the model are:

$$C_\mu = 0,09 , \quad \sigma_k = 1,0 , \quad \sigma_\varepsilon = 1,33 , \quad C_{\varepsilon 1} = 1,44 , \quad C_{\varepsilon 2} = 1,92 .$$

G denotes the *production rate* of turbulent kinetic energy defined by

$$G = \mu_t \left(\frac{\partial \overline{v}_i}{\partial x_j} + \frac{\partial \overline{v}_j}{\partial x_i} \right) \frac{\partial \overline{v}_i}{\partial x_j} .$$

Inserting the relation (11.4) into the momentum equation (11.3) and defining $\tilde{p} = \bar{p} + 2k/3$, the following system of partial differential equations results, which has to be solved for the unknowns $\tilde{p}$, $\bar{v}_i$, k, and ε:

$$\frac{\partial \bar{v}_i}{\partial x_i} = 0, \tag{11.9}$$

$$\frac{\partial(\rho \bar{v}_i)}{\partial t} + \frac{\partial}{\partial x_i}\left[\rho \bar{v}_i \bar{v}_j - (\mu + \mu_t)\left(\frac{\partial \bar{v}_i}{\partial x_j} + \frac{\partial \bar{v}_j}{\partial x_i}\right)\right] = -\frac{\partial \tilde{p}}{\partial x_i} + \rho f_i, \tag{11.10}$$

$$\frac{\partial(\rho k)}{\partial t} + \frac{\partial}{\partial x_j}\left[\rho \bar{u}_j k - \left(\mu + \frac{\mu_t}{\sigma_k}\right)\frac{\partial k}{\partial x_j}\right] = G - \rho\varepsilon, \tag{11.11}$$

$$\frac{\partial(\rho\varepsilon)}{\partial t} + \frac{\partial}{\partial x_j}\left[\rho \bar{u}_j \varepsilon - \left(\mu + \frac{\mu_t}{\sigma_\varepsilon}\right)\frac{\partial\varepsilon}{\partial x_j}\right] = C_{\epsilon1}G\frac{\varepsilon}{k} - C_{\epsilon2}\rho\frac{\varepsilon^2}{k} \tag{11.12}$$

with μ_t according to (11.5). By employing $\tilde{p}$ instead of $\bar{p}$ in the momentum equation (11.10) the derivative of k does not appear explicitly ($\bar{p}$ can be computed afterwards from $\tilde{p}$ and k).

11.2.2 Boundary Conditions

An important issue when using statistical turbulence models for the computation of turbulent flows is the prescription of "reasonable" boundary conditions. We will discuss this topic for the k-ε model, but note that analogous considerations have to be taken into account for other models as well.

At an inflow boundary $\bar{v}_i$, k, and ε have to be prescribed (often based on experimental data). The prescription of the dissipation rate ε poses a particular problem, since this usually cannot be measured directly. Alternatively, an estimated value for the turbulent length scale l introduced in the preceding section, which represents a physically interpretable quantity, can be prescribed. From this ε can be computed according to $\varepsilon = k^{3/2}/l$. Usually, near a wall l grows linearly with the wall distance δ as

$$l = \frac{\kappa}{C_\mu^{3/4}}\delta, \tag{11.13}$$

where $\kappa = 0,41$ is the *Kármán constant*. In areas far from walls l is constant. If no better information is available, the inflow values for l can be estimated by means of these values. Often, no precise inflow data are available for k as well. From experiments or experience sometimes only the *turbulence degree* T_v is known. With this k can be estimated as follows:

$$k = \frac{1}{2}T_v^2 \bar{v}_i^2.$$

If T_v is also not known, just a "small" inflow value for k can be used, e.g., $k = 10^{-4}\bar{v}_i^2$.

In most cases inaccuracies in the inflow values of k and ε are not that critical. This is because in the equations for k and ε frequently the source terms dominate so that the production rate downstream is relatively large and thus the influence of the inflow values becomes small. However, in any case it is recommended to perform an investigation on the influence of the estimated inflow values on the downstream results (by just comparing results for computations with different values).

At an outflow boundary for k as well as for ε a vanishing normal derivative can be assumed. For the same reasons as above, the influence of this assumption to the upstream results is usually minor. The same condition also can be used for k and ε at symmetry boundaries.

The most critical problem consists in the treatment of wall boundaries – a matter that principally applies to *all* turbulence models. The reason is that in the area close to the wall a "thin" laminar layer (viscous sublayer) with very steep velocity gradients exists (see Fig. 11.4) in which the assumptions of the turbulence models are no longer valid. There are basically two possibilities to tackle this problem:

- The layer is resolved by a sufficiently fine grid accompanied by an adaption of the turbulence model in the near wall range (known as *low-Re* modification).
- The layer is not resolved, but modeled by using special *wall functions*.

Both approaches will be discussed briefly in the following.

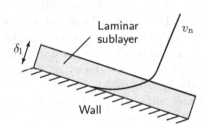

Fig. 11.4. Laminar sublayer at wall boundary

Resolving the layer by the grid means that at least a few (say 5) CVs are located in normal wall direction within the layer. In this case at the wall the boundary conditions

$$\overline{v}_i = 0, \quad k = 0, \quad \text{and} \quad \frac{\partial \varepsilon}{\partial x_i} n_i = 0$$

can be chosen. The necessary modification of the model in the near wall range can be accomplished by controling the turbulence model effects, for instance, in the transport equation for ε and the relation for the eddy-viscosity μ_t by special damping functions:

$$\frac{\partial (\rho \tilde{\varepsilon})}{\partial t} + \frac{\partial (\rho \overline{v}_j \tilde{\varepsilon})}{\partial x_j} = \frac{\partial}{\partial x_j}\left[\left(\mu + \frac{\mu_t}{\sigma_\varepsilon}\right)\frac{\partial \tilde{\varepsilon}}{\partial x_j}\right] + C_{\varepsilon 1}f_1 P\frac{\tilde{\varepsilon}}{k} - C_{\varepsilon 2}\rho f_2 \frac{\tilde{\varepsilon}^2}{k} + E,$$

$$\mu_t = C_\mu \rho f_\mu \frac{k^2}{\tilde{\varepsilon}},$$

where the quantity $\tilde{\varepsilon}$ is related to the dissipation rate ε by

$$\varepsilon = \tilde{\varepsilon} + 2\frac{\mu k}{\rho \delta^2} \tag{11.14}$$

with the wall distance δ. The corresponding models, which have been proposed in numerous variants, are known as *low-Re turbulence models*. Utilizing, for instance, the Chien approach as a typical representative of such a model, the damping functions f_1, f_2, and E are chosen to be

$$f_1 = 1, \quad f_2 = 1 - 0.22e^{-\left(\rho k^2/6\tilde{\varepsilon}\mu\right)^2}, \quad f_\mu = 1 - e^{-0.0115y^+}, \quad E = -2\frac{\mu\tilde{\varepsilon}}{\delta^2}e^{-0.5y^+},$$

where y^+ denotes a normalized (dimensionless) distance to the nearest wall point defined by

$$y^+ = \frac{\rho u_\tau \delta}{\mu}$$

with the *wall shear stress velocity*

$$u_\tau = \sqrt{\frac{\tau_w}{\rho}},$$

where

$$\tau_w = \mu\frac{\partial \overline{v}_t}{\partial x_i}n_i$$

denotes the *wall shear stress*. $\overline{v}_t$ is the tangential component of the mean velocity. The model constants in this case change to $C_\mu = 0.09$, $C_{\varepsilon 1} = 1.35$, and $C_{\varepsilon 2} = 1.80$. A survey on low-Re models can be found, for instance, in [28].

The thickness δ_l of the laminar layer decreases with the Reynolds number according to

$$\delta_l \sim \frac{1}{\sqrt{Re}}.$$

Therefore, for larger Reynolds numbers its resolution by the numerical grid becomes critical because this would result in very high numbers of grid points. As an alternative, *wall functions* can be employed which in some sense can "bridge" the laminar layer. The physical background of this approach is that in a fully developed turbulent flow a *logarithmic wall law* is valid, i.e., the velocity beyond the laminar layer logarithmically increases in a certain range (see also Fig. 11.5):

$$v^+ = \frac{1}{\kappa} \ln y^+ + B .$$ (11.15)

Here, $B = 5.2$ is a further model constant and v^+ is a normalized quantity for the tangential velocity $\bar{v}_t$ defined by

$$v^+ = \frac{\bar{v}_t}{u_\tau} .$$

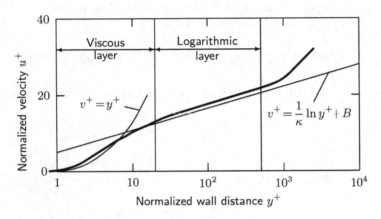

Fig. 11.5. Velocity distribution in turbulent flow near wall (logarithmic wall law)

Under the assumption of a local equilibrium of production and dissipation of turbulent kinetic energy and constant turbulent stresses, for the wall shear stress velocity u_τ the expression

$$u_\tau = C_\mu^{1/4} \sqrt{k}$$

can be derived, which is valid approximately in the range

$$30 \leq y^+ \leq 300 .$$

Using the wall law (11.15) one gets:

$$\tau_w = \rho u_\tau^2 = \rho u_\tau C_\mu^{1/4} \sqrt{k} = \frac{\bar{v}_t}{u^+} \rho C_\mu^{1/4} \sqrt{k} = \frac{\bar{v}_t \kappa \rho C_\mu^{1/4} \sqrt{k}}{\ln y^+ + \kappa B} .$$ (11.16)

This relation can be employed as a boundary condition for the momentum equations. On the basis of the last term in (11.16) the wall shear stress can be approximated, for instance, by:

$$\tau_w = -\frac{\bar{v}_P}{\delta_P} \underbrace{\frac{\kappa C_\mu^{1/4}\sqrt{k_P}\rho\delta_P}{\ln y_P^+ + \kappa B}}_{\mu_w} , \qquad (11.17)$$

where the index P denotes the midpoint of the boundary CV. By introducing δ in the nominator and denominator on the right hand side of (11.17) the discretization near the wall can be handled analogously as in the laminar case. Instead of μ in (11.17) just the marked quantity μ_w has to be used.

In addition to the above boundary condition in the transport equation (11.11) for k, the expression for the production rate G near the wall must be modified because the usual linear interpolation for the computation of the gradients of the mean velocity components would yield errornous resaults. Since in G the derivative of $\bar{v}_t$ in the direction to the normal dominates, the following approximation can be used:

$$G_P = \tau_w \left(\frac{\partial \bar{v}_t}{\partial x_i}\right)_P n_i = \tau_w \frac{C_\mu^{1/4}\sqrt{k_P}}{\kappa\delta_P} .$$

For the dissipation rate ε normally no boundary condition in the usual form is employed. The transport equation (11.12) for ε is "suspended" in the CV nearest to the wall and ε is computed in the point P from the corresponding value of k using the relations (11.6) and (11.13):

$$\varepsilon_P = \frac{C_\mu^{3/4}k_P^{3/2}}{\kappa\delta_P} .$$

Note that for scalar quantities, like temperature or concentrations, there also are corresponding wall laws so that similar modifications as described above for the velocities can be incorporated also for the corresponding boundary conditions for these quantities.

When using wall functions particular attention must be given to the y^+-values for the midpoints of the wall boundary CVs to be located in the range $30 \leq y^+ \leq 300$. Since y^+ is usually not exactly known in advance (it depends on the unknown solution) one can proceed as follows:

(1) determine a "rough" solution by a computation with a "test grid",
(2) compute from these results the y^+-values for the wall boundary CVs,
(3) carry out the actual computation with a correspondingly adapted grid.

It might be necessary to repeat this procedure. In particular, in connection with an estimation of the discretization error by a systematic grid refinement (see Sect. 8.2) the variation of the values of y^+ should be carefully taken into account.

11.2.3 Discretization and Solution Methods

The discretization of the equations (11.9)-(11.12), each having the form of the general scalar transport equation, can be done as in the laminar case. The only issue that should be mentioned in this respect concerns the treatment of the source terms in the equations for k and ε. Here, it is very helpful with respect to an improved convergence behavior of the iterative solution method to split the source terms in the following way:

$$- \int_V (\rho\varepsilon - G)\, dV \approx \underbrace{-\rho\delta V \frac{\varepsilon_P}{k_P^*}\, k_P}_{a_k} + \underbrace{G_P\delta V}_{b_k},$$

$$- \int_V \left(C_{\varepsilon 2}\rho\frac{\varepsilon^2}{k} - C_{\varepsilon 1}G\frac{\varepsilon}{k}\right) dV \approx \underbrace{-C_{\varepsilon 2}\rho\delta V \frac{\varepsilon_P^*}{k_P}\, \varepsilon_P}_{a_\varepsilon} + \underbrace{C_{\varepsilon 1}G_P\delta V \frac{\varepsilon_P^*}{k_P}}_{b_\varepsilon}.$$

The values marked with "*" can be treated explicitly within the iteration process (i.e., using values from the preceding iteration). In the term with a_k the quantity k is introduced "artificially" (multiplication and division by k_P), which does not change the equation, but, owing to the additional positive contribution to the main diagonal of the corresponding system matrix, its diagonal dominance is enhanced.

The solution algorithm for the discrete coupled equation system (11.9)-(11.12) can also be derived in a manner similar to the laminar case. The course of the pressure-correction method described in Sect. 10.3.1 for the turbulent case is illustrated schematically in Fig. 11.6. Again an under-relaxation is necessary, where usually also the equations for k and ε have to be under-relaxed. A typical combination of under-relaxation parameters is:

$$\alpha_{\overline{v}_i} = \alpha_k = \alpha_\varepsilon = 0.7\,, \quad \alpha_{\tilde{p}} = 0.3\,.$$

Additionally, the changes of μ_t also can be under-relaxed with a factor α_{μ_t} by combining the "new" values with a portion of the "old" ones:

$$\mu_t^{\text{new}} = \alpha_{\mu_t} C_\mu \rho \frac{k^2}{\varepsilon} + (1 - \alpha_{\mu_t})\mu_t^{\text{old}}\,.$$

Due to the analogies in the equation structure and the course of the solution process in the turbulent and laminar cases, both can be easily integrated into a single program. In the laminar case μ_t is just set to zero, and the equations for k and ε are not solved.

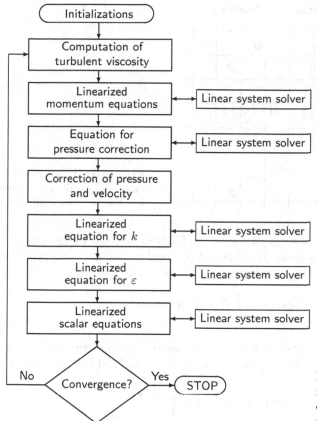

Fig. 11.6. Pressure-correction method for flow computations with k-ε model

11.3 Large Eddy Simulation

The large eddy simulation in some way represents an intermediate approach between DNS and RANS models. The basic idea of LES is to directly compute the large scale turbulence structures, which can be resolved with the actual numerical grid, and to suitably model the small scale structures (subgrid-scales, SGS) for which the actual grid is too coarse (see Fig. 11.7). Compared to a statistic modeling one has the advantage that the small scale structures (for sufficiently fine grids) are easier to model, and that the model error – just like the numerical error – decreases when the grid becomes finer.

For the mathematical formulation of the LES it is necessary to decompose the flow quantities, which we again denote generally with ϕ, into large and small (in a suitable relation to the grid) scale portions $\overline{\phi}$ and ϕ', respectively:

$$\phi(\mathbf{x}, t) = \overline{\phi}(\mathbf{x}, t) + \phi'(\mathbf{x}, t).$$

Computation   Modeling

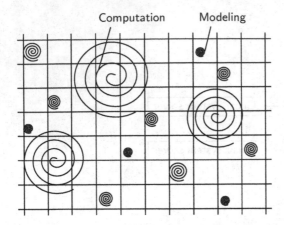

Fig. 11.7. Treatment of
large and small scale turbulence structures with LES

According to the formal similarity with the RANS approach we use the same
notations here, but emphasize the different physical meanings of the quantities
for LES. In the LES context $\overline{\phi}$ has to be defined by a *filtering*

$$\overline{\phi}(\mathbf{x},t) = \int\limits_V G(\mathbf{x},\mathbf{y})\phi(\mathbf{x},t)\,\mathrm{d}\mathbf{y}$$

with a (low-pass) *filter function* G. For G several choices are possible (see,
e.g., [18]). For instance, the *top-hat* (or *box*) *filter* is defined by (see Fig. 11.8)

$$G(\mathbf{x},\mathbf{y}) = \prod_{i=1}^{3} G_i(x_i,y_i) \quad \text{with} \quad G_i(x_i,y_i) = \begin{cases} 1/\Delta_i & \text{for } |x_i - y_i| < \Delta_i/2, \\ 0 & \text{otherwise}, \end{cases}$$

where Δ_i is the filter width in x_i-direction.

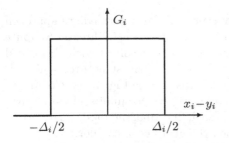

Fig. 11.8. Box filter in one dimension

Filtering the continuity and momentum equations (2.70) and (2.71) yields:

$$\frac{\partial \overline{v}_i}{\partial x_i} = 0, \tag{11.18}$$

$$\frac{\partial(\rho\overline{v}_i)}{\partial t} + \frac{\partial(\rho\overline{v}_i\overline{v}_j)}{\partial x_j} = \frac{\partial}{\partial x_j}\left[\mu\left(\frac{\partial \overline{v}_i}{\partial x_j} + \frac{\partial \overline{v}_j}{\partial x_i}\right)\right] - \frac{\partial \tau_{ij}^{\text{sgs}}}{\partial x_j} - \frac{\partial \overline{p}}{\partial x_i}, \tag{11.19}$$

where

$$\tau_{ij}^{\mathrm{sgs}} = \overline{v_i v_j} - \overline{v}_i \overline{v}_j \qquad (11.20)$$

is the *subgrid-scale stress tensor*, which – as for the Reynolds stress tensor in the case of RANS models – has to be modeled by a so called *subgrid-scale model* to close the problem formulation.

There are a variety of subgrid-scale models available – similar to RANS models – starting with zero-equation models and ending up with Reynolds stress models. The most important property of the subgrid-scale models is that they simulate energy transfer between the resolved scales and the subgrid-scales at a roughly correct magnitude. Since only the small scales have to be modeled, the models used in LES are more universal and simpler than the RANS models.

The most frequently used subgrid-scale models are eddy-viscosity models of the form

$$-\tau_{ij}^{\mathrm{sgs}} = 2\mu_t \overline{S}_{ij} - \frac{1}{3}\delta_{ij}\tau_{kk}^{\mathrm{sgs}}\,, \qquad (11.21)$$

where the subgrid-scale stresses τ_{ij}^{sgs} are proportional to the large scale strain-rate tensor

$$\overline{S}_{ij} = \frac{1}{2}\left(\frac{\partial \overline{v}_i}{\partial x_j} + \frac{\partial \overline{v}_j}{\partial x_i}\right). \qquad (11.22)$$

Since the trace of the resolved strain-rate tensor is zero in incompressible flows, only the traceless part of the subgrid-scale tensor has to be modeled.

One of the most popular eddy-viscosity subgrid-scale models is the *Smagorinsky model*. Here, the eddy viscosity is defined as

$$\mu_t = \rho l^2 |\overline{S}_{ij}| \quad \text{with} \quad |\overline{S}_{ij}| = \sqrt{2\overline{S}_{ij}\overline{S}_{ij}}\,,$$

where the length scale l is related to the filter width $\Delta = (\Delta_1 \Delta_2 \Delta_3)^{1/3}$ by

$$l = C_{\mathrm{s}}\Delta$$

with the Smagorinsky constant C_{s}. The theoretical value of the Smagorinsky constant for a homogeneous isotropic flow is $C_{\mathrm{s}} \approx 0.16$ (e.g., [18]).

Germano proposed a procedure that allows the determination of the Smagorinsky constant dynamically from the results of the LES, such that it is no longer a constant, but – being more realistic – a function of space and time: $C_{\mathrm{s}} = C_{\mathrm{s}}(\mathbf{x}, t)$. The dynamical computation of C_{s} is accomplished by defining a test filter with width $\widehat{\Delta}$ which is larger than the grid filter width Δ (e.g., $\widehat{\Delta} = 2\Delta$ has proven to be a good choice).

Applying the test filter to the filtered momentum equation (11.19), the *subtest-scale stresses* $\tau_{ij}^{\mathrm{test}}$ can be written as

$$\tau_{ij}^{\mathrm{test}} = \widehat{\overline{v_i v_j}} - \widehat{\overline{v}}_i \widehat{\overline{v}}_j\,. \qquad (11.23)$$

The subgrid-scale and subtest-scale stresses are approximated using the Smagorinsky model as

$$\tau_{ij}^{\text{sgs}} - \frac{1}{3}\delta_{ij}\tau_{kk}^{\text{sgs}} = -2C_{\text{g}}\Delta^2|\overline{S}|\overline{S}_{ij} =: -2C_{\text{g}}\alpha_{ij}^{\text{sgs}}, \qquad (11.24)$$

$$\tau_{ij}^{\text{test}} - \frac{1}{3}\delta_{ij}\tau_{kk}^{\text{test}} = -2C_{\text{g}}\widehat{\Delta}^2|\widehat{\overline{S}}|\widehat{\overline{S}}_{ij} =: -2C_{\text{g}}\alpha_{ij}^{\text{test}}, \qquad (11.25)$$

where the model parameter is defined as $C_{\text{g}} = C_{\text{s}}^2$.

The resolved turbulent stresses are defined by

$$L_{ij} = \widehat{\overline{v}_i\overline{v}_j} - \widehat{\overline{v}}_i\widehat{\overline{v}}_j, \qquad (11.26)$$

which represent the scales with the length between the grid-filter length and the test-filter length. Inserting the approximations (11.24) and (11.25) into the *Germano identity* (e.g., [18])

$$L_{ij} = \tau_{ij}^{\text{test}} - \widehat{\tau_{ij}^{\text{sgs}}} \qquad (11.27)$$

yields

$$L_{ij} = -2C_{\text{g}}\alpha_{ij}^{\text{test}} + 2\widehat{C_{\text{g}}\alpha_{ij}^{\text{sgs}}}. \qquad (11.28)$$

Employing the approximation

$$\widehat{C_{\text{g}}\alpha_{ij}^{\text{sgs}}} \approx C_{\text{g}}\widehat{\alpha_{ij}^{\text{sgs}}}, \qquad (11.29)$$

which means that C_{g} is assumed to be constant over the test filter width, one gets

$$L_{ij} = 2C_{\text{g}}\left(\widehat{\alpha_{ij}^{\text{sgs}}} - \alpha_{ij}^{\text{test}}\right) =: 2C_{\text{g}}M_{ij}. \qquad (11.30)$$

Since both sides of (11.30) are symmetric and the traces are zero, there are five independent equations and one unknown parameter C_{g}. Lilly (1992) proposed to applying the least-squares method to minimize the square of the error

$$E = L_{ij} - 2C_{\text{g}}M_{ij} \qquad (11.31)$$

yielding

$$\frac{\partial\left(E^2\right)}{\partial C_{\text{g}}} = 4\left(L_{ij} - 2C_{\text{g}}M_{ij}\right)M_{ij} = 0, \qquad (11.32)$$

such that the model parameter finally becomes

$$C_{\text{g}}(\mathbf{x}, t) = \frac{L_{ij}M_{ij}}{2M_{ij}M_{ij}}. \qquad (11.33)$$

The eddy viscosity then is evaluated by

$$\mu_t = \rho C_{\mathrm{g}} \Delta^2 |\overline{S}_{ij}|. \tag{11.34}$$

The parameter C_{g} may take also negative values, which can cause numerical instabilities. To overcome this problem, different proposals have been made. For instance, an averaging in time and locally in space can be applied or simply a clipping of negative values can be employed:

$$C_{\mathrm{g}}(\mathbf{x}, t) = \max\left\{ \frac{L_{ij}M_{ij}}{2M_{ij}M_{ij}}, 0 \right\}, \tag{11.35}$$

The dynamic procedure can be used also together with other models than the Smagorinsky model.

As for RANS models, LES also has the problem with the proper treatment of the laminar sublayer at wall boundaries. Here, again, wall functions can be applied. For this and for another problem with LES, i.e., the prescription of suitable values at inflow boundaries, we refer to the corresponding literature (e.g., [18]).

11.4 Comparison of Approaches

Comparing the three different approaches for the computation of turbulent flows, i.e., RANS models, LES, and DNS, one can state that the numerical effort from the simplest methods of statistical turbulence modeling to a fully resolved DNS increases dramatically. However, with the numerical effort the generality and modeling quality of the methods increases (see Fig. 11.9).

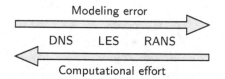

Fig. 11.9. Relation of model quality and computational effort for DNS, LES, and RANS

As an example, in Table 11.2 the computing power and memory requirements for the different methods for aerodynamic computations of a complete airplane is given. The Flop rate is based on the assumption that the "desired" computing time is *one* hour in each case. Considering the numerical effort for the different approaches, it becomes apparent, why nowadays almost only RANS models are employed for practical applications (mostly two-equation models).

In actual flow computation codes usually a variety of RANS models of different complexity are available and it is up to the user to decide which model is best suited for a specific application (usually a non-trivial task). The assumptions made in the derivation of simpler RANS models, as for instance in the k-ε model, provide a good description of the physical situation in highly

Table 11.2. Effort for computing turbulent flow around an airplane within one hour CPU time

Method	Computing power (Flops)	Memory (Byte)
RANS	10^9 - 10^{11}	10^9 - 10^{10}
LES	10^{13} - 10^{17}	10^{12} - 10^{14}
DNS	10^{19} - 10^{23}	10^{16} - 10^{18}

turbulent flows with isotropic turbulence (e.g., channel flows, pipe flows, ...). Problems arise, in particular, in the following situations:

- flows with separation,
- bouyancy-driven flows,
- flows along curvilinear surfaces,
- flows in rotating systems,
- flows with sudden change of mean strain rate.

In such cases, with more advanced RANS models, like Reynolds stress models or even LES, a significantly better modeling quality can be achieved so that it can be worth accepting the higher computational effort.

12

Acceleration of Computations

For complex practical problems the numerical simulation of the corresponding continuum mechanical model equations usually is highly demanding with respect to the efficiency of the numerical solution methods as well as to the performance of the computers. In order to achieve sufficiently accurate numerical solutions, in particular for flow simulations, in many practically relevant cases a very fine resolution is required and consequently results in a high computational effort and high memory requirements. Thus, in recent years intensive efforts have been undertaken to develop techniques to improve the efficiency of the computations. For the acceleration basically two major directions are possible:

- the usage of improved algorithms,
- the usage of computers with better performance.

With respect to both aspects in recent years tremendous progress has been achieved. Concerning the algorithms, adaptivity and multigrid methods represent important acceleration techniques, and concerning the computers, in particular, the usage of parallel computers is one of the key issues. In this chapter we will address the major ideas related to these aspects.

12.1 Adaptivity

A key issue when using numerical methods is the question how to select the numerical parameters (i.e., grid, time step size, ...) so that on the one hand a desired accuracy is reached and on the other hand the computational effort is as low as possible. Since the exact solution is not known, a proper answer to this question is rather difficult. Adaptive methods try to deal with these issues iteratively during the solution process on the basis of information provided by the actual numerical solution.

The principle procedure with adaptive methods is illustrated schematically in Fig. 12.1. First, with some initial choice of numerical parameters a preliminary "rough" solution is computed, which then is evaluated with respect to its accuracy and efficiency. Based on this information the numerical parameters are adjusted and the solution is recomputed. This process is repeated until a prescribed tolerance is reached. The adaptation process, in principle, can apply to any numerical parameters of the underlying scheme that influence the accuracy and efficiency, i.e., relaxation factors, number of grid points and time steps, and order of the discretization (see Sect. 8.4). One of the most important numercial parameters in this respect is the local properties of the numerical grids to which we will now turn.

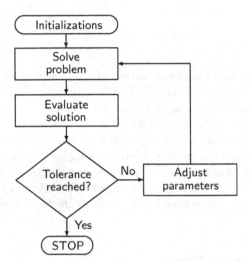

Fig. 12.1. Schematic representation of adaptive solution strategy

Important ingredients of a mesh adaptive solution strategy are the techniques for adjusting the grid and for evaluating the accuracy of the numerical solution. We will briefly address these issues in the following sections.

12.1.1 Refinement Strategies

For a local mesh adaptation one basically can distinguish between three different refinement strategies:

- *r-refinement*: The general idea of this approach is to move the nodal variables in the problem domain to locations such that – for a given fixed number of nodes – the error is minimized. On structured grids this can be achieved by the use of control functions or variational approaches (see [5]). A simple clustering technique has already been presented in Sect. 3.3.1.
- *h-refinement*: In this approach the number and size of elements or CVs is locally adapted such that the error is equidistributed over the problem domain. Thus at critical regions, i.e., with high gradients in the solution,

more elements or CVs are placed. An example of a locally conformingly h-refined mesh can be seen in Fig. 9.9.

- *p-refinement*: This approach consists in an increase of the order of approximation in critical regions with high gradients.

Combinations of the methods are also possible (and sensible). As an example, in Fig. 12.2 the three strategies are illustrated schematically for quadrilateral finite elements. While r- and h-refinement can be used similarily for both finite-element and finite-volume methods, the p-refinement is best suited for the context of finite-element approximations.

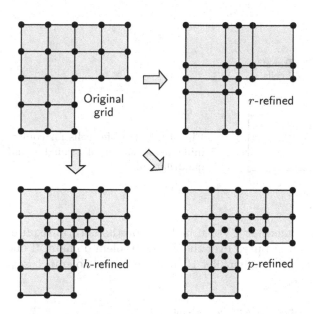

Original grid r-refined

h-refined p-refined

Fig. 12.2. Schematic representation of grid refinement strategies

In the case of a local h-refinement the interfaces between refined and non-refined regions need special attention. In Fig. 12.3 examples of *conforming* and *non-conforming* refinements of triangles and quadrilaterals are indicated. In the non-conforming case the quality of the elements or CVs is better, but hanging nodes appear. Here, either special *transfer cells* can be employed to handle the transition conformingly (see Fig. 12.4) or the hanging nodes must be taken into account directly in the discretization, e.g., by special ansatz functions. The same applies for p-refinement where special ansatz functions can be employed to handle the mixed approximation in interfacial elements.

Note that for time dependent problems also a *mesh unrefinement* is sensible, i.e., in regions where a refined mesh is no longer needed, owing to the temporal behavior of the solution the mesh can be derefined again.

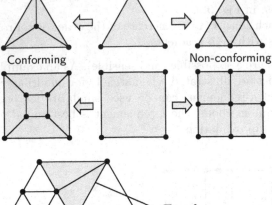

Fig. 12.3. Conforming and non-conforming h-refinement of triangles and quadrilaterals

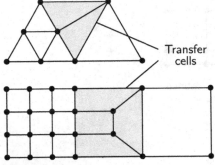

Fig. 12.4. Transfer cells for conforming h-refinement of triangles and quadrilaterals

12.1.2 Error Indicators

The decision to refine an element or CV can be based on the natural requirement that the error is equidistributed over the whole problem domain. If one requires that the global error e_h is smaller than a prescribed tolerance ε_{tol}, i.e.,

$$e_h = \|\phi - \phi_h\| \leq \varepsilon_{\text{tol}},$$

a criterion to refine the i-th element or CV can be

$$e_h^i = \|\phi^i - \phi_h^i\| \geq \frac{\varepsilon_{\text{tol}}}{N},$$

where e_h^i is the local error in the i-th element or CV (for $i = 1, \ldots, N$) and N is the total number of elements or CVs.

The question that remains is how an estimation of the local errors e_h^i in the individual elements or CVs can be obtained. For this, so-called *a posteriori error indicators* are employed, which usually work with ratios or rates of changes of gradients of the actual solution. There is an elaborate theory on this subject, especially in the context of finite-element methods, and a variety of approaches are available, i.e., residual based methods, projection methods, hierarchical methods, dual methods, or averaging methods. We will not discuss this matter in detail here (see, e.g., [29]) and, as examples, just give two simple possibilities in the framework of finite-volume methods.

From a computational point of view, one of the simplest criteria is provided by the so-called *jump indicator* E^i_{jump}, which for a two-dimensional quadrilateral CV is defined by

$$E^i_{\text{jump}} = \max\left\{\left|\phi^i_e - \phi^i_w\right|, \left|\phi^i_n - \phi^i_s\right|\right\} . \tag{12.1}$$

E^i_{jump} measures the variation of ϕ in the directions of the local coordinate system ξ_i defined by opposite cell face midpoints (see Fig. 12.5) and it is proportional to the gradient of ϕ weighted with the characteristic size h of the corresponding CV:

$$E^i_{\text{jump}} \sim \max_{k=1,2}\left\{h\left|\left(\frac{\partial\phi}{\partial x_k}\right)_P\right|\right\} . \tag{12.2}$$

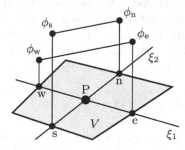

Fig. 12.5. Illustration of jump error indicator

Another simple error indicator, which conveniently can be employed together with multigrid methods described in the next section, is based on the difference of solutions at two different grid levels, i.e., ϕ_h and ϕ_{2h}:

$$E^i_{\text{grid}} = \left|\phi^i_{h,P} - \left(\mathcal{I}^h_{2h}\phi_{2h}\right)^i_P\right| , \tag{12.3}$$

where $\mathcal{I}^h_{2h}$ is a suitable interpolation operator (see Sect. 12.2.3, in particular Fig. 12.10).

12.2 Multi-Grid Methods

Conventional iterative solution methods – such as the Jacobi or Gauß-Seidel methods (see Sect. 7.1.2) – for linear equation sytems, which result from a discretization of differential equations, converge slower the finer the numerical grid is.

In general, with this kind of methods the number of required iterations to reach a certain accuracy increases with the number of grid points (unknowns). Since for an increasing number of unknowns also the number of arithmetic operations per iteration increases, the total computing time increases disproportionately with the number of unknowns (quadratically for the Jacobi or Gauß-Seidel methods, see Sect. 7.1.7). Using multigrid methods it is possible, to keep the required number of iterations mostly independent from the grid spacing, with the consequence that the computing time only increases proportionally with the number of grid points.

12.2.1 Principle of Multi-Grid Method

The idea of multigrid methods is based on the fact that an iterative solution algorithm just eliminates efficiently those error components of an approximate solution whose wavelengths correspond to the grid spacing, whereas errors with a larger wavelength with such a method can be reduced only slowly. The reason for this is that via the discretization scheme for each grid point only local neighboring relations are set up, which has the consequence that the global information exchange (e.g., the propagation of boundary values into the interior of the solution domain) with iteration methods only happens very slowly. To illustrate this issue we consider as the simplest example the one-dimensional diffusion problem

$$\frac{\partial^2 \phi}{\partial x^2} = 0 \quad \text{for } 0 < x < 1 \quad \text{and} \quad \phi(0) = \phi(1) = 0 \,,$$

which obviously has the analytical solution $\phi = 0$. Using a central difference discretization on an equidistant grid with $N - 1$ inner grid points, for this problem one obtains the discrete equations

$$\phi_{i+1} - 2\phi_i + \phi_{i-1} = 0 \quad \text{for } i = 1, \dots, N - 1 \,.$$

The iteration procedure for the Jacobi method for the solution of this tridiagonal equation system reads ($k = 0, 1, \dots$):

$$\phi_i^{k+1} = \frac{\phi_{i+1}^k + \phi_{i-1}^k}{2} \quad \text{for } i = 1, \dots, N - 1 \quad \text{and} \quad \phi_0^{k+1} = \phi_N^{k+1} = 0 \,.$$

Assuming that the two initial solutions ϕ^0 as shown in Fig. 12.6 (top) for the indicated grid are given, after one Jacobi iteration the approximative solutions shown in Fig. 12.6 (bottom) result.

One can observe the different error reduction behavior for the two different starting values:

- In case (a) the iterative algorithm is very efficient. The correct solution is obtained after just one iteration.
- In case (b) the improvement is small. Many iterations are required to get an accurate solution.

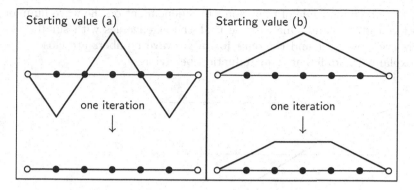

Fig. 12.6. Error reduction with the Jacobi method for solution of one-dimensional diffusion problem with different starting values

Normally, a starting value contains many different error components that are superposed. The high frequency components are reduced rapidly, while the low frequency components are reduced very slowly.

Quantitative assertions about the convergence behavior of classical iterative methods can be derived by Fourier analysis (at least for model problems). For a one-dimensional problem the error e_{ih} at the location x_i can be expressed as a Fourier series as follows:

$$e_{ih} = \sum_{k=1}^{N-1} a_k \sin(ik\pi/N).$$

Here, k is the so-called *wave number*, h is the grid spacing, $N = 1/h$ is the number of nodal values, and a_k are the Fourier coefficients. The components with $k \leq N/2$ and $k > N/2$ are denoted as *low* and *high frequency errors*, respectively. The absolute value of the eigenvalue λ_k of the iteration matrix (see Sect. 7.1) of the employed iterative method determines the reduction of the corresponding error component:

- "good" reduction, if $|\lambda_k|$ is "close to" 0,
- "bad" reduction, if $|\lambda_k|$ is "close to" 1.

Let us consider as an example the *damped Jacobi method* with the iteration matrix

$$\mathbf{C} = \mathbf{I} - \frac{1}{2}\mathbf{A}_{\mathrm{D}}^{-1}\mathbf{A}. \tag{12.4}$$

Using this method for the solution of the above diffusion problem with $N-1$ inner grid points results in an iteration matrix $\mathbf{C}$ for which the eigenvalues are given analytically as

$$\lambda_k = 1 - \frac{1}{2}\left[1 - \cos(k\pi/N)\right] \quad \text{for} \quad k = 1, \dots, N-1.$$

In Fig. 12.7 the magnitude of the eigenvalues dependent on the wave number is illustrated graphically. One can see that the eigenvalues with small wave numbers are close to 1 and the ones for large wave numbers are close to 0, which explains the different error reduction behavior.

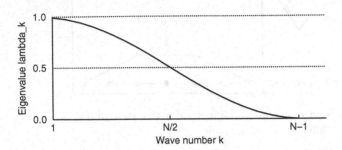

Fig. 12.7. Eigenvalue distribution of iteration matrix of damped Jacobi method for one-dimensional diffusion problem

The idea of multigrid methods now is to involve a hierarchy of successively coarsened grids into the iteration process in order to reduce there the low frequency error components. A multigrid algorithm transfers the computation after some fine grid iterations, where the error function afterwards is "smooth" (i.e., free of high frequency components), to a coarse grid, which, for instance, only involves every second grid point in each spatial direction. Smooth functions can be represented on coarser grids without a big loss of information. On the coarser grids the low frequency error components from the fine grid – relative to the grid spacing – look more high frequency and thus can be reduced more efficiently there. Heuristically, this can be interpreted by the faster global information exchange on coarser grids (see Fig. 12.8).

The efficiency of multigrid methods is due to the fact that on the one hand a significantly more efficient error reduction is achieved and on the other hand the additional effort for the computations on coarser grids is relatively small owing to the lower number of grid points.

12.2.2 Two-Grid Method

We will outline the procedure for multigrid methods first by means of a two-grid method and afterwards show how this can be extended to a multigrid method. In order to reduce the error of the fine grid solution on a coarser grid an error equation (*defect or correction equation*) has to be defined there. Here one has to distinguish between linear and nonlinear problems.

We first consider the linear case. Let

$$\mathbf{A}_h \phi_h = \mathbf{b}_h \tag{12.5}$$

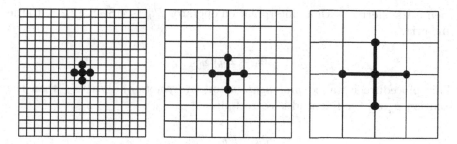

Fig. 12.8. Relation between information exchange and grid size due to neighboring relations of discretization schemes

be the linear equation system resulting from a discretization on a grid with grid spacing h. Starting from an initial value ϕ_h^0 after some iterations with an iteration procedure $\mathcal{S}_h$ a *smooth* approximation $\tilde{\phi}_h$ is obtained, which only contains low frequency error components:

$$\tilde{\phi}_h \leftarrow \mathcal{S}_h(\phi_h^0, \mathbf{A}_h, \mathbf{b}_h).$$

$\tilde{\phi}_h$ fulfils the original equation (12.5) only up to a residual $\mathbf{r}_h$:

$$\mathbf{A}_h \tilde{\phi}_h = \mathbf{b}_h - \mathbf{r}_h.$$

Subtracting this equation from (12.5) yields the fine grid error equation:

$$\mathbf{A}_h \mathbf{e}_h = \mathbf{r}_h$$

with the error $\mathbf{e}_h = \phi_h - \tilde{\phi}_h$ as the unknown quantity. For the further treatment of the error on a coarse grid (e.g., with grid spacing $2h$) the matrix $\mathbf{A}_h$ and the residual $\mathbf{r}_h$ have to be transferred to the coarse grid:

$$\mathbf{A}_{2h} = \mathcal{I}_h^{2h} \mathbf{A}_h \quad \text{and} \quad \mathbf{r}_{2h} = \mathcal{I}_h^{2h} \mathbf{r}_h.$$

This procedure is called *restriction*. $\mathcal{I}_h^{2h}$ is a restriction operator, which we will specify in more detail in Sect. 12.2.3. In this way an equation for the error $\mathbf{e}_{2h}$ on the coarse grid is obtained:

$$\mathbf{A}_{2h} \mathbf{e}_{2h} = \mathbf{r}_{2h}. \tag{12.6}$$

The solution of this equation can be done with the same iteration scheme as on the fine grid:

$$\tilde{\mathbf{e}}_{2h} \leftarrow \mathcal{S}_{2h}(0, \mathbf{A}_{2h}, \mathbf{r}_{2h}).$$

As initial value in this case $\mathbf{e}_{2h}^0 = 0$ can be taken, since the solution represents an error which should vanish in case of convergence. The coarse grid error $\tilde{\mathbf{e}}_{2h}$

then is transferred with an interpolation operator $\mathcal{I}_{2h}^h$ (see Sect. 12.2.3) to the fine grid:

$$\tilde{\mathbf{e}}_h = \mathcal{I}_{2h}^h \tilde{\mathbf{e}}_{2h} \,.$$

This procedure is called *prolongation* (or *interpolation*). With $\tilde{\mathbf{e}}_h$ then the solution $\tilde{\phi}_h$ on the fine grid is corrected:

$$\phi_h^* = \tilde{\phi}_h + \tilde{\mathbf{e}}_h.$$

Afterwards some iterations on the fine grid are carried out (with initial value ϕ_h^*), in order to damp high frequency error components that might arise due to the interpolation:

$$\tilde{\phi}_h^* \leftarrow \mathcal{S}_h(\phi_h^*, \mathbf{A}_h, \mathbf{b}_h) \,.$$

The described procedure is repeated until the residual on the fine grid, i.e.,

$$\tilde{\mathbf{r}}_h^* = \mathbf{b}_h - \mathbf{A}_h \tilde{\phi}_h^* \,,$$

fulfils a given convergence criterion. The described procedure in the literature is also known as *correction scheme (CS)*.

Now let us turn to the nonlinear case with the problem equation

$$\mathbf{A}_h(\phi_h) = \mathbf{b}_h \,. \tag{12.7}$$

For the application of multigrid methods to nonlinear problems there exist two principal approaches:

- linearization of the problem (e.g., with Newton method or successive iteration, see Sect. 7.2) and application of a linear multigrid method in each iteration.
- Direct application of a nonlinear multigrid method.

In many cases a nonlinear multigrid method, the so-called *full approximation scheme (FAS)*, has turned out to be advantageous, and we will therefore briefly describe this next.

After some iterations with a solution method for the nonlinear system (12.7) (e.g., the Newton method or the SIMPLE method for flow problems) one obtains an approximative solution $\tilde{\phi}_h$ fulfilling:

$$\mathbf{A}_h(\tilde{\phi}_h) = \mathbf{b}_h - \mathbf{r}_h \,.$$

The starting point for the linear two-grid method was the error equation $\mathbf{A}_h \mathbf{e}_h = \mathbf{r}_h$. For nonlinear problems this makes no sense, since the superposition principle is not valid, i.e., in general it is

$$\mathbf{A}_h(\phi_h + \psi_h) \neq \mathbf{A}_h(\phi_h) + \mathbf{A}_h(\psi_h) \,.$$

Therefore, for a nonlinear multigrid method a nonlinear error equation has to be defined, which can be obtained by a linearization of $\mathbf{A}_h$:

$$\mathbf{A}_h(\tilde{\phi}_h + \mathbf{e}_h) - \mathbf{A}_h(\tilde{\phi}_h) = \mathbf{r}_h \quad \text{with} \quad \mathbf{e}_h = \phi_h - \tilde{\phi}_h. \qquad (12.8)$$

The nonlinear error equation (12.8) is now the basis for the coarse grid equation, which is defined as

$$\mathbf{A}_{2h}(\mathcal{I}_h^{2h}\tilde{\phi}_h + \mathbf{e}_{2h}) - \mathbf{A}_{2h}(\mathcal{I}_h^{2h}\tilde{\phi}_h) = \mathcal{I}_h^{2h}\mathbf{r}_h.$$

Thus, to set up the coarse grid equation $\mathbf{A}_h$, $\tilde{\phi}_h$, and $\mathbf{r}_h$ have to be restricted to the coarse grid ($\mathcal{I}_h^{2h}$ is the restriction operator). As coarse grid variable $\phi_{2h} := \mathcal{I}_h^{2h}\tilde{\phi}_h + \mathbf{e}_{2h}$ can be used, such that the following coarse grid problem results:

$$\mathbf{A}_{2h}(\phi_{2h}) = \mathbf{b}_{2h} \quad \text{with} \quad \mathbf{b}_{2h} = \mathbf{A}_{2h}(\mathcal{I}_h^{2h}\tilde{\phi}_h) + \mathcal{I}_h^{2h}\mathbf{r}_h. \qquad (12.9)$$

$\mathcal{I}_h^{2h}\tilde{\phi}_h$ can be employed as initial value for the solution iterations for this equation. After the solution of the coarse grid equation (the solution is denoted by $\tilde{\phi}_{2h}$), as in the linear case, the *error* (only this is smooth) is transferred to the fine grid and is used to correct the fine grid solution:

$$\phi_h^* = \tilde{\phi}_h + \tilde{\mathbf{e}}_h \quad \text{with} \quad \tilde{\mathbf{e}}_h = \mathcal{I}_{2h}^h(\tilde{\phi}_{2h} - \mathcal{I}_h^{2h}\tilde{\phi}_h).$$

Note that the quantities ϕ_{2h} and $\mathbf{b}_{2h}$ do *not* correspond to the solution and the right hand side, which would be obtained from a discretization of the continuous problem on the coarse grid. ϕ_{2h} is an approximation of the *fine grid solution*, hence the name full approximation scheme. In the case of convergence, all coarse grid solutions (where they are defined) are identical to the fine grid solution.

12.2.3 Grid Transfers

The considerations so far have been mostly independent from the actual discretization method employed. Multigrid methods can be defined in an analogous way for finite difference, finite-volume, and finite-element methods, where, however, in particular, for interpolation and restriction the corresponding specifics of the discretization have to be taken into account. As an example, we will briefly discuss these grid transfer operations for a finite-volume discretization.

For finite-volume methods it is convenient to perform a CV oriented grid coarsening, such that one coarse grid CV is formed from 2^d fine grid CVs, where d denotes the spatial dimension (see Fig. 12.9 for the two-dimensional case). For the grid transfers the interpolation and restriction operators $\mathcal{I}_h^{2h}$ and $\mathcal{I}_{2h}^h$ have to be defined, respectively. Also the coarse grid equation should be based on the conservation principles underlying the finite-volume method.

As outlined in Chap. 4 the matrix coefficients are formed from convective and diffusive parts. The mass fluxes for the convective parts simply can be determined by adding the corresponding fine grid fluxes. The diffusive parts usually are newly computed on the coarse grid. The coarse grid residuals result as the sum of the corresponding fine grid residuals. This is possible because the coarse grid equation can be interpreted as the sum of the fine grid equations (conservation principle). The variable values can be transferred to the coarse grid, for instance by bilinear interpolation (see Fig. 12.9).

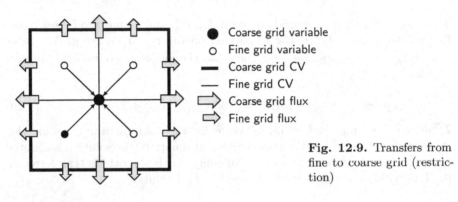

Fig. 12.9. Transfers from fine to coarse grid (restriction)

The interpolation from the coarse to the fine grid must be consistent with the order of the underlying discretization scheme. For a second-order discretization, for instance, again a bilinear interpolation can be employed (see Fig. 12.10).

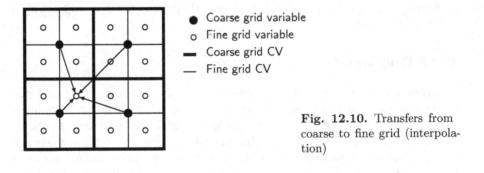

Fig. 12.10. Transfers from coarse to fine grid (interpolation)

12.2.4 Multigrid Cycles

The solution of the coarse grid problem in the above two-grid method for fine grids still can be very costly. The coarse grid problem, i.e., equation (12.6) in the linear case or equation (12.9) in the nonlinear case, again can be solved by a two-grid method. In this way from a two-grid method a multigrid method can be defined recursively.

The choice of the coarsest grid is problem dependent and usually is determined by the problem geometry, which should be described sufficiently accurately also by the coarsest grid (typical values are about 4-5 grid levels for two-dimensional and 3-4 grid levels for three-dimensional problems). For the cycling through the different grid levels several strategies exists. The most common ones are the so-called *V-cycles* and *W-cycles*, which are illustrated in Fig. 12.11. Using W-cycles the effort per cycle is higher than with V-cycles, but usually a lower number of cycles are required to reach a certain convergence criterion. Depending on the underlying problem there can be advantages for the one or the other variant, but usually these are not very significant.

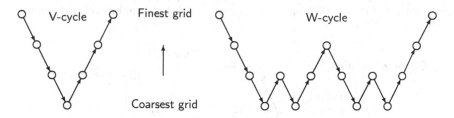

Fig. 12.11. Schematic course of V-cycle and W-cycle

In contrast to classical iterative methods the convergence rate of multigrid methods is mostly *independent* of the grid spacing. For model problems it can be proven that the solution with the multigrid method requires an asymptotic effort which is proportional to $N \log N$, where N is the number of grid points. Numerically, this convergence behavior can be proven also for many other (more general) problems.

For the estimation of the effort of multigrid cycles we denote the effort for one iteration on the finest grid by W, and by k the number of fine grid iterations within one cycle. For the effort of the V-cycle one obtains:

$$\text{2-D: } W_{\mathrm{MG}} = (k+1)W \left[1 + \frac{1}{4} + \frac{1}{16} + \cdots\right] \le \frac{4}{3}(k+1)W,$$

$$\text{3-D: } W_{\mathrm{MG}} = (k+1)W \left[1 + \frac{1}{8} + \frac{1}{64} + \cdots\right] \le \frac{8}{7}(k+1)W.$$

For $k = 4$, which is a typical value, one V-cycle requires in the two-dimensional case only about as much time as 7 iterations on the finest grid (approximately 6 in the three-dimensional case). However, the error reduction is better by orders of magnitude than with single-grid methods (see examples in Sect. 12.2.5).

For a further acceleration, for steady problems multigrid methods can be used in combination with the method of *nested iteration*, which serves for the improvement of the initial solution on the finest grid by using solutions for coarser grids as initial guesses for the finer ones. The computation is started on the coarsest grid. The converged solution obtained there is extrapolated to the next finer grid serving as initial solution for a two-grid cycle there. The procedure is continued until the finest grid is reached. This combination of a nested iteration with a multigrid method is called *full multigrid method (FMG method)*. The procedure is illustrated in Fig. 12.12 for the case when combined with V-cycles. The computing time spent for the solutions on the coarser grids is saved on the finest grid because there the iteration process can be started with a comparably good starting value. Thus, with relatively low effort with the FMG method a further acceleration of the solution process can be achieved. One obtains an asymptotically optimal method, where the computational effort only increases *linearly* with the number of grid points.

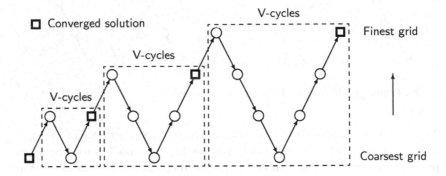

Fig. 12.12. Schematic course of full multigrid method

An additional aspect of the FMG method, which is very important in practice, is that at the end of the computation converged solutions for all grid levels involved are available, which directly can be used for an estimation of the discretization errors (see Sect. 8.4).

12.2.5 Examples of Computations

An example of the acceleration by multigrid methods for the solution of linear problems already has been given in Sect. 7.1.7 (see Table 7.2). As an example for the multigrid efficiency for nonlinear problems, we consider the computation of a laminar natural convection flow in a square cavity with a complex obstacle (see Fig. 12.13). The cavity walls and the obstacle possess constant temperatures T_C and T_H, respectively, where $T_C < T_H$. For the computations a second-order finite-volume method with a SIMPLE method on a colocated grid (with selective interpolation) is employed. Here, the SIMPLE method acts as the smoother for a nonlinear multigrid method with V-cycles and bilinear interpolation for the grid transfers.

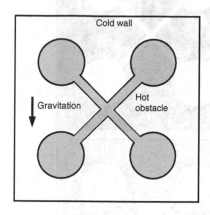

Fig. 12.13. Problem configuration for natural convection flow in square cavity with complex obstacle

For the computation up to 6 grid levels (from 64 CVs to 65 536 CVs) are employed. The coarsest and the finest grid as well as the corresponding computed velocity fields are represented in Fig. 12.14. One can see, in particular, that the coarsest grid does not exactly model the geometry and also the velocity field computed there is relatively far away from the "correct" result. However, as the following results will show this does not have an unfavorable influence on the efficiency of the multigrid method.

In Table 12.1 a comparison of the numbers of required fine grid iterations for the single-grid and multigrid methods each with and without nested iteration is given. The corresponding computing times are summarized in Table 12.2.

One can observe the enormous acceleration, which is achieved on the finer grids with the multigrid method due to the nearly constant iteration number. The necessity to use such fine grids with respect to the numerical accuracy can be seen from Table 8.1: the numerical error in the Nußelt number, for

Table 12.1. Number of fine grid iterations for single-grid (SG) and multigrid (MG) with and without nested iteration (NI) for cavity with complex obstacle

| Method | Control volumes | | | | | |
	64	256	1 024	4 096	16 384	65 536
SG	52	42	128	459	1 755	4 625
SG+NI	52	36	79	269	987	3 550
MG	52	31	41	51	51	51
MG+NI	52	31	31	31	31	31

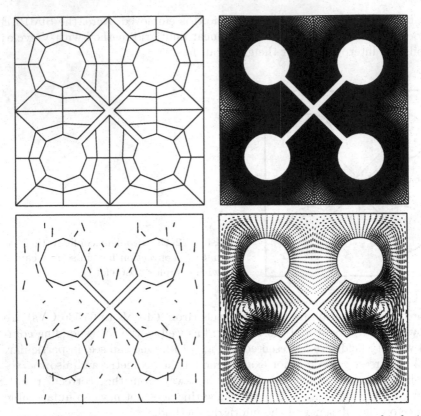

Fig. 12.14. Coarsest and finest grids and corresponding computed velocity fields for cavity with complex obstacle

Table 12.2. Computing times for single-grid (SG) and multigrid (MG) with and without nested iteration (NI) for cavity with complex obstacle

| Method | Control volumes | | | | | |
	64	256	1 024	4 096	16 384	65 536
SG	3	7	70	902	13 003	198 039
SG+NI	3	10	54	590	8 075	110 628
MG	3	7	34	144	546	2 096
MG+NI	3	11	36	124	451	1 720

instance, on the 40×40 grid still is about 7%. Also the additional acceleration effect due to the nested iteration becomes apparent. Although this also applies to the single-grid method the increase in the number of iterations with the number of grid points cannot be avoided this way. In Fig. 12.15 the acceleration effect of the multigrid method with nested iteration compared to the single-grid method (without nested iteration) depending on the grid size is given graphically in a double logarithmic representation. One clearly can observe the quadratic dependence of the effort for the single-grid method (line with slope 2) in contrast to the linear dependence in the case of the multigrid method (line with slope 1).

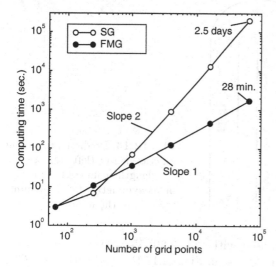

Fig. 12.15. Computing times versus number of grid points for single-grid and multigrid methods for cavity with complex obstacle

In general, the acceleration factors, which can be achieved with multigrid methods, strongly depend on the problem. In Table 12.3 typical acceleration factors for steady and unsteady laminar flow computations in the two- and three-dimensional cases are given (each for a grid with around 100 000 CVs).

Table 12.3. Typical acceleration factors with multigrid methods for laminar flow computations (with around 100 000 CVs)

Flow	2-d (5 grids)	3-d (3 grids)
Steady	80-120	40-60
Unsteady	20-40	5-20

Also for the computation of turbulent flows with RANS models or LES (see Chap. 11) significant accelerations can be achieved when using multigrid methods. However, the acceleration factors are (still) lower.

Let us consider as an example the turbulent flow in an axisymmetric bend, which consists of a circular cross-section entrance followed by an annulus in the opposite direction connected by a curved section of 180°. Figure 12.16 shows the configuration together with the predicted turbulent kinetic energy and turbulent length scale when using the standard k-ϵ model with wall functions (see Sects. 11.2.1 and 11.2.2). The Reynolds number based on the block inlet velocity and the entrance radius is Re = 286 000.

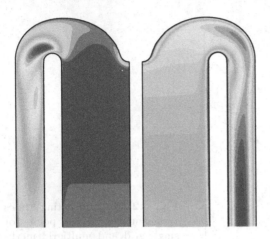

Fig. 12.16. Predicted turbulent kinetic energy (left) and turbulent length scale (right) for flow in axisymmetric bend (symmetry axis in the middle).

The multigrid procedure is used with (20,20,20)-V-cycles with a coarsest grid of 256 CVs for up to 5 grid levels (in Fig. 12.17 the grid with 4 096 CVs is shown).

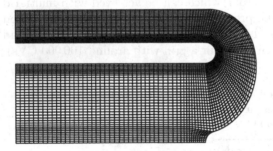

Fig. 12.17. Numerical grid for flow in axisymmetric bend (the bottom line is the symmetry axis)

In Fig. 12.18, the computing times and the numbers of fine grid iterations are given for the single-grid and multigrid methods, each with and without nested iteration, for different grid sizes. Although still significant, at least for finer grids, the multigrid acceleration is lower than in comparable laminar cases. The acceleration factors increase nearly linearly with the grid level. A general experience is that the acceleration effect decreases with the complexity of the model. Here further research appears to be necessary.

Another experience which is worth noting is that, in general, the multigrid method stabilizes the computations, i.e., the method is less sensitive with

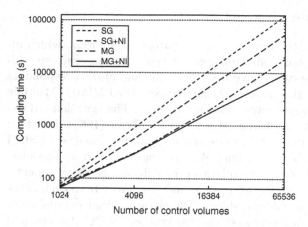

Fig. 12.18. CPU times for single-grid (SG) and multigrid (MG) methods with and without nested iteration (NI) versus number of CVs for turbulent bend flow

respect to numerical parameters (e.g., under-relaxation factors or grid distortions) than is the corresponding single-grid method.

12.3 Parallelization of Computations

Despite the high efficiency of the numerical methods that could be achieved by improvements of the solution algorithms in recent years (e.g., with the techniques described in the preceding sections), many practical computations, in particular flow simulations, still are very demanding with respect to computing power and memory capacity. Due to the complexity of the underlying problems the number of arithmetic operations per variable and time step cannot fall below a certain number. A further acceleration of the computations can be achieved by the use of computers with better performance.

Due to the enormous advances that could be achieved concerning the processor speeds (clock rates), the operating time per floating point operation drastically could be reduced. This way a tremendous reduction in the total computing times could be achieved where, in particular, the usage of vectorization and cache techniques has to be emphazised. To go beyond the principal physical limitations for the acceleration of the individual processors (mainly defined by the speed of light), the possibilities of parallel computing can be exploited, i.e., a computational task is accomplished by several processors simultaneously, resulting in a further significant reduction of the computing time.

In this section the most important aspects for using parallel computers for continuum mechanical computations are discussed and typical effects arising for concrete applications are shown by means of example computations. A detailed discussion of the subject from the computer science point of view can be found in [21].

12.3.1 Parallel Computer Systems

While at the beginning of the development of parallel computers, which already date back to the 1960s, quite different concepts for the concrete realization of corresponding systems have been pursued (and also corresponding systems were available on the market). Meanwhile so called MIMD (Multiple Instruction Multiple Data) systems clearly dominate. The notation MIMD goes back to a classification scheme introduced by Flynn (1966), in which classical sequential computers (PCs, workstations) can be classified as SISD (Single Instruction Single Data) systems. We will not detail further this classification scheme which has lost its relevance in light of the developments. In a MIMD system all processors can operate independently from each other (different instructions with different data). All relevant actual parallel computer systems, like multiprocessor systems, workstation or PC clusters, and also high-performance vector computers, which nowadays all possess multiple vector processors, can be grouped into this class.

An important classification attribute of parallel computing systems for the continuum mechanical computations of interest here is the way of memory access. Here, mainly two concepts are realized (see Fig. 12.19):

- *Shared memory systems*: each processor can access directly the whole memory via a network.
- *Distributed memory systems*: each processor only has direct access to its own local memory.

Typical shared memory computers are high-performance vector computers, while PC clusters are typical representatives for distributed memory systems. Shared memory systems with identical individual processors are also known as *symmetric multiprocessor (SMP) systems*. A very popular architecture nowadays are clusters of SMP systems, which somehow represent a compromise between shared and distributed memory systems.

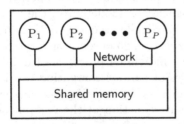 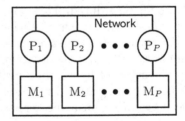

Fig. 12.19. Assignment of processors $P_1, \ldots, P_P$ and memory for parallel computers with shared memory (left) and distributed memory $M_1, \ldots, M_P$ (right)

For the programming of parallel computers there exist different programming models, the functionality of which, in particular, also is mainly determined by the possibilities of the memory access:

- *Parallelizing compilers*: The (sequential) program is parallelized automatically on the basis of an analysis of data dependencies of the program (maybe supported by compiler directives). This works – at least fairly adequately – only at the loop level for shared-memory systems and for relatively small processor numbers. One does not expect a really fully automatic *efficiently* parallelizing compiler to be available sometime.
- *Virtual shared memory*: The operating system or the hardware simulates a global shared memory on systems with (physically) distributed memory. This way, an automatic, semi-automatic, or user-directed parallelization is possible, where the efficiency of the resulting program increases in the same sequence. Such a concept is implemented by an extension of programming languages (usually Fortran or C) by an array syntax and compiler directives and the parallelization can be done by a generation of threads (controlled by directives) that are distributed on the different processors.
- *Message passing*: The data exchange between the individual processors is performed solely by sending and receiving of messages, where corresponding communication routines are made available via standardized library calls. The programs must be parallelized "manually", perhaps with the help of supporting tools (e.g., FORGE or MIMDIZER). However, this way also the best efficiency can be achieved. Meanwhile, as quasi-standards for the message passing some systems such as *Parallel Virtual Machine (PVM)* or *Message Passing Interface (MPI)*, have been established and are available for all relevant systems.

Note that programs that are parallelized on the basis of message passing can be used efficiently also on shared-memory systems (the corresponding communication libraries also are available on such systems). The converse usually is not the case. In this sense, the message passing concept can be viewed as the most general approach for the parallelization of continuum mechanical computations. Thus, all subsequent considerations relate to this concept and also the given numerical examples are realized on this basis.

12.3.2 Parallelization Strategies

For the parallelization of continuum mechanical computations almost exclusively *data decomposition techniques* are applied, which decompose the data space into certain partitions that are distributed to the different processors and treated there sequentially and locally. If required, a data transfer to the other partitions is carried out. The most important concepts for a concrete realization of such a data decomposition are: *grid partitioning, domain decomposition, time parallelization*, and *combination methods*. Grid partitioning techniques are by far most frequently applied in practice, and we will therefore address this in more detail.

Grid partitioning techniques are based on a decomposition of the (spatial) problem domain into non-overlapping subdomains, for which certain portions of the computations can be performed by different processors simultaneously. The coupling of the subdomains is handled by a data exchange between adjacent subdomain boundaries. For the exemplification of the procedure, we restrict ourselves to the case of (two-dimensional) block-structured grids, which provide a natural attempt for a grid partitioning (the principle – with corresponding additional effort – can be realized in an analogous way also for unstructured grids).

The starting point for the block-structured grid partitioning is the geometric block structure of the numerical grid. For the generation of the partitioning two cases have to be distinguished. If the number of processors P is larger than the number of geometrical blocks, the latter are further decomposed such that a new (parallel) block structure results, for which the number of blocks equals the number of processors. These blocks then can be assigned to the individual processors (see Fig. 12.20). If the number of processors is smaller than the number of geometrical blocks, the latter are suitably grouped together such that the number of groups equals the number of processors. These groups then can be assigned to the individual processors (see Fig. 12.21).

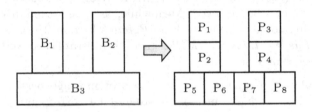

Fig. 12.20. Assignment of blocks to processors (more processors than blocks)

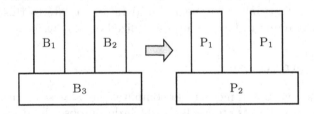

Fig. 12.21. Assignment of blocks to processors (more blocks than processors)

For the generation of the parallel block structure or the grouping of the blocks several strategies with different underlying criteria are possible. The simplest and most frequently employed approach is to take the number of grid points, which are assigned to each processor, as the sole criterion. If the numbers of grid points per processor are nearly the same, a good load

balancing on the parallel computer can be achieved, but other criteria such as
the number of neighboring subdomains or the length of adjacent subdomain
boundaries are not considered this way.

In order to keep the communication effort between the processors as small
as possible, usually along adjacent subdomain boundaries additional auxiliary
CVs are introduced, which correspond to adjacent CVs of the neighboring domain (see Fig. 12.22). When in the course of an iterative solution procedure
the variable values in the auxiliary CVs are actualized at suitable points in
time, the computation of the coefficients and source terms of the equation
systems in the individual subdomains can be carried out fully independently
from each other. Due to the locality of the discretization schemes only values of
neighboring CVs are involved in these computations, which are then available
to the individual processors in the auxiliary CVs. For higher-order methods,
for which also farther neighboring points are involved in the discretization,
corresponding additional "layers" of auxiliary CVs can be introduced. In general, this way the computations for the assembling of the equation systems do
not differ from those in the serial case.

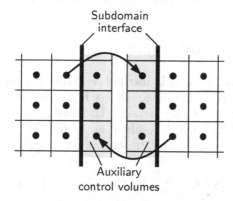

Fig. 12.22. Auxiliary CVs for data exchange along subdomain interfaces

Usually, the only issue, where – from the numerical point of view – a parallel algorithm differs from a corresponding serial one, is the solver for the linear
equation systems. For instance, ILU or SOR solvers are organized strongly recursively (a fact that greatly contributes to their high efficiency), such that
a direct parallelization is related to a very high communication effort and in
most cases does not turn out to be efficient. For such solvers it is advantageous
to partially break up the recursivity by considering the grid partitioning that
leads to an algorithmic modification of the solver. Since this plays an important role with respect to the efficiency of a parallel implementation, we will
explain this procedure briefly.

If the nodal values are numbered subdomain by subdomain, the equation
system that has to be solved gets a block structure corresponding to the grid
partitioning:

$$
\underbrace{\begin{bmatrix} \mathbf{A}_{1,1} & \mathbf{A}_{1,2} & \cdots & \mathbf{A}_{1,P} \\ \mathbf{A}_{2,1} & \mathbf{A}_{2,2} & & \cdot \\ \vdots & \cdot & \cdots & \cdot \\ \mathbf{A}_{P,1} & & \cdots & \mathbf{A}_{P,P} \end{bmatrix}}_{\mathbf{A}} \underbrace{\begin{bmatrix} \phi_1 \\ \phi_2 \\ \vdots \\ \phi_P \end{bmatrix}}_{\phi} = \underbrace{\begin{bmatrix} \mathbf{b}_1 \\ \mathbf{b}_2 \\ \vdots \\ \mathbf{b}_P \end{bmatrix}}_{\mathbf{b}}, \tag{12.10}
$$

where P denotes the number of processors (subdomains). In the vector ϕ_i ($i = 1, \ldots, P$) the unknowns of the subdomain i are summarized and in $\mathbf{b}_i$ the corresponding right hand sides are summarized. The matrices $\mathbf{A}_{i,i}$ are the principal matrices of the subdomains and have the same structure as the corresponding matrices in the serial case. The matrices outside the diagonals describe the coupling of the corresponding subdomains. Thus, the matrix $\mathbf{A}_{i,j}$ represents the coupling between the subdomains i and j (via the corresponding coefficients a_c). If the subdomains i and j are not connected to each other by a common interface $\mathbf{A}_{i,j}$ is a zero matrix.

Assuming that in the serial case the linear system solver is defined by an iteration process of the form (see Sect. 7.1)

$$
\phi^{k+1} = \phi^k - \mathbf{B}^{-1}\left(\mathbf{A}\phi^k - \mathbf{b}\right),
$$

for instance, with $\mathbf{B} = \mathbf{LU}$ in the case of an ILU method. For the parallel version of the method, instead of $\mathbf{B}$ the corresponding matrices $\mathbf{B}_i$ are used for the individual subdomains. For an ILU solver, for instance, $\mathbf{B}_i$ is the product of the lower and upper triangular matrices, which result from the incomplete LU decomposition of $\mathbf{A}_{i,i}$. The corresponding iteration scheme is defined by

$$
\phi_i^{k+1} = \phi_i^k - \mathbf{B}_i^{-1}(\sum_{j=1}^{p} \mathbf{A}_{i,j}\phi_j^k - \mathbf{b}_i). \tag{12.11}
$$

The computation can be carried out simultaneously for all $i = 1, \ldots, P$ by the individual processors, if ϕ^k is available in the auxiliary CVs for the computation of $\mathbf{A}_{i,j}\phi_j^k$ for $i \neq j$. In order to achieve this, the corresponding values at the subdomain boundaries have to be exchanged (updated) after each iteration.

Thus, with the described variant of the solver each subdomain during an iteration is treated if it were a self-contained solution domain and the coupling is done at the end of each iteration by a data transfer along all subdomain interfaces. This strategy results in a high numerical efficiency, because the subdomains are closely coupled, but requires a relatively high communication effort. An alternative would be to perform the exchange of the interface data not after each iteration, but after a certain number of iterations. This reduces the communication effort, but results in a deterioration of the convergence rate of the solver and thus in a decrease in the numerical efficiency of the method. Which variant finally is the better one strongly depends on the ratio of communication and computing power of the parallel computer employed.

Besides the local processor communication for the exchange of the subdomain interface data, a parallel computation always also requires some global communication over all subdomains. For instance, for the determination of global residuals for the checking of convergence criteria, it is necessary – after the local residuals are computed within the subdomains by the corresponding processors – to add these over all processors, such that the global residual is available to all processors and in case the convergence criterion is fulfilled all processors can finish the iteration process. As further communication processes, at the beginning of the computation the required data for its subdomain have to be provided to each processor (e.g., size of subdomain, grid coordinates, boundary conditions,...). For instance, these can be read by one processor, which then sends it to the other ones. At the end of the computation the result data for the further processing (e.g., for graphical purposes) must be suitably merged and stored, which again requires global communication.

In Fig. 12.23, as an example, a flow diagram for a flow computation with a parallel pressure-correction method is shown, which illustrates the parallel course of outer and inner iterations with the necessary local and global communication processes. Such a computation usually is realized following the *Single Program Multiple Data (SPMD) concept*, i.e., on all processors the same program is loaded and run, but with different data.

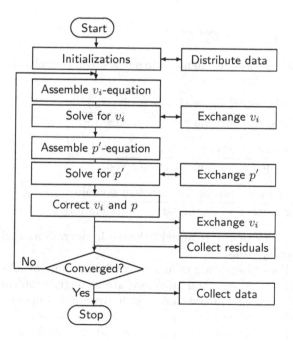

Fig. 12.23. Flow diagram of parallel pressure-correction method for flow computation

12.3.3 Efficieny Considerations and Example Computations

For the assessment of the performance of parallel computations usually the *speed-up* S_P and the *efficiency* E_P are defined:

$$S_P = \frac{T_1}{T_P} \quad \text{and} \quad E_P = \frac{T_1}{P T_P}, \tag{12.12}$$

where T_P denotes the computing time for the solution of the full problem with P processors. The ideal case, i.e., $S_P = P$ or $E_P = 100\%$, due to the additional effort in the parallel implementation, usually cannot be reached (exceptions from this can occur, for instance, by the influence of cache effects). Of course, a major objective of the parallelization must be to keep this additional effort as small as possible. For computations parallelized with a grid partitioning technique the loss factor can be split into the following three portions:

- The processor communication for the data transfers (local and global).
- The increase in the number of required arithmetic operations to achieve convergence by introduction of additional (artificial) inner boundaries, which are treated explicitly (modification of the solver).
- The unbalanced distribution of the computing load to the processors, e.g., when the number of grid points per processor is not the same or when there are different numbers of boundary grid points in the individual sub-domains.

With respect to a distinction of these portions the efficieny E_P can be split according to

$$E_P = E_P^{\text{num}} E_P^{\text{par}} E_P^{\text{load}}$$

with the parallel, numerical, and load balancing efficiencies, respectively, defined as

$$E_P^{\text{par}} = \frac{\text{CT(parallel algorithm with } one \text{ processor})}{P \cdot \text{CT(parallel algorithm with } P \text{ processors})},$$

$$E_P^{\text{num}} = \frac{\text{OP(best serial algorithm on } one \text{ processor})}{P \cdot \text{OP(parallel algorithm on } P \text{ processors})},$$

$$E_P^{\text{load}} = \frac{\text{CT(one iteration on the full problem domain)}}{P \cdot \text{CT(one iteration on the largest subdomain)}}.$$

Here $\text{OP}(\cdot)$ denotes the number of the required arithmetic operations and $\text{CT}(\cdot)$ is the required computing time.

The numerical and parallel efficiency are influenced by the number of sub-domains and their topology (coupling), and therefore are strongly problem dependent. For the parallel efficiency, in addition, also hardware and operating system data of the parallel computer are important influence factors. The load balancing efficiency only depends on the grid data and the partitioning into the subdomains and, if the grid size is chosen appropriately in relation to the processor number, it is relatively easy to achieve here a value close to 100%.

In order to illustrate the corresponding influences on the parallel computation, in Fig. 12.24 the efficiencies for different grid sizes and numbers of processors are given for the computation of a typical flow problem (again the natural convection flow with complex obstacle from Sect. 12.2.5). Some effects can be observed that generally apply for such computations. For constant grid size the efficiency decreases with increasing numbers of processors, because the portion of communication in the computation increases. For constant numbers of processors the efficiency increases with increasing grid size, because the communication portion becomes smaller.

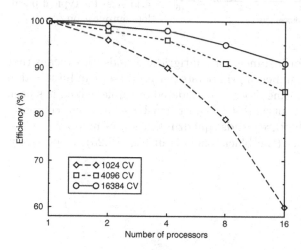

Fig. 12.24. Dependence of efficiency on number of processors for different grid sizes for typical parallel computation

The impact of the efficiency behavior on the corresponding computing times – only these are interesting for the user, the efficiency considerations are only a tool – are illustrated in Fig. 12.25, where also the ideal cases (i.e., efficiency $E_P = 100\%$) are indicated. For a given grid an increase in the number of processors results in an increasing deviation from the ideal case, which is larger the coarser the grid is.

The above considerations have the consequence that from a certain number of processors on (for a fixed grid) the total computing time increases. Thus, for a given problem size there is a maximum number of processors P_{max} which still leads to an acceleration of the computation (see Fig. 12.26). This maximum number increases with the size of the problem. So, from a certain number of processors on, the advantage of the usage of parallel computers no longer is to solve the same problem in a shorter computing time, but to solve a larger problem in a "not much longer" time (e.g., to achieve a better accuracy). If the convergence rate of the solver does not depend on the number of the

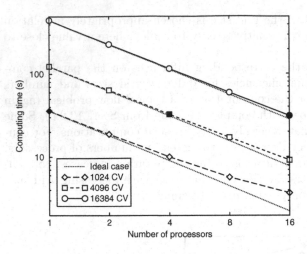

Fig. 12.25. Dependence of computing time on number of processors for different grid sizes for typical parallel computation

processors, as is the case, for instance, for multigrid methods, this means that problems with a constant number of grid points per processor can be solved in nearly the same computing time. For the considered example, which has been computed with a multigrid method, this effect can be seen if one compares the computing times for the three grids (quadruplication of number of CVs) with 1, 4, and 16 processors (the black symbols in Fig. 12.25), respectively, which nearly are identical.

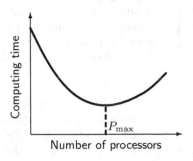

Fig. 12.26. Maximum sensible number of processors

For a concrete computation a corresponding "sensible" number of processors can be estimated – at least roughly – in advance by simple preliminary considerations taking into account the parallel and load balancing efficiencies (a mutual influence of the numerical efficiency usually is very difficult due to the complexity of the problems). The load balancing efficiency can simply be estimated by considering the numbers of grid points assigned to the individual processors. For the parallel efficiency the time T_K needed for a communication has to be taken into account:

$$T_K = T_L + \frac{N_B}{R_T},$$

where T_L is the *latency time* (or *set up time*) for a communication process, R_T is the *data transfer rate*, and N_B is the number of bytes that has to be transferred. For a concrete computer system all these parameters usually are known. For a specific solution algorithm the communication processes per iteration can be counted and a model equation for the parallel efficiency can be derived. While the times for the global communication (besides the dependency on T_K) strongly depend on the total number of processors P (the larger P, the more costly), this is not the case for the local communication, which can be done parallelly, and therefore mostly independent of P.

While the above described effects qualitatively do not depend on the actual computer employed, of course, quantitatively they are strongly determined by specific hardware and software parameters of the parallel system. In particular, the ratio of communication and arithmetic performance plays an important role in this respect. The larger this ratio, the lower the efficiency. With respect to the parallel efficiency, in particular, a high latency time has a rather negative influence. Parallel computers with fast processors can only be used efficiently for parallel continuum mechanical computations – at least for larger numbers of processors – together with a communication system of corresponding performance.

As already mentioned, for the user in the first instance the computing time for a computation is the relevant quantity (and not the efficiency!). In this context the performance of the underlying numerical algorithms play an important role. In order to point this out in Fig. 12.27 the computing times for a single-grid and a multigrid method versus the number of processors is shown (again for the natural convection flow with complex obstacle). The deviation from the ideal case with increasing number of processors for the multigrid method is much larger, i.e., the efficiencies for the multigrid method are lower, because on the coarser grids the ratio of required communications to the arithmetic operations that have to be performed is larger. However, one can see that the computing times with the parallel multigrid method in total still are significantly lower than with the parallel single-grid method. This is an aspect which generally applies: simpler, numerically less efficient methods can be parallelized relatively easily and efficiently, but usually with respect to the computing time they are inferior to parallel methods which are characterized also by a high numerical efficiency. By the way, the same applies to the vectorization of computations, which, however, we will not discuss here.

From the considerations above several requirements for parallel computer systems become apparent, in order for them to be efficiently used for continuum mechanical computations. In order to keep the losses in efficiency as small as possible the ratio of communication and computing performance should be "balanced". In particular, the ratio of the latency time and the time for a floating point operation should not be "too large" (i.e., smaller than 200). Since, in general, the portion of the communication time on the total computing time increases with the number of processors, a given computing power should be achieved with as few processors as possible. In order for these processors to

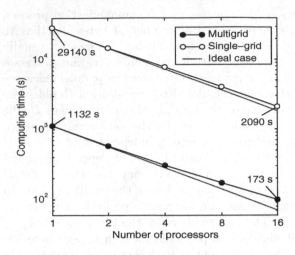

Fig. 12.27. Computing times for single-grid and multigrid methods versus number of processors (for 16 384 CVs)

be optimally be utilized, they should have sufficient memory capacity (i.e., at least 1 GByte per GFlops). Furthermore, the computer architecture conceptually should be suited for future increased demands with respect to memory, computing, and communication capacity, without larger modifications on the software side becoming necessary.

With respect to the above requirements, MIMD systems with local memory appear to be a well suited parallel computer architecture. They are – at least theoretically– arbitrarily scalable, and possess the most flexibility possible with respect to arithmetic and communication operations. They can be realized with high processor performance (with commercial standard processors), with balanced ratio of communication and computing power, and with reasonable price-performance ratio.

Exercises for Chap. 12

Exercise 12.1. Discretize the bar equation (2.38) with the boundary conditions (2.39) with a second-order finite-volume method for an equidistant grid with 4 CVs. The problem data are $L = 4\,\mathrm{m}$, $A = 1\,\mathrm{m}^2$, $u_0 = 0$, and $k_L = 2\,\mathrm{N}$. Formulate a two-grid method (two coarse grid CVs) each with one damped Jacobi iteration (12.4) for smoothing and compute one cycle with zero starting value.

Exercise 12.2. Let the bar problem from Exercise 12.1 be partitioned with the grid partitioning strategy from Sect. 12.3.2 into two equally sized subdomains. Perform two iterations with the Gauß-Seidel method parallelized according to (12.11). Compare the result with that for the Gauß-Seidel method without partitioning.

List of Symbols

In the following the meaning of the important symbols in the text with the corresponding physical units are listed. Some letters are multiply used (however, only in different contexts), in order to keep as far as possible the standard notations as they commonly appear in the literature.

Matrices, dyads, higher order tensors

$\mathbf{A}$		general system matrix
$\mathbf{B}$		procedure matrix for iterative methods
$\mathbf{C}$		iteration matrix for iterative methods
$\mathbf{C}, C_{ij}$	N/m^2	material matrix
$\mathbf{E}, E_{ijkl}$	N/m^2	elasticity tensor
$\mathbf{G}, G_{ij}$		Green-Lagrange strain tensor
$\mathbf{L}$		general lower triangular matrix
$\mathbf{I}$		unit matrix
$\mathbf{J}, J_{ij}$		Jacobi matrix
$\mathbf{P}$		preconditioning matrix
$\mathbf{P}, P_{ij}$	N/m^2	2nd Piola-Kirchhoff stress tensor
$\mathbf{S}, S_{ij}$		strain rate tensor
$\mathbf{S}, S_{ij}$		stiffness matrix
$\mathbf{S}^e, S_{ij}^e$		unit element stiffness matrix
$\mathbf{S}^k, S_{ij}^k$		element stiffness matrix
$\mathbf{T}, T_{ij}, \tilde{T}_{ij}$	N/m^2	Cauchy stress tensor
$\mathbf{U}$		general upper triangular matrix
δ_{ij}		Kronecker symbol
ϵ, ϵ_{ij}		Green-Cauchy strain tensor
ϵ_{ijk}		permutation symbol
τ_{ij}^{sgs}	N/m^2	subgrid-scale stress tensor
τ_{ij}^{test}	N/m^2	subtest-scale stress tensor

Vectors

$\mathbf{a}, a_i$	m	material coordinates
$\mathbf{b}, b_i$		load vector
$\mathbf{b}, \tilde{b}_i$	N/kg	volume forces per unit mass
$\mathbf{b}^e, b_i^e$		unit element load vector
$\mathbf{b}^k, b_i^k$		element load vector
$\mathbf{c}$	m/s	translating velocity vector
$\mathbf{d}, d_i$	Nms	moment of momentum vector
$\mathbf{e}_i, e_{ij}$		Cartesian unit basis vectors
$\mathbf{f}, f_i$	N/kg	volume forces per mass unit
$\mathbf{h}, h_i$	N/ms	heat flux vector
$\mathbf{j}, j_i$	kg/m^{2}s	mass flux vector
$\mathbf{n}, n_i$		unit normal vector
$\mathbf{p}, p_i$	Ns	momentum vector
$\mathbf{t}, t_i$		unit tangent vector
$\mathbf{t}, t_i$	N/m^2	stress vector
$\mathbf{u}, u_i$	m	displacement vector
$\mathbf{v}, v_i$	m/s	velocity vector
$\mathbf{v}^{\mathrm{g}}, v_i^{rmg}$	m/s	grid velocity
$\mathbf{w}, \tilde{w}_i$	m/s	relative velocity vector
$\mathbf{x}, x_i$	m	spatial coordinates
$\boldsymbol{\varphi}, \varphi_i$		test function vector
$\boldsymbol{\omega}, \omega_i$	1/s	angular velocity vector

Scalars (latin upper case letters)

A	m^2	cross sectional area
B_i^n		Bernstein polynomial of degree n
B	Nm^2	flexural stiffness
C		Courant number
C_{s}		Smagorinsky constant
C_{g}		dynamic Germano parameter
D		diffusion number
D	kg/ms	diffusion coefficient
D_0	m^2	unit triangle
D_i	m^2	general triangle
G, G_i	1/m^3, 1/m	filter function
E	N/m^2	elasticity modulus
$E_{\mathrm{grid}}^i, E_{\mathrm{jump}}^i$		error indicators
E_P		efficiency for P processors
E_P^{par}		parallel efficiency for P processors
E_P^{num}		numerical efficiency for P processors
E_P^{last}		load balancing efficiency for P processors

F_c		flux through face S_c
F_c^C		convective flux through face S_c
F_c^D		diffusive flux through face S_c
G	N/m^2s	production rate of turbulent kinetic energy
H	m	height
I	m^4	axial angular impulse
J		Jacobi determinant
K	Nm	plate stiffness
L	m	length
M	Nm	bending moment
N_B	s	data transfer time
N_j^e		shape function in unit element
N_j^i		local shape function
N_j		global shape function
N_{it}		number of iterations
P	Nm	potential energy
P_a	Nm/s	power of external forces
Q_0	m^2	unit square
Q_i	m^2	general quadrilateral
Q	Nm/s	power of heat supply
Q	N	transverse force
R	kg/m^3s	mass source
R_T	1/s	data transfer rate
R	Nm/kgK	specific gas constant
S	m^2 bzw. m	surface or boundary curve
S_c		control volume face
S_P		speed-up for P processors
T	K	temperature
$\tilde{T}$	K	reference temperature
T_L	s	latency time for data transfer
T_K	s	data transfer time
T_v		turbulence degree
T_H		higher order terms
V, V_i	m^3 bzw. m^2	(control) volume or (control) area
V_0	m^3	reference volume
W	Nm	total energy of a body
W	N/m^2	strain energy density function

Scalars (latin lower case letters)

a	m/s	speed of sound
c		species concentration
c_p	Nm/kgK	specific heat capacity at constant pressure
c_v	Nm/kgK	specific heat capacity at constant volume

d	m	plate thickness
e	Nm/kg	specific internal energy
e_P^n		total numerical error at point P and time t_n
f		general source term
f	N/m^3	force density
g		scalar source term
g	m/s^2	acceleration of gravity
h	m	measure for grid spacing
f_l	N/m	longitudinal load
f_q	N/m	lateral load
k	Nm/kg	turbulent kinetic energy
k_L	N	boundary force (bar)
l	m	turbulent length scale
m	kg	mass
$\dot{m}_c$	kg/s	mass flux through face S_c
p	N/m^2	pressure
q	Nm/skg	heat source
p', p''	N/m^2	pressure correction
s	Nm/kgK	specific internal entropy
t	s	time
u	m/s	velocity component in x-direction
u_τ	m/s	wall shear stress velocity
u^+		normalized tangential velocity
v	m/s	velocity component in y-direction
v_n	m/s	normal component of velocity
v_t	m/s	tangential component of velocity
$\bar{v}$	m/s	characteristic velocity
w	m	deflection
w_i		weights for Gauß quadrature
x	m	spatial coordinate
y	m	spatial coordinate
y^+		normalized wall distance

Scalars (greek letter)

α		general diffusion coefficient
α_ϕ		under-relaxation factor for ϕ
α	Nm/kg	thermal expansion coefficient
α_{num}		numerical (artificial) diffusion
$\tilde{\alpha}$	N/Kms	heat transfer coefficient
β		flux-blending parameter
β_c		artificial compressibility parameter
Γ		domain boundary
γ		interpolation factor

δ	m	wall distance
ϵ	Nm/s	dissipation of turbulent kinetic energy
ε_{tol}		error tolerance
$\eta, \tilde{\eta}$	m	spatial coordinate
θ	K	temperature deviation
θ		control parameter for θ-method
κ	N/Ks	heat conductivity
κ		condition number of a matrix
κ		Kármán constant
λ	N/m^2	Lamé constant
λ_P		aspect ratio of grid cell
λ_{max}		spectral radius
μ	N/m^2	Lamé constant
μ_t	kg/ms	turbulent viscosity
μ	kg/ms	dynamic viscosity
ν		Poisson number
ν	m^2/s	kinematic viscosity
Ω		problem domain
$\xi, \tilde{\xi}$	m	spatial coordinate
ξ_c		grid expansion ratio
Π	Nm	strain energy
ρ	kg/m^3	density
ρ_0	kg/m^3	reference density
τ	N/m^2	stiffness
τ_w	N/m^2	wall shear stress
τ_P^n		truncation error at point P and time t_n
ϕ		scalar transport quantity
$\overline{\phi}$		filtered or averaged quantity ϕ
ϕ'		small scale portion or fluctuation of ϕ
φ		virtual displacement
ψ	N/m^{2}s	specific dissipation function
ψ		general conservation quantity
ψ	m^2/s	velocity potential
ω		relaxation parameter for SOR method

Others

Ma		Mach number
Re		Reynolds number
Nu		Nußelt number
Pe		Peclet number
Pe$_h$		grid Peclet number
δS_c	m or m^2	length or area of control volume face S_c
δV	m^3 or m^2	volume or area of V

Δt	s	time step size
Δx	m	spatial grid spacing
Δy	m	spatial grid spacing
$\mathcal{F}$		discretization rule
$\mathcal{H}$		function space for test functions
$\mathcal{I}_{2h}^{h}$		interpolation operator
$\mathcal{I}_{h}^{2h}$		restriction operator
$\mathcal{L}$		spatial discretization operator
$\mathcal{S}$		iteration method

References

1. O. Axelsson und V.A. Barker
   Finite Element Solution of Boundary Value Problems
   Academic Press, Orlando, 1984 (for Chap. 7)
2. K.-J. Bathe
   Finite-Element Procedures
   Prentice Hall, New Jersey, 1995 (for Chaps. 5 and 9)
3. D. Braess
   Finite Elements
   2nd edition, University Press, Cambridge, 2001 (for Chaps. 5 and 9)
4. W. Briggs
   Multi-Grid Tutorial
   2nd edition, SIAM, Philadelphia, 2000 (for Chap. 12)
5. T.J. Chung
   Computational Fluid Mechanics
   Cambridge University Press, 2002 (for Chaps. 3, 4, 6, 8, 10, 11, and 12)
6. H. Eschenauer, N. Olhoff, and W. Schnell
   Applied Structural Mechanics
   Springer, Berlin, 1997 (for Chaps. 2, 5, and 9)
7. G.E. Farin
   Curves and Surfaces for Computer Aided Geometric Design: A Practical Guide
   5th edition, Academic Press, London, 2001 (for Chap. 3)
8. J. Ferziger und M. Perić
   Computational Methods for Fluid Dynamics
   3rd edition, Springer, Berlin, 2001 (for Chaps. 4, 6, 8, 10, and 11)
9. C.A.J. Fletcher
   Computational Techniques for Fluid Dynamics (Vol. 1, 2)
   Springer, Berlin, 1988 (for Chaps. 4, 6, and 10)
10. W. Hackbusch
    Multi-Grid Methods and Applications
    Springer, Berlin, 1985 (for Chap. 12)
11. W. Hackbusch
    Iterative Solution of Large Sparse Systems of Equations
    Springer, Berlin, 1998 (for Chap. 7)

12. C. Hirsch
    Numerical Computation of Internal and External Flows (Vol. 1, 2)
    Wiley, Chichester, 1988 (for Chaps. 4, 6, 7, 8, and 10)
13. K.A. Hoffmann und S.T. Chang
    Computational Fluid Dynamics for Engineers I, II
    Engineering Education System, Wichita, 1993 (for Chaps. 3, 6, and 8)
14. G.A. Holzapfel
    Nonlinear Solid Mechanics
    Wiley, Chichester, 2000 (for Chap. 2)
15. P. Knupp und S. Steinberg
    Fundamentals of Grid Generation
    CRC Press, Boca Raton, 1994 (for Chap. 3)
16. R. Peyret (Editor)
    Handbook of Computational Fluid Mechanics
    Academic Press, London, 1996 (for Chaps. 10 and 11)
17. S.B. Pope
    Turbulent Flows
    University Press, Cambridge, 2000 (for Chap. 11)
18. P. Sagaut
    Large Eddy Simulation for Incompressible Flows
    2nd edition, Springer, Berlin, 2003 (for Chap. 11)
19. J. Salençon
    Handbook of Continuum Mechanics
    Springer, Berlin, 2001 (for Chap. 2)
20. H.R. Schwarz
    Finite Element Methods
    Academic Press, London, 1988 (for Chaps. 5 and 9)
21. L.R. Scott, T. Clark, and B. Bagheri
    Scientific Parallel Computing
    Princeton University Press, 2005 (for Chap. 12)
22. R. Siegel and J.R. Howell
    Thermal Radiation Heat Transfer
    4th edition, Taylor & Francis, New York, 2002 (for Chap. 2)
23. J.H. Spurk
    Fluid Mechanics
    Springer, Berlin, 1997 (for Chap. 2)
24. J. Stoer und R. Bulirsch
    Introduction to Numerical Analysis
    3rd edition, Springer, Berlin, 2002 (for Chaps. 6 and 7)
25. S. Timoschenko und J.N. Goodier
    Theory of Elasticity
    McGraw Hill, New York, 1970 (for Chap. 2)
26. J.F. Thompson, B.K. Soni, N.P. Weatherhill (Editors)
    Handbook of Grid Generation
    CRC Press, Boca Raton, 1998 (for Chaps. 3 and 4)
27. M. van Dyke
    An Album of Fluid Motion
    Parabolic Press, Stanford, 1988 (for Chaps. 2 and 11)

28. D. Wilcox
    Turbulence Modeling for CFD
    DCW Industries, La Cañada, 1993 (for Chap. 11)
29. O.C. Zienkiewicz, R.L. Taylor, and J.Z. Zhu
    The Finite-Element-Method (Vol. 1, 2, 3)
    6th edition, Elsevier Butterworth-Heinemann, Oxford, 2005 (for Chaps. 5, 9, and 12)

Index